WHITE MATTER IN COGNITIVE NEUROSCIENCE

ADVANCES IN DIFFUSION TENSOR IMAGING AND ITS APPLICATIONS

ANNALS OF THE NEW YORK ACADEMY OF SCIENCES
Volume 1064

WHITE MATTER IN COGNITIVE NEUROSCIENCE

ADVANCES IN DIFFUSION TENSOR IMAGING AND ITS APPLICATIONS

Edited by John L. Ulmer, Lawrence Parsons, Michael Moseley, and John Gabrieli

The New York Academy of Sciences
New York, New York
2005

Library of Congress Cataloging-in-Publication Data

White matter in cognitive neuroscience: advances in diffusion tensor imaging and its applications / edited by John L. Ulmer ... [*et al.*].
p.; cm. — (Annals of the New York Academy of Sciences; v. 1064)
"Result of a workshop entitled White Matter in the Cognitive Neurosciences: Advances in Diffusion Tensor Imaging and Its Applications, held August 19–20, 2004, at the New York Academy of Sciences in New York City"—Table of contents.
Includes bibliographical references and index.
ISBN 1-57331-545-1 (cloth: alk. paper) — ISBN 1-57331-546-X (pbk.: alk. paper)
1. Brain—Magnetic resonance imaging—Congresses. 2. Diffusion tensor imaging—Congresses. 3. Myelinated neurofibrils—Congresses. I. Ulmer, John L. II. Series.
[DNLM: 1. Brain—Congresses. 2. Diffusion Magnetic Resonance Imaging—Congresses. 3. Neurosciences—Congresses. W1 AN626YL v.1064 2005 / WL 300 W586 2005]
Q11.N5 vol. 1064
[RC386.6.M34]
500 s—dc22
[616.8'047548]

2005027371

GYAT / PCP

ISBN 1-57331-545-1 (cloth)
ISBN 1-57331-546-X (paper)
ISSN 0077-8923

ANNALS OF THE NEW YORK ACADEMY OF SCIENCES

Volume 1064
December 2005

WHITE MATTER IN COGNITIVE NEUROSCIENCE

ADVANCES IN DIFFUSION TENSOR IMAGING AND ITS APPLICATIONS

Editors

JOHN L. ULMER, LAWRENCE PARSONS, MICHAEL MOSELEY, AND JOHN GABRIELI

This volume is the result of a workshop entitled **White Matter in the Cognitive Neurosciences: Advances in Diffusion Tensor Imaging and Its Applications**, held on August 19–20, 2004, at the New York Academy of Sciences in New York City.

CONTENTS

Financial assistance was received from:

- GENERAL ELECTRIC HEALTH CARE SYSTEMS
- MUSHETT FAMILY FOUNDATION
- NATIONAL INSTITUTES OF HEALTH
- NATIONAL SCIENCE FOUNDATION
- NEW YORK ACADEMY OF SCIENCES
- SIEMENS MEDICAL

Preface

LEIGHTON P. MARK AND JOHN L. ULMER

Medical College of Wisconsin, Milwaukee, Wisconsin, USA

Diffusion tensor imaging (DTI), a new and evolving MRI technique, has the potential to enhance and enrich our insights about the functional connectivity of the human brain. Given that we do not know as much about the functional implications of white matter structures as we do about cortical structures, we sometimes run the risk of oversimplifying the concept of brain activity. It is not enough to consider only structures that encode sensory information and command movements. We must be careful not to perceive white matter fibers as mere conduits for the appropriate gray matter centers involved in sensory and motor functions. To make this mistake is to reduce most brain functions to one large reflex arc.

While motor and sensory regions account for only a fraction (approximately 20%) of the cerebral cortex, most of the brain consists of the so-called association cortices that enable diverse functions collectively referred to as "cognition". Some of these functions include awareness of physical and social circumstances (consciousness), the ability to have thoughts and emotions, sexual attraction, expressions of these thoughts with language, emotional memory, etc. It can be argued that cognitive abilities represent the most complex, important, and intriguing cerebral functions. In other words, these are the very psychological and neurological processes that help to define ourselves and our lives.

A closer look at association white matter tracts illuminates the complexity of the neuronal network and readily dispels the notion of a simple one-to-one connection from one cortical neuron to another. The signals that these fibers project to other association cortices via the thalamus have already been processed in the primary motor and sensory areas and are fed back to the association regions for further processing. The information, therefore, is a relay from other cortical areas rather than of primary motor or peripheral sensory signals. This type of corticocortical connection explains the observed enrichment or multiplication of input fibers from other cortical areas to any one particular association area. The functional implications for this increasingly complex network foster the speculation about subcortical processing of complex behavior.

Most of the evidence that supports our anatomic understanding of these white matter tracts is derived from anatomic tracing studies in nonhuman primates supplemented by limited pathway tracings done in postmortem human brain tissue. The inferred functional association of this information then depends on critical correlation with clinical observations of patients with cortical lesions. The ability to demon-

Address for correspondence: John L. Ulmer, M.D., Department of Radiology, Medical College of Wisconsin, 9200 W. Wisconsin Avenue, Milwaukee, WI 53226. Voice: 414-805-3122; fax: 414-259-9290.

julmer@mcw.edu

Ann. N.Y. Acad. Sci. 1064: vii–ix (2005). © 2005 New York Academy of Sciences.
doi: 10.1196/annals.1340.001

strate these tracts *in vivo*, therefore, represents a huge advantage in direct observation and may even generate new observations and conclusions that can challenge assumptions built upon the previous indirect data.

The potential for DTI to dramatically impact the clinical neurosciences is clear. In today's clinical practice, most routine MR imaging (including functional MR imaging) used to determine the functional implications of cerebral disease largely relies on an analysis of the relationships and functional imaging correlates to deep and superficial gray matter structures. This is an understandable natural consequence of readily identifiable gray matter landmarks such as sulcal and gyral anatomy as well as many well-defined deep gray matter margins. The underlying white matter tracts are more opaque to routine imaging and clinical evaluation, owing to less visible margins and more complex functional associations. With the addition of DTI, the promise that *in vivo* depiction of underlying neuronal infrastructure will provide greater understanding of the ways that disease and treatments can affect brain function is certainly appealing. There is hope that DTI will also allow us to begin to visualize the underpinnings of more profound neurological disorders affecting consciousness, attention, and awareness. The technique, when applied to the spinal cord and brain stem, may enable us to better characterize the impact of disease on connectivity between the brain and other organ systems as well.

This overall enthusiasm for DTI, however, is for its potential and not the current reality. The technique at its current stage of development effectively represents only an improved *in vivo* method of producing white matter contrast. This is a method that, so far, only allows the visualization of the directional orientation of fibers in the white matter based on eigenvector constructs, which is not the same as depicting functionally specific white matter tracts. Depicting particularly tortuous white matter tracts, for example, that temporarily merge with or lie close to various other white matter tracts at different points throughout their course could be quite challenging. The seemingly reasonable connection between directional vectors in different voxels based on existing anatomic information is an unproven assumption that may be vulnerable to specious reasoning in future developments. The major eigenvector also only represents the longest axis of a diffusion ellipsoid and may not be representative of the actual path of the particular tract one wishes to map.

Fiber tracking techniques based in DTI have the potential to further enhance specificity in characterizing brain connectivity in the normal and disease human brains. However, even the most sophisticated fiber tracing algorithms are likely to be challenged by anatomic constraints, such as parallel but functionally distinct tracts and acutely angled white matter pathways. Also, disease-specific biosystems can alter the very structural processes necessary to construct diffusion ellipsoids. Nonetheless, the initial information produced by DTI techniques has immediate value and applications. The technical challenge for the near future is to clearly distinguish tracts that are defined by their common functionality. The combined use of cortical and white matter mapping has the potential to see this to fruition.

This proceeding contains some of the topics discussed at a diffusion tensor imaging workshop sponsored in part by the New York Academy of Sciences in late August of 2004. The workshop had four major goals: (i) to explore advanced uses of DTI for modeling white matter neural connectivity and tractography; (ii) to discuss methods for integrating DTI of white matter into cognitive and clinical neuroscience data and models; (iii) to discuss how to promote new advances in DTI techniques for applications relevant to cognitive and clinical neuroscience; and (iv) to discuss how to implement new advances in DTI in readily accessible software that can be distributed to the cognitive and clinical neuroscience communities.

There has been an explosion of DTI research in recent years, exploring both technical issues and its potential to elucidate normal brain connectivity, aging, neurodegenerative diseases, psychiatric disorders, learning disabilities, and pretreatment surgical and radiation planning. The material covered in the workshop and contained in this volume is broad in scope, but represents only a fraction of current and planned work on the topic. Rather, the presentations are meant to stimulate interest and investigations in the field of white matter mapping and to foster interdisciplinary approaches to the refinement of emerging DTI techniques specifically for the purposes of analyzing white matter networks noninvasively, both in normal brains and in disease.

Combining Functional and Diffusion Tensor MRI

DAE-SHIK KIM[a] AND MINA KIM[b]

[a]*Center for Biomedical Imaging, Department of Anatomy and Neurobiology, Boston University School of Medicine, Boston, Massachusetts, USA*

[b]*Center for Magnetic Resonance Research, Department of Radiology, University of Minnesota Medical School, Minneapolis, Minnesota, USA*

ABSTRACT: Functional magnetic resonance imaging (fMRI) of the perceptual, motor, and cognitive capacities in humans is of increasing importance for basic and clinical neurosciences. The explanatory power of current fMRI techniques could be greatly expanded, however, if the pattern of the neuronal connections between the active cortical areas could likewise be visualized. In this study, we acquired blood-oxygenation level dependent (BOLD) fMRI signals during the stimulation of subjects with a set of localizer stimuli for cortical visual areas. Subsequently, diffusion tensor imaging (DTI) data from the same subjects were obtained, and the activation areas identified through fMRI were utilized as seeding points for 3D DTI fiber reconstruction algorithms. The methods developed in this study have the potential to lay a foundation for *in vivo* neuroanatomy and the ability for noninvasive longitudinal studies of brain development.

KEYWORDS: diffusion tensor imaging (DTI); functional magnetic resonance imaging (fMRI); blood-oxygenation level dependent (BOLD); cortical; visual

INTRODUCTION

The rapid progress of blood-oxygenation level dependent (BOLD) functional magnetic resonance imaging (fMRI) in recent years[1–3] has raised the hope that the functional architecture of the human brain can be studied directly in a noninvasive manner. The BOLD technique is based on the use of deoxyhemoglobin as nature's own intravascular paramagnetic contrast agent.[4–6] When placed in a magnetic field, deoxyhemoglobin alters the magnetic field in its vicinity, particularly when it is compartmentalized as it is within red blood cells and vasculature. The effect increases as the concentration of deoxyhemoglobin increases. At concentrations found in venous blood vessels, a detectable local distortion of the magnetic field surrounding the red blood cells and surrounding blood vessels is produced. This affects the magnetic resonance behavior of the water proton nuclei within and surrounding the vessels, which in turn results in *decreases* in the transverse relaxation times (T_2 and

Address for correspondence: Dae-Shik Kim, Ph.D., Center for Biomedical Imaging, Department of Anatomy and Neurobiology, Boston University School of Medicine, 715 Albany Street, L-1004, Boston, MA 02118. Voice: 617-414-2361; fax: 617-414-2362.
dskim@bu.edu

Ann. N.Y. Acad. Sci. 1064: 1–15 (2005).
doi: 10.1196/annals.1340.005

T_2*).[4,6] During the activation of the brain, this process is reduced: increase in neuronal and metabolic activity results in a *reduction* of the relative deoxyhemoglobin concentration due to an increase of blood flow (and hence increased supply of fresh oxyhemoglobin) that follows. Consequently, in conventional BOLD fMRI, brain "activity" can be measured as an *increase* in T_2- or T_2*-weighted MR signals.[1–3] Since its introduction about 10 years ago, BOLD fMRI was successfully applied, among numerous other examples, to precisely localize the cognitive,[7] motor,[8] and perceptual[9–11] function of the human *cortex cerebri*. The explanatory utility of BOLD fMRI has been further strengthened in recent years through the introduction of high (~3 T) and ultrahigh (~7 T) MRI scanners.[12] Stronger magnetic field not only increases the fMRI signal per se, but in addition it will specifically enhance the signal components originating from parenchymal capillary tissue, thus enhancing the spatial specificity of BOLD fMRI.

While the aforementioned fMRI-based neuroimaging techniques provide detailed information about the spatial location of the functionally active cortical areas, the question of functional interdependency between the cortical areas remains elusive. This is of particular importance for higher cortical functions as they require a well-coordinated balancing of cortical computations across multiple brain areas. This is most evident in the mammalian visual system. Here, the visual information is represented and processed across a multitude of visual areas. The number of visual areas range from 10–15 for cats[13] to 10–30 for humans and macaques,[14] respectively. Furthermore, similar to other primates, human visual areas are clustered along two "streams" diverging from the occipital pole: the ventrotemporal "what or perception" stream and the dorsal "where or action" stream.[15,16] While the areas in the dorsal stream are tuned for visual stimuli and tasks related to stimulus location and/or action, the ventral stream consists of a web of exquisitely category-selective areas. For example, a region in the lateral occipital cortex (LOC) extending anteriorly into the temporal cortex responds strongly to a variety of complex-shaped objects such as polygonal figures, chairs, gloves, etc.[17,18] On the other hand, in the so-called fusiform face area (FFA)[19] located within the fusiform gyrus, cells are tuned to faces and facial stimuli in a way comparable to the receptive field properties of face-selective neurons in primate inferotemporal cortex (IT).[20,21] Further down the temporal cortex, in the so-called parahippocampal place area (PPA),[22] maximum functional response can be obtained using scenic or place types of stimuli. The description of highly specialized areas such as FFA and PPA raises the question of how many category-selective regions of cortex exist in the human visual system and, more generally, how the ventral temporal cortex is organized. Hypotheses range from the assumption that there are a few specialized processing modules, that is, for faces, places, letters, and human body parts,[19,22] up to the proposal of widely distributed and overlapping cortical object representations.[23–25] Effects of category-related expertise[26,27] and, more recently, different category-related resolution needs[17] have also been proposed to explain the topology of the human "what"-pathway. FIGURE 1 displays the retinotopic and nonretinotopic ventral visual areas that have been identified to exist in the human occipitoventral visual system. FIGURE 1 also signifies the major aim of the present study: given the multitude of the ventral visual areas, what is the overall organizing scheme that ties the individual areas together? It is the working hypothesis of this study that no such organizing principle could be understood without the knowledge about the pattern of axonal connections between the occipitoventral areas.

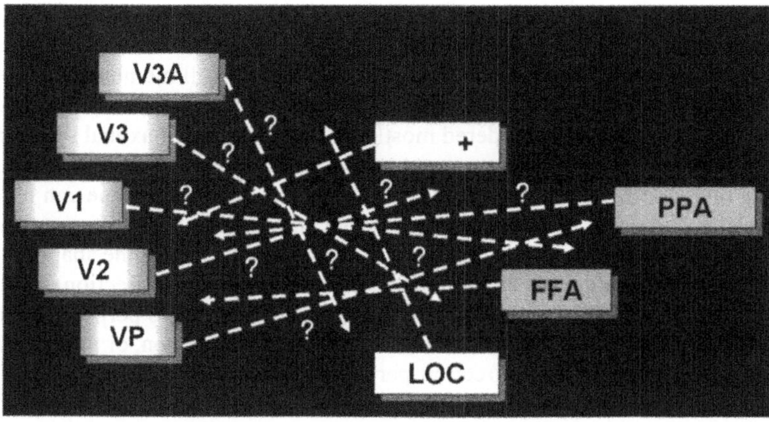

FIGURE 1. The retinotopic and nonretinotopic ventral visual areas that have been identified to exist in the human occipitoventral visual system. While the patterns of occipitoventral connectivities in animal models are relatively well established, obtaining homologous data in humans has been hampered due to the invasive nature of most neuroanatomical techniques. The *dashed lines* (and *question marks*) indicate the elusive nature of the axonal connections across the human ventral stream.

The importance of large-scale, functional neuroanatomical information notwithstanding, labeling axonal connections in humans is greatly hampered by methodological limitations of existing techniques. In general, thalamocortical, corticofugal, and cortico-cortical connections between neurons and neuronal populations can be studied using a variety of techniques. The earliest developed techniques include postmortem methods, such as dissection of white matter, strychnine neuronography,[28] and the Nauta[29] methods of tracing neuronal degeneration after localized lesions. More recently, implantation of carbocyanine dyes, such as DiI and DiA,[30] has replaced the older "degeneration methods". However, besides the long duration needed for passive labeling of the axonal fibers (often several months), the aforementioned postmortem methods inherently fail to yield a correlation to the foci of functional activation. Here, several modern labeling methods using transneuronal markers [e.g., horseradish peroxidase (HRP), rhodamine- and fluorescein-conjugated latex microspheres,[31] biocytin,[32] and biotinylated dextran amine (BDA)] can be used to track the pattern of neuronal connectivity "*in vivo*". However, while the aforelisted neurotracing techniques that utilize actual, physical tracers will always remain the gold standard for defining the pattern of neuronal connections, they pose several limitations: (a) low number of labeled tracts; (b) short tracing distance (mostly, across 2–3 synapses only); (c) the need for invasive procedure, such as inserting a glass probe; and most importantly (d) the need to sacrifice the animal before the labeling pattern can be visualized. Naturally, such limitations invalidate the use of existing "*in vivo*" tracing methods for human studies.

The newly developed diffusion tensor imaging (DTI) technique based on well-established diffusion-weighted imaging (DWI) has the potential to successfully overcome some of these problems. DTI is a powerful MRI technique that enables us to translate the self-diffusion or microscopic motion of water molecules in tissue into an MRI measure of tissue integrity and structure. Namely, the spatial characteristic of water diffusion highly depends on the barriers imposed on the water molecule motion, those barriers being the elements of tissue such as cell membranes, myelin sheath, intracellular microorganelles, and others. Specifically, in white matter, water self-diffusion is restricted, or hindered mostly by the intracellular axonal space, and by the interstitial, extracellular space among the well-packed axons in the fiber tract. By taking several diffusion-weighted images in several dimensions, one can reconstruct the so-called diffusion tensor for each image unit, or pixel. The diffusion tensor gives a 3D representation of the preferred direction of diffusion, in the shape of the 3D ellipsoid. The main axis of the ellipsoid gives the "preferred" direction of diffusion, typically parallel to the axonal tract, and the two minor axes perpendicular to the main direction typically provide a measure to "lateral" diffusion, or the ability of water molecules to move in a direction perpendicular to the main direction. This "lateral" diffusion proves to be a critical parameter in evaluating the integrity of white matter structure as it is possibly related to increase in water mobility in the interstitial space due to, for example, myelin breakdown.

Recently, DTI techniques in combination with a variety of 3D fiber reconstruction algorithms were used to generate spectacular images of the axonal connectivity pattern *in vivo* in humans,[33–35] rodents,[36] and (recently) cats.[37] The differences in detailed fiber reconstruction algorithms notwithstanding, a key in making DTI into an outstanding tool for cognitive neurosciences is to develop selection criteria to determine the seeding region of interest (ROI) for DTI fiber tracing. In the majority of fiber-tracking algorithms, tracking starts at a user-defined seeding point or ROI. Such "seeding points" are selected either based on the quality of the underlying DWI or based on a priori anatomical criteria that are known from postmortem studies. Such anatomically motivated tracking strategies are of greatest importance for testing for anatomical irregularities *in vivo*. However, in normal subjects, strictly anatomically defined DTI fiber reconstructions have the tendency to simply revalidate what has been already known from conventional anatomical and histological techniques, thus resulting in partially tautological statements. An alternative way of DTI fiber reconstruction therefore is to use the foci of functional activity—such as obtained with BOLD contrast—as the "initial" and "termination" ROIs. This is a more natural choice for most questions in cognitive neurosciences as the main interest here is to elucidate the pattern of neuronal circuitry underlying the observed functional activation for a particular task.

MATERIALS AND METHODS

Subject Preparation

All studies were performed with the approval of the Institutional Review Board (IRB) of the University of Minnesota Medical School and Boston University School of Medicine. Following the proper instruction of the subjects, the subjects were asked to lay on the MRI table inside the magnet, and view visual stimuli. Paddings and foams were used to maintain the subject's head in a stable position. Earplugs and headphones were employed to reduce the noise due to the switching gradients.

Visual Stimulation

Visual stimuli were generated on a PC using custom-written MATLAB (The Mathworks Inc., Natick, MA) software utilizing functions provided by PsychToolbox.[38] Stimuli were presented binocularly with a video projector on a rear projection screen. Conventional checkerboard stimuli that consisted of four triangular wedges for the upper/lower and left/right visual fields, and four segmented expanding rings for foveal representation were used for mapping retinotopic areas. Localizer stimuli known to activate the respective areas within the human ventral stream were used to identify the FFA (conventional/scrambled faces),[39] PPA (buildings and scenes), and LOC (set of complex objects). Within each all-novel epoch, subjects saw four categories of pictures that each contained 30 different photographs. Within multiple-repeat epochs (four repetitions), subjects saw different photographs from the same category. A localizer was used to identify hMT+ (motion) as a part of the human dorsal stream.

MRI Acquisition

High-resolution fMRI and T_1-weighted anatomical images were obtained at 3 tesla (Siemens Trio or Philips Intera). The imaging parameters for the Siemens scanner were as follows: **T1 MPRAGE** (nonselective IR)—no. of slices: 144; slice thickness: 1 mm; FoV: 256 mm × 256 mm; matrix: 256 × 256; TR: 2100 ms; TE: 3.93 ms; TI: 1100 ms; flip angle: 15°; 1 NEX; **fMRI: gradient echo EPI**—no. of slices: 30; slice thickness: 2 mm; FoV: 256 mm × 256 mm; matrix: 128 × 128; no. of volumes: 132; TR: 3000 ms; TE: 40 ms. Parameters for the Philips scanner were as follows: **T1 MPRAGE** (NS-IR)—no. of slices: 144; slice thickness: 1 mm; FoV: 230 mm × 230 mm; matrix: 256 × 256; TR: 2100 ms; TE: 4.6 ms; TI: 1100 ms; flip angle: 8°; 1 NEX; **fMRI: gradient echo EPI**—no. of slices: 30; slice thickness: 2 mm; FoV: 230 mm × 230 mm; matrix: 128 × 128; no. of volumes: 132; TR: 3000 ms; TE: 40 ms.

Diffusion-Weighted MRI

Conventional methods for DWI were used in order to calculate the voxel-based diffusion tensors. Diffusion imaging parameters for the Siemens scanner were as follows: **DTI: spin echo EPI**—no. of slices: 64; slice thickness: 2 mm; FoV: 256 mm × 256 mm; matrix: 128 × 128; no. of directions: 12; TR: 11,500 ms; TE: 111 ms; 3 NEX. Parameters for the Philips scanner were as follows: **DTI: spin echo EPI**—no. of slices: 73; slice thickness: 1.5 mm; FoV: 230 mm × 230 mm; matrix: 256 × 256; no. of directions: 15; TR: 10,646 ms; TE: 91 ms; 1 NEX.

Data Analysis

Functional imaging scans were used to localize areas hMT+, LOC, FFA, and PPA, as well as retinotopic areas (V1, V2, V3, V3A, VP, and V4v). fMRI data were analyzed using BrainVoyager (Brain Innovation, Maastricht, the Netherlands). Each area was segmented by mapping borders in the flattened representation of cortex, and reconstructed as 3D volume-rendered ROIs (see FIG. 2). ROIs were then imported into a custom-written DTI reconstruction software. Diffusion tensors, fractional anisotropy (FA), and fiber tracts were calculated using custom-written MATLAB (The Mathworks Inc., Natick, MA) software. Standard methods for FA exclusion and

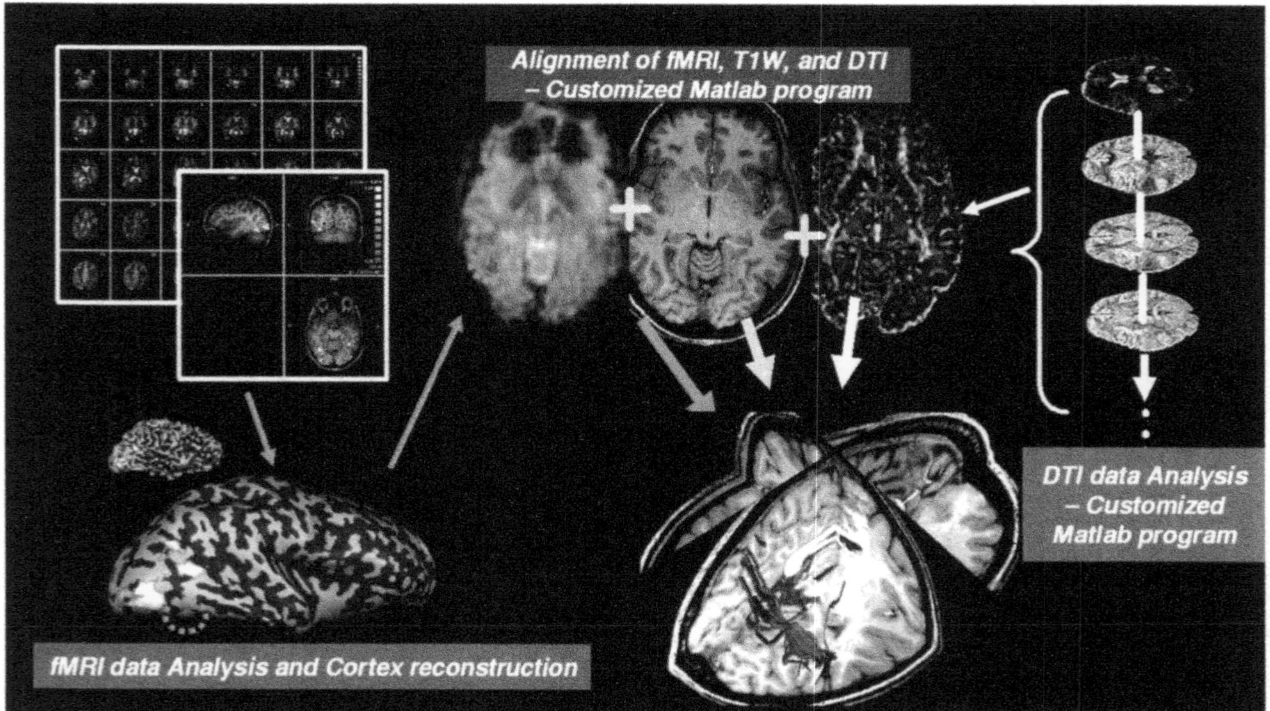

FIGURE 2. Methods involved in combined fMRI and DTI measurements. Localizer stimuli were used to label areas V1, V2, V3, V3A, VP, V4v, mediotemporal (hMT+), LOC, FFA, and PPA. These areas were then used as seeding ROIs for DTI-based fiber reconstructions. See text for further details.

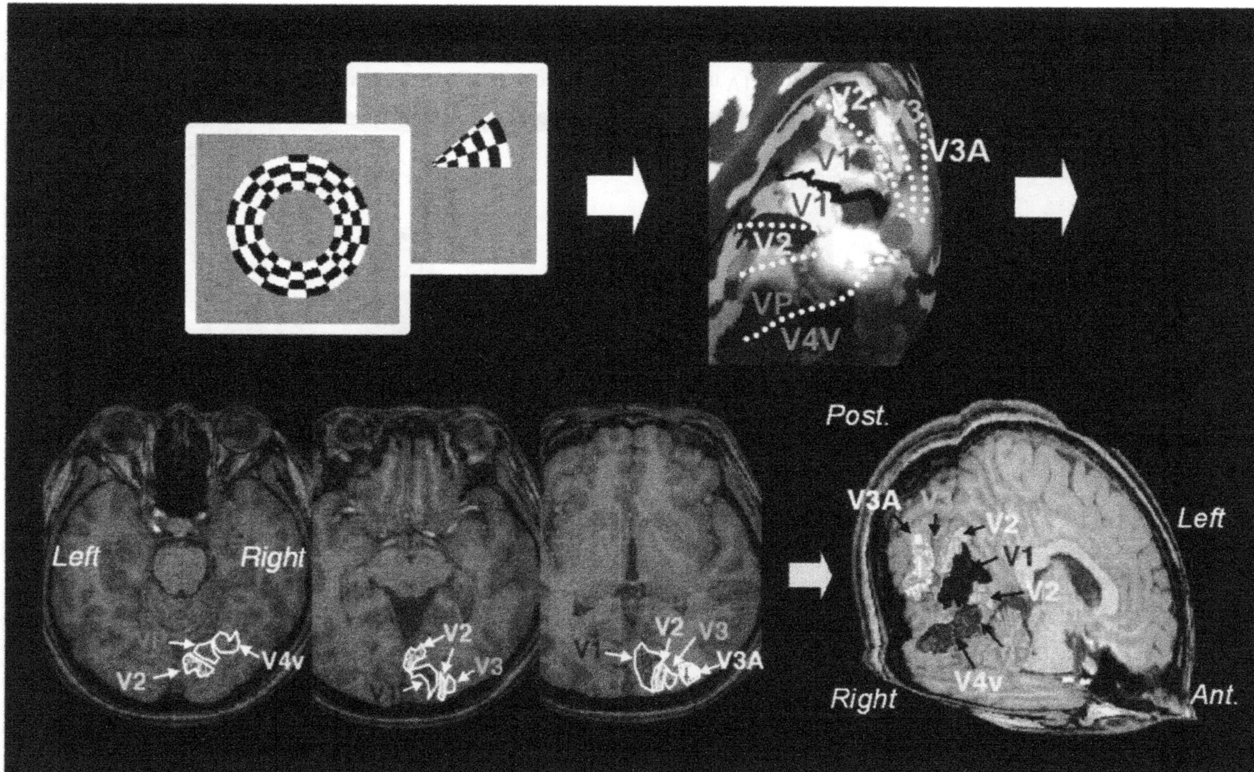

FIGURE 3. Volume-rendered representation of visual areas and the experimental protocol for labeling retinotopic areas. Cortical areas responding to isoeccentricity and polar angles were used in order to determine the areal boundaries between the retinotopic areas V1, V2, V3, V3A, VP, and V4v. The surface tangents of the individual areas were volume rendered to yield 3D reconstruction of the respective visual areas. See text for further details.

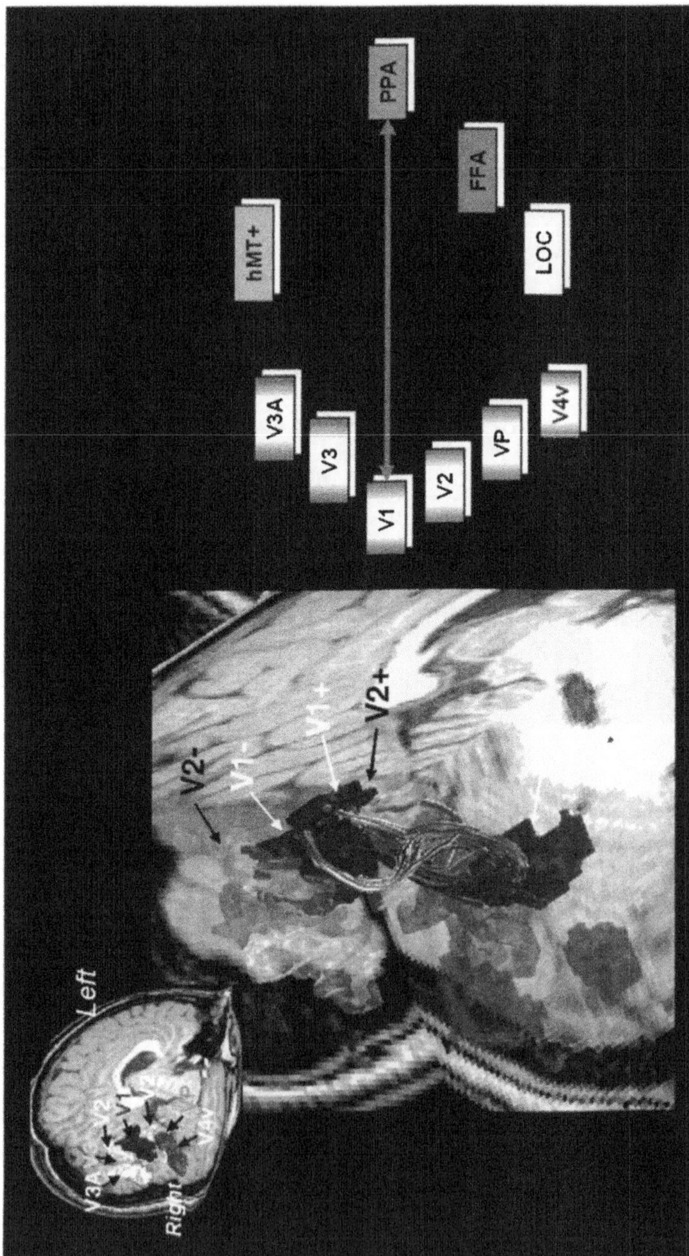

FIGURE 4. Combined fMRI and DTI data for labeling fiber connections between V1 and parahippocampal place area (PPA, Talairach coordinate: 28, −39, −6). Following the initial localization, voxels from retinotopic and nonretinotopic isofunctional areas were used as "seeding" points for DTI-based fiber reconstructions. Note the highly specific nature of the observed connections between these areas.

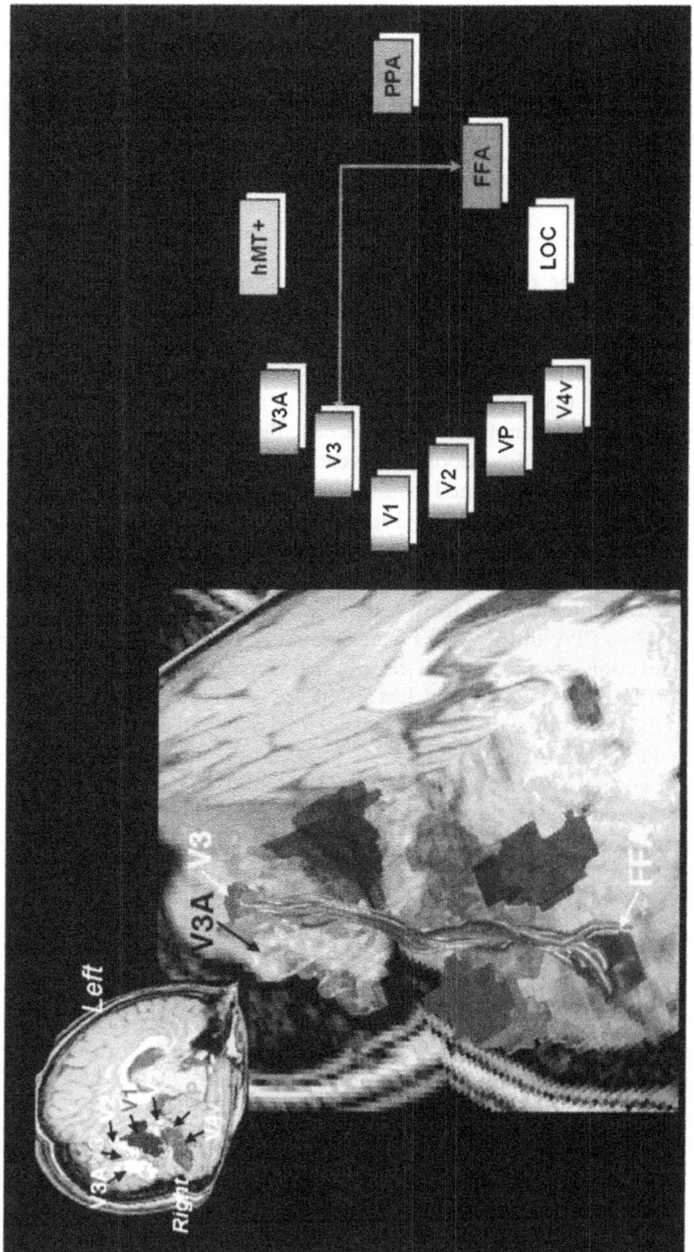

FIGURE 5. Connections between V3 and fusiform face area (FFA, Talairach coordinate: 36, –48, –16). See text for further details.

tracking algorithms were used:[34] that is, minimum FA of 0.2 and 60° of maximum angle. Since fMRI activation is limited to the gray/white matter border, we chose to include at least one more voxel (beyond the gray/white matter boundary) into our seeding ROI. To investigate the connectivity pattern, tracing was performed between two selected ROIs (i.e., V1 and PPA pair, V2 and FFA pair, etc.) for each individual area. Subsequently, corresponding fMRI and DTI tracing data were superimposed on 3D anatomical images for visualization.

RESULTS

FIGURE 3 shows the result of our initial retinotopic mapping studies. Cortical areas responding to isoeccentricity and polar angles were used in order to determine the areal boundaries between the retinotopic areas V1, V2, V3, V3A, VP, and V4v. The surface tangents of the individual areas were volume rendered to yield 3D reconstruction of the respective visual areas.

For high-order nonretinotopic visual areas, the respective "localizer" stimuli were used. Talairach coordinates of the investigated areas were found to be in good agreement with previous published results—hMT+: (46, –58, 4); LOC: (46, –63, 8); FFA: (36, –48, –16); and PPA: (28, –39, –6).

Following the initial localization, voxels from retinotopic and nonretinotopic isofunctional areas were used as "seeding" points for DTI-based fiber reconstructions. FIGURES 4 and 5 demonstrate, from the same subject, the pattern of connections between V1/V2 ↔ PPA and V3 ↔ FFA, respectively. Note the highly specific nature of the observed connections between these areas. Overall, of the functionally defined areas tested (FFA, PPA, LOC, and hMT), PPA was found to be the most highly connected to the retinotopically defined visual areas. Strong connections were found between the PPA and areas VP and V4v. In addition, we also found pronounced connections between PPA and visual areas V1 and V2. In contrast, a significant connection between primary visual areas (V1/V2) and FFA was observed in only one subject (out of four subjects used for this type of analysis). On the other hand, connections to/from FFA originated (or ended) mostly in V3, V4v, and V3A (observed in three out of four subjects analyzed for this study). The most pronounced connection for area LOC was with area V3A (three out of four). hMT was also found to have significant connections to V3A in addition to area V3.

FIGURE 6 shows a cartoon diagram of the observed pattern of connectivities. Note that hMT+ connects with FFA, and FFA in turn connects with LOC. Finally, LOC seems to connect with hMT+, thus completing a looplike pattern of connectivity between these three areas. Most surprisingly, we found that PPA was not a part of this ventral-stream loop between FFA, LOC, and hMT+ (see FIG. 6). Rather, PPA directly connects with retinotopic areas, such as V1/V2, VP, and V4v.

DISCUSSION

The results of our study suggest that high-resolution BOLD MRI and DTI can be obtained from the same cortical tissue *in vivo* at 3-tesla magnetic fields. Furthermore, in our study, the foci of fMRI activation were successfully utilized as seeding

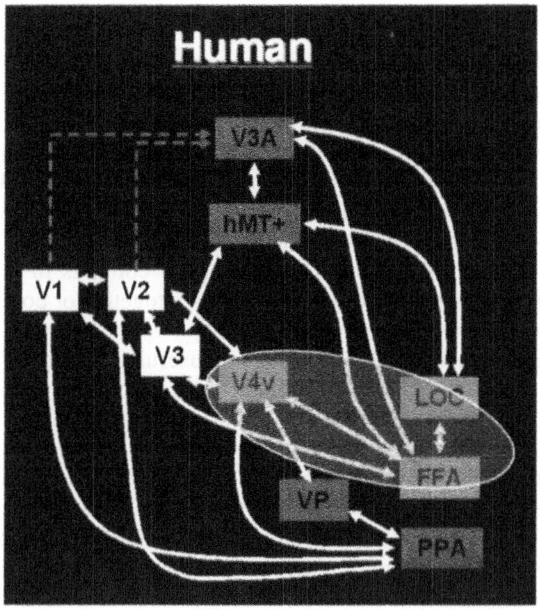

FIGURE 6. Cartoon illustration of the connectivity pattern observed in this study. The diagram shows the reconstructed fiber connections between retinotopic areas and nonretinotopic areas, as well as intrinsic connections between the ventrotemporal areas (FFA, PPA, LOC, and hMT+).

points for 3D DTI fiber reconstruction algorithms, thus providing the map of the axonal circuitry between neuronal populations participating in occipitoventral visual information processing. The results of our preliminary study suggest that the functional organization of the human occipitoventral stream is governed by a distinct pattern of inter-areal connectivities. The areas hMT+, LOC, and FFA are tightly interconnected, thus forming a loop for visual information processing. The principal connections to/from this ventral loop are provided by areas V3 and V3A, while the influence of the primary visual areas V1 and V2 to/from this processing loop is limited. The observed connections between V3/V3A and FFA are consistent with the hierarchical pattern of visual area known from macaques.[40] Consistent with classical tractography work done for the visual system of rhesus macaques, our fMRI/DTI data suggest a linear connectivity relationship between association visual cortex of the occipital lobe and association cortex of the temporal lobe.

Overall, our combined fMRI/DTI studies resulted in an occipitoventral stream connectivity pattern that is highly specific and comparable to homologous data obtained in macaques using well-established neuroanatomical techniques. However, before this new technique can be utilized for addressing clinical and basic neuroscience questions *de novo*, major interpretative issues will have to be addressed. Some of the critical issues are discussed below.

Can We Trust DTI-Based Fiber Reconstructions?

Despite intense research, the structural correlate of DTI remains elusive. For example, the precise contribution of the underlying fiber density and myelination on the anisotropy index has not been completely understood. Thus, it is not clear to what degree the results of DTI correspond to actual density and orientation of the local axonal fiber bundles. It is also important to understand how white matter is, in general, organized. How often do fiber pathways cross and, when they do, what happens? Can fibers change direction sharply in deep white matter? How often do individual fibers branch? When axons descend from columns in gray matter into white matter, what route do they take? Do they remain together or do they intermingle with fibers from more distant regions of cortex?

In discussing the veracity of DTI fibers, it is useful to distinguish between the reliability and the validity of DTI. "Reliability" tells us something about the "goodness" of DTI as a methodological tool, while "validity" is a measure about what we are actually measuring. Obviously, the relevant issues for DTI reliability are the test/retest stability of the measurements, as well as the statistical power of the technique. Unfortunately, there is no accepted statistical test for DTI fiber reconstructions, and thus no p-value that we can attach to the reconstructed fibers. Fiber uncertainty measures[41] and/or generating normalized statistical DTI maps with subsequent probabilistic fiber reconstructions have been suggested as viable ways to increase the reliability of DTI fiber reconstructions. However, unlike conventional fMRI data, averaging multiple DTI data is notoriously difficult as it has to be clarified as to the level that the averaging should be performed: at the level of DWI, tensors, FAs, or resulting fiber reconstructions?

Addressing the issue of DTI validity is even more serious and complicated: not unlike BOLD fMRI until a few years ago, we do not have a clear understanding about what DTI is measuring, nor what is the degree of coregistration between the reconstructed fibers and the underlying actual fiber bundles. Here, combining DTI studies with conventional (e.g., DiI) and/or paramagnetic (e.g., Mn^{2++}) neurotracer injections would be of great benefit.

Selection Criteria for DTI Data

In addition to the reliability and validity problems related to DTI methodologies per se, the combination of DTI with fMRI scans raises a set of unique challenges and problems. The relevant issues are:

(i) How do we extract compact seeding ROIs based on "patchy" BOLD functional activity? What is the appropriate statistical threshold for this? Should all fMRI voxels (above certain threshold) be treated as homologous DTI seeding ROIs regardless of their p-values and/or cross-correlation coefficients?

(ii) Areas of high BOLD activity are located within the cortical gray matter. The cortical gray matter, however, is characterized by low FA (because of the presence of multidirectional fibers/fiber bundles within the imaging voxels in the gray matter). Therefore, most reconstructed fibers will terminate at the gray/white matter boundary. If seeding ROIs were placed exclusively within the fMRI voxels, conventional DTI algorithms will

result in zero or low number of reconstructed fibers. We will thus need to develop an automatic "search" algorithm (this is currently done manually) for the closest fiber termination points that have to be included as part of the seeding ROI.

(iii) We will need to address the question of how the experimentally acquired voxel size should be related to the density of the seeding points. For example, a given $1 \times 1 \times 1$ mm^3 voxel in the BOLD fMRI study may contain hundreds of thousands of physical axonal fibers. For the subsequent DTI fiber reconstruction, should the same voxel be then treated as *one* seeding point? Or should the imaging voxel be "regrided" into a multitude of subseeding points?

In summary, the BOLD-based DTI fiber reconstruction method described in this study allows the local orientation of fiber bundles in the white matter to be determined in an absolutely noninvasive manner, thus enabling *in vivo* neuroanatomy in both animals and humans. The methods developed in this study have the potential to lay a foundation for *in vivo* neuroanatomy and the ability for noninvasive longitudinal studies of brain development.

ACKNOWLEDGMENTS

We thank Mathieu Ducros, Itamar Ronen, Susumu Mori, and Rainer Goebel for their helpful discussions. This work was supported by the NIH (Grant No. NS44825) and the Human Frontiers Science Program.

REFERENCES

1. Ogawa, S. *et al.* 1992. Intrinsic signal changes accompanying sensory stimulation: functional brain mapping with magnetic resonance imaging. Proc. Natl. Acad. Sci. USA **89:** 5951–5955.
2. Bandettini, P.A., E.C. Wong, R.S. Hinks *et al.* 1992. Time course EPI of human brain function during task activation. Magn. Reson. Med. **25:** 390–397.
3. Kwong, K.K. *et al.* 1992. Dynamic magnetic resonance imaging of human brain activity during primary sensory stimulation. Proc. Natl. Acad. Sci. USA **89:** 5675–5679.
4. Ogawa, S., T.M. Lee, A.S. Nayak & P. Glynn. 1990. Oxygenation-sensitive contrast in magnetic resonance image of rodent brain at high magnetic fields. Magn. Reson. Med. **14:** 68–78.
5. Pauling, L. & C.D. Coryell. 1936. The magnetic properties and structures of hemoglobin, oxyhemoglobin, and carbonmonoxyhemoglobin. Proc. Natl. Acad. Sci. USA **22:** 210–216.
6. Thulborn, K.R., J.C. Waterton, P.M. Mattews & G.K. Radda. 1982. Dependence of the transverse relaxation time of water protons in whole blood at high field. Biochim. Biophys. Acta **714:** 265–270.
7. Wagner, A.D. *et al.* 1998. Building memories: remembering and forgetting of verbal experiences as predicted by brain activity. Science **281:** 1188–1191.
8. Kim, S.G. *et al.* 1993. Functional magnetic resonance imaging of motor cortex: hemispheric asymmetry and handedness. Science **261:** 615–617.
9. Sereno, M.I. *et al.* 1995. Borders of multiple visual areas in humans revealed by functional magnetic resonance imaging. Science **268:** 889–893.
10. Tootell, R.B. *et al.* 1997. Functional analysis of V3A and related areas in human visual cortex. J. Neurosci. **17:** 7060–7078.

11. ENGEL, S.A., G.H. GLOVER & B.A. WANDELL. 1997. Retinotopic organization in human visual cortex and the spatial precision of functional MRI. Cereb. Cortex **7:** 181–192.
12. UGURBIL, K., L. TOTH & D-S. KIM. 2003. How accurate is magnetic resonance imaging of brain function? Trends Neurosci. **26:** 108–114.
13. PAYNE, B.R. & A. PETERS. 2002. The Cat Primary Visual Cortex. Academic Press. San Diego.
14. FELLEMAN, D.J. & D.C. VAN ESSEN. 1991. Distributed hierarchical processing in the primate cerebral cortex. Cereb. Cortex **1:** 1–47.
15. UNGERLEIDER, L.G. & J.V. HAXBY. 1994. "What" and "where" in the human brain. Curr. Opin. Neurobiol. **4:** 157–165.
16. GOODALE, M.A. & A.D. MILNER. 1992. Separate visual pathways for perception and action. Trends Neurosci. **15:** 20–25.
17. MALACH, R., I. LEVY & U. HASSON. 2002. The topography of high-order human object areas. Trends Cogn. Sci. **6:** 176–184.
18. GRILL-SPECTOR, K. 2003. The neural basis of object perception. Curr. Opin. Neurobiol. **13:** 159–166.
19. KANWISHER, N., J. MCDERMOTT & M.M. CHUN. 1997. The fusiform face area: a module in human extrastriate cortex specialized for face perception. J. Neurosci. **17:** 4302–4311.
20. TANAKA, K., H. SAITO, Y. FUKADA & M. MORIYA. 1991. Coding visual images of objects in the inferotemporal cortex of the macaque monkey. J. Neurophysiol. **66:** 170–189.
21. DESIMONE, R., T.D. ALBRIGHT, C.G. GROSS & C. BRUCE. 1984. Stimulus-selective properties of inferior temporal neurons in the macaque. J. Neurosci. **4:** 2051–2062.
22. EPSTEIN, R. & N. KANWISHER. 1998. A cortical representation of the local visual environment. Nature **392:** 598–601.
23. HAXBY, J.V. *et al.* 2001. Distributed and overlapping representations of faces and objects in ventral temporal cortex. Science **293:** 2425–2430.
24. ISHAI, A., L.G. UNGERLEIDER, A. MARTIN *et al.* 1999. Distributed representation of objects in the human ventral visual pathway. Proc. Natl. Acad. Sci. USA **96:** 9379–9384.
25. ISHAI, A., L.G. UNGERLEIDER, A. MARTIN & J.V. HAXBY. 2000. The representation of objects in the human occipital and temporal cortex. J. Cogn. Neurosci. **12**(suppl. 2): 35–51.
26. GAUTHIER, I., M.J. TARR, A.W. ANDERSON *et al.* 1999. Activation of the middle fusiform "face area" increases with expertise in recognizing novel objects. Nat. Neurosci. **2:** 568–573.
27. TARR, M.J. & I. GAUTHIER. 2000. FFA: a flexible fusiform area for subordinate-level visual processing automatized by expertise. Nat. Neurosci. **3:** 764–769.
28. PRIBRAM, K. & P. MACLEAN. 1953. Neuronographic analysis of medial and basal cerebral cortex. J. Neurophysiol. **16:** 324–340.
29. WHITLOCK, D.G. & W.J.H. NAUTA. 1956. Subcortical projections from temporal neocortex in *Macaca mulatta*. J. Comp. Neurol. **106:** 183–212.
30. GALUSKE, R.A., W. SCHLOTE, H. BRATZKE & W. SINGER. 2000. Interhemispheric asymmetries of the modular structure in human temporal cortex. Science **289:** 1946–1949.
31. KATZ, L.C. & D.M. IAROVICI. 1990. Green fluorescent latex microspheres: a new retrograde tracer. Neuroscience **34:** 511–520.
32. KISVARDAY, Z.F., D-S. KIM, U.T. EYSEL & T. BONHOEFFER. 1994. Relationship between lateral inhibitory connections and the topography of the orientation map in cat visual cortex. Eur. J. Neurosci. **6:** 1619–1632.
33. CONTURO, T.E. *et al.* 1999. Tracking neuronal fiber pathways in the living human brain. Proc. Natl. Acad. Sci. USA **96:** 10422–10427.
34. BASSER, P.J., S. PAJEVIC, C. PIERPAOLI *et al.* 2000. *In vivo* fiber tractography using DT-MRI data. Magn. Reson. Med. **44:** 625–632.
35. BASSER, P.J., J. MATTIELLO & D. LEBIHAN. 1994. MR diffusion tensor spectroscopy and imaging. Biophys. J. **66:** 259–267.
36. MORI, S. & P.B. BARKER. 1999. Diffusion magnetic resonance imaging: its principle and applications. Anat. Rec. **257:** 102–109.
37. KIM, D-S. *et al.* 2001. Soc. Neurosci. Abstr. 783.3.
38. BRAINARD, D.H. 1997. The Psychophysics Toolbox. Spat. Vision **10:** 433–436.

39. BLANZ, V. & T. VETTER. 1999. *In* Computer Graphics Proceedings SIGGRAPH, pp. 187–194 (Los Angeles).
40. VAN ESSEN, D.C., C.H. ANDERSON & D.J. FELLEMAN. 1992. Information processing in the primate visual system: an integrated systems perspective. Science **255:** 419–423.
41. JONES, D.K. 2003. Determining and visualizing uncertainty in estimates of fiber orientation from diffusion tensor MRI. Magn. Reson. Med. **49:** 7–12.

Investigating the Functional Role of Callosal Connections with Dynamic Causal Models

KLAAS E. STEPHAN,[a] WILL D. PENNY,[a] JOHN C. MARSHALL,[b] GEREON R. FINK,[c,d] AND KARL J. FRISTON[a]

[a]*Wellcome Department of Imaging Neuroscience, Institute of Neurology, University College London, London WC1N 3BG, United Kingdom*

[b]*Neuropsychology Unit, University Department of Clinical Neurology, Radcliffe Infirmary, Oxford OX2 6HE, United Kingdom*

[c]*Institute of Medicine (IME), Research Center Jülich, 52425 Jülich, Germany*

[d]*Department of Neurology/Cognitive Neurology, University of Aachen, 52074 Aachen, Germany*

ABSTRACT: The anatomy of the corpus callosum has been described in considerable detail. Tracing studies in animals and human postmortem experiments are currently complemented by diffusion-weighted imaging, which enables noninvasive investigations of callosal connectivity to be conducted. In contrast to the wealth of anatomical data, little is known about the principles by which interhemispheric integration is mediated by callosal connections. Most importantly, we lack insights into the mechanisms that determine the functional role of callosal connections in a context-dependent fashion. These mechanisms can now be disclosed by models of effective connectivity that explain neuroimaging data from paradigms that manipulate interhemispheric interactions. In this article, we demonstrate that dynamic causal modeling (DCM), in conjunction with Bayesian model selection (BMS), is a powerful approach to disentangling the various factors that determine the functional role of callosal connections. We first review the theoretical foundations of DCM and BMS before demonstrating the application of these techniques to empirical data from a single subject.

KEYWORDS: fMRI; DTI; DCM; effective connectivity; corpus callosum; interhemispheric integration

INTRODUCTION

Ever since the description of localizable lesion and excitation effects in the nineteenth century, modern neuroscience has revolved around the twin themes of functional specialization and functional integration.[1,2] Functional specialization refers to the notion that local neural units are specialized in certain aspects of information processing, for example, the processing of particular stimulus properties. Traditional

Address for correspondence: Klaas E. Stephan, Wellcome Department of Imaging Neuroscience, Institute of Neurology, University College London, 12 Queen Square, London WC1N 3BG, United Kingdom. Voice: +44-207-8337481; fax: +44-207-8131420.
k.stephan@fil.ion.ucl.ac.uk

Ann. N.Y. Acad. Sci. 1064: 16–36 (2005). © 2005 New York Academy of Sciences.
doi: 10.1196/annals.1340.008

methods for investigating functional specialization include invasive recordings from animals and neuropsychological investigations of patients with brain lesions. More recently, functional neuroimaging techniques like functional magnetic resonance imaging (fMRI) have made it possible to investigate functional specialization across the whole brain in a noninvasive manner. In contrast, functional integration refers to the causal interactions among distinct neural units. Critically, the form of these causal interactions, which mediate complex cognitive processes, is constrained by the anatomical connections between the neural units. Consequently, in order to understand the basis of functional integration, much effort has been invested in characterizing anatomical connectivity in the mammalian brain. For example, thousands of tract tracing experiments have been performed in several species over the last decades. These techniques require the *in vivo* injection of specific dyes into particular areas. Depending on its biophysical properties, the dye is taken up by neuronal somata or axonal terminals and transported in an anterograde or retrograde direction, respectively.[3] Subsequent histological processing of the brain can then reveal the regions that receive connections from (anterograde tracer) or send connections to (retrograde tracer) the injected area. For the macaque monkey alone, more than 36,000 individual experimental findings from tracing experiments are described by the connectivity database CoCoMac[4] (see www.cocomac.org).

Unfortunately, for the human brain, we have considerably less knowledge about its anatomical connections. This is because tract tracing procedures, as the gold-standard method to reveal anatomical connections, are too invasive to use in the human brain. There has been an intensive search for postmortem methods as an alternative to investigate human brain connectivity, but these methods are either restricted to very large fiber bundles[5] or limited to very short intracortical connections.[6]

Recently, the advent of noninvasive diffusion-weighted imaging (DWI) techniques, for example, diffusion tensor imaging (DTI), has raised great hopes that we may be able to obtain a complete picture of human brain connectivity in the not too distant future. However, the resolution of current DWI approaches is still too coarse to allow for characterizations that are comparable to those obtained from tracing techniques, and fundamental problems like intravoxel fiber crossings still need to be solved convincingly. Using improved acquisition schemes, probabilistic approaches,[7,8] and models that do not make strong a priori assumptions about the shape of the spatial distribution of diffusion,[9] these technical limitations might eventually be overcome. Should this indeed be possible, would this mean that we have all the information we need to understand principles of functional integration in the human brain? The answer is, unfortunately, no. Even if we knew everything about the anatomical connectivity of a particular neural system, we cannot directly derive its dynamics from its connectional structure.[10] Further knowledge is required, for example, of the time constants of activity propagation or the strength of individual connections and how these change as a function of cognitive context (task requirements, learning, etc.).[11] These time constants and connection strengths are parameters that have to be estimated from empirical observations. In conclusion, knowledge of anatomical connectivity is a necessary, but not sufficient, condition to build dynamical models of brain function.

One good example is the corpus callosum. This massive fiber bundle, which contains a huge number of axons that link and functionally integrate the two hemispheres, has been investigated in great detail. We know the source and target laminae of neurons projecting through the corpus callosum,[12] the spatial distribution of axonal diameters across the callosum,[13] and the topography of individual callosal projections.[14,15] Neuropsychological studies have demonstrated the involvement of very restricted parts of the corpus callosum in specific cognitive processes.[16] More recently, the corpus callosum has been studied intensively by DWI studies that have investigated its connectivity in both healthy subjects[17] and patients.[18] Yet, in spite of all these data, we still lack any powerful theory of callosal function and how it underlies interhemispheric integration. Banich and colleagues have recently advanced a useful framework that relates the functional role of the corpus callosum to the complexity of cognitive tasks and attentional processing,[19] but this theory is neither quantitative nor directly embedded into a precise neurobiological model.

An important starting point for more precise theories of callosal function would be to investigate neurobiologically plausible models of interhemispheric integration in order to (i) identify those cognitive factors that determine the functional role of specific callosal connections, (ii) determine quantitatively the strength of callosal connections as a function of experimentally controlled cognitive context, and (iii) analyze the pattern of context-dependent connection strengths with regard to interesting features, for example, directional asymmetries or differences in modulation by different cognitive factors. Models of effective connectivity, which are based on empirical neuroimaging data and model the modulation of connection strengths by experimentally controlled changes in context, would be ideal to do this. Previously available techniques for studying effective connectivity, for example, structural equation modeling (SEM), are not well suited for models of high connectional complexity, for example, multiple reciprocal connections and loops, due to potential problems of identifiability.[20] This is a particular problem for models of interhemispheric integration because callosal connections appear to be generally reciprocal between homotopic regions.[21,22] Furthermore, SEM operates at the level of measured hemodynamic responses and does not offer a model of the underlying neural processes.

In this article, we demonstrate how a novel method to study effective connectivity, dynamic causal modeling (DCM[11]), can be combined with Bayesian model selection (BMS[23]) to address quite complex questions about callosal function. First, we briefly review the theoretical foundations of both DCM and BMS. Subsequently, we apply DCM to fMRI data from a single subject who performed a task that manipulated interhemispheric interactions. Specifically, we focus on the question of how competing hypotheses about the functional role of callosal connections in a specific cognitive context can be disambiguated using BMS.

METHODS

Dynamic Causal Modeling (DCM)

DCM is a method to make inferences about neural processes that underlie measured time series—in our case, fMRI data. The general idea is to estimate the parameters of a reasonably realistic neuronal system model such that the predicted blood oxygen level–dependent (BOLD) signal, which results from converting the modeled neural dynamics into hemodynamic responses, corresponds as closely as possible to the

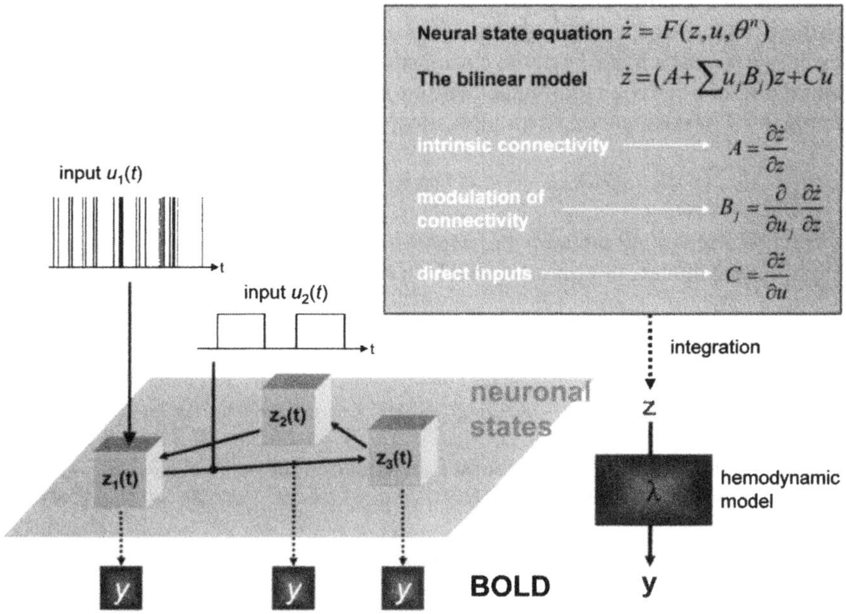

FIGURE 1. Schematic summary of the conceptual basis of DCM. The dynamics in a system of interacting neuronal populations (*left lower panel*), which are not directly observable by fMRI, is modeled using a bilinear state equation (*right upper panel*). Integrating the state equation gives predicted neural dynamics (z) that enter a model of the hemodynamic response (λ) to give predicted BOLD responses (y) (*right lower panel*). The parameters at both neural and hemodynamic levels are adjusted such that the differences between predicted and measured BOLD series are minimized. Critically, the neural dynamics are determined by experimental manipulations. These enter the model in the form of external inputs (*left upper panel*). Driving inputs (u_1; e.g., sensory stimuli) elicit local responses directly that are propagated through the system according to the intrinsic connections. The strengths of these connections can be changed by modulatory inputs (u_2; e.g., changes in cognitive set, attention, or learning). In this figure, the structure of the system and the scaling of the inputs are arbitrary.

observed BOLD time series. As in state-space models, two distinct levels constitute a DCM (see FIG. 1). The hidden level, which cannot be directly observed using fMRI, represents a simple model of neural dynamics in a system of k coupled brain regions. Each system element i is represented by a single state variable z_i, and the dynamics of the system is described by the change of the neural state vector $z = [z_1, \ldots, z_k]^T$ over time. The neural state variables do not correspond directly to any common neurophysiological measurement (such as spiking rates or local field potentials), but represent a summary index of neural population dynamics in the respective regions. Importantly, DCM models how the neural dynamics are driven by external perturbations that result from experimentally controlled manipulations. These perturbations are described by means of external inputs u that enter the model in two different ways: they can elicit responses through direct influences on specific regions ("driving" inputs, e.g., evoked responses in early sensory areas) or they can change the strength of coupling among regions ("modulatory" inputs, e.g., during learning or attention). Overall, DCM models the temporal evolution of the neural state vector, that is, $\dot{z} = dz/dt$, as a function of the current state, the inputs u, and some parameters θ^n that define the functional architecture and interactions among brain regions at a neuronal level (n is not an exponent, but simply denotes "neural"):

$$[\dot{z}_1 \ldots \dot{z}_k] = \dot{z} = F(z, u, \theta^n). \tag{1}$$

In this neural state equation, the state z and the inputs u are time-dependent, whereas the parameters are time-invariant. In DCM, F has the bilinear form

$$\dot{z} = Az + \sum_{j=1}^{m} u_j B_j z + Cu. \tag{2}$$

The parameters of this bilinear neural state equation, $\theta^n = \{A, B_1, \ldots, B_m, C\}$, can be expressed as partial derivatives of F:

$$\begin{aligned} A &= \partial F/\partial z = \partial \dot{z}/\partial z \\ B_j &= \partial^2 F/\partial z \partial u_j = (\partial/\partial u_j)(\partial \dot{z}/\partial z) \\ C &= \partial F/\partial u. \end{aligned} \tag{3}$$

These parameter matrices describe the nature of the three causal components that underlie the modeled neural dynamics: (i) context-independent effective connectivity among brain regions, mediated by anatomical connections ($k \times k$ matrix A); (ii) context-dependent changes in effective connectivity induced by the j-th input u_j ($k \times k$ matrices $B_1, \ldots, B_m$); and (iii) direct inputs into the system that drive regional activity ($k \times m$ matrix C). As will be demonstrated below, the posterior distributions of these parameters can inform us about the impact that different mechanisms have on determining the dynamics of the model. Notably, the distinction between "driving" and "modulatory" is neurobiologically relevant: driving inputs exert their effects through direct synaptic responses in the target area, whereas modulatory inputs change synaptic responses in the target area in response to inputs from another area. This distinction represents an analogy, at the level of large neural populations, to the concept of driving and modulatory afferents in studies of single neurons.[24]

DCM combines this model of neural dynamics with a biophysically plausible and experimentally validated hemodynamic model that describes the transformation of neuronal activity into a BOLD response. This so-called "balloon model" was initially formulated by Buxton and colleagues[25] and later extended by Friston *et al.*[26,27] Briefly summarized, it consists of a set of differential equations that describe the relations between four hemodynamic state variables, using five parameters (θ^h). More specifically, changes in neural activity elicit a vasodilatory signal that leads to increases in blood flow and subsequently to changes in blood volume v and deoxyhemoglobin content q. The predicted BOLD signal y is a nonlinear function of blood volume and deoxyhemoglobin content: $y = \lambda(v, q)$. Details of the hemodynamic model can be found in other publications.[11,26,27]

By combining the neural and hemodynamic states into a joint state vector x and the neural and hemodynamic parameters into a joint parameter vector $\theta = [\theta^n \ \theta^h]^T$, we obtain the full forward model that is defined by the neural and hemodynamic state equations

$$\begin{aligned} \dot{x} &= F(x, u, \theta) \\ y &= \lambda(x). \end{aligned} \tag{4}$$

For any given set of parameters θ and inputs u, the joint state equation can be integrated and passed through the output nonlinearity λ to give a predicted BOLD response $h(u, \theta)$. This can be extended to an observation model that includes observation error ε and confounding effects X (e.g., scanner-related low-frequency drifts):

$$y = h(u, \theta) + X\beta + \varepsilon. \tag{5}$$

This formulation is the basis for estimating the neural and hemodynamic parameters from the measured BOLD data, using a fully Bayesian approach with empirical priors for the hemodynamic parameters and conservative shrinkage priors for the neural coupling parameters. Details of the parameter estimation scheme, which rests on a Gauss-Newton gradient ascent embedded in an expectation maximization (EM) algorithm, can be found elsewhere.[11] In brief, under Gaussian assumptions about the posterior distributions (Laplace approximation), this scheme returns the posterior expectations (= maximum a posteriori [MAP] estimates) $\eta_{\theta|y}$ and posterior covariance $C_{\theta|y}$ for the parameters as well as hyperparameters for the covariance of the observation noise, C_ε.

After fitting the model to measured BOLD data, the posterior distributions of the parameters can be used to test hypotheses about the size and nature of effects at the neural level. Although inferences could be made about any of the parameters in the model, hypothesis testing usually concerns context-dependent changes in coupling (i.e., specific parameters from the B matrices). As will be demonstrated below, at the single-subject level, these inferences concern the question of how certain one can be that a particular parameter or, more generally, a contrast of parameters, $c^T\eta_{\theta|y}$, exceeds a particular threshold γ (e.g., zero; see FIG. 6 later). Under the assumptions of the Laplace approximation, this is easy to test (ϕ_N denotes the cumulative normal distribution):

$$p(c^T\eta_{\theta|y} > \gamma) = \phi_N[(c^T\eta_{\theta|y} - \gamma)/(\sqrt{c^T C_{\theta|y} c})]. \tag{6}$$

For example, for the special case $c^T\eta_{\theta|y} = \gamma$, the probability is $p(c^T\eta_{\theta|y} > \gamma) = 50\%$, that is, it is equally likely that the parameter is smaller or larger than the chosen threshold γ.

We conclude this section on the theoretical foundations of DCM by noting that the parameters can be understood as rate constants (units: 1/s = Hz) of neural population responses that have an exponential nature (see also figure 1 in ref. 28). This is easily understood if one considers that the solution to a linear ordinary differential equation of the form $\dot{z} = Az$ is an exponential function (compare the state equation in equation 2).

Bayesian Model Selection (BMS)

A generic problem encountered by any kind of modeling approach is the question of model selection: given some observed data, which of several alternative models is the optimal one? This problem is not trivial because the decision cannot be made solely by comparing the relative fit of the competing models. One also needs to take into account the relative complexity of the models as expressed, for example, by the number of free parameters in each model. Model complexity is important to consider because there is a trade-off between model fit and generalizability (i.e., how well the model explains different data sets that were all generated from the same underlying process). As the number of free parameters is increased, model fit increases monotonically, whereas beyond a certain point model generalizability decreases. The reason for this is "overfitting": an increasingly complex model will, at some point, start to fit noise that is specific to one data set and thus become less generalizable across multiple realizations of the same underlying generative process. (Generally, in addition to the number of free parameters, the complexity of a model also depends on its functional form.[29] This is not an issue for DCM, however, because here all possible models have the same functional form.)

Therefore, the question, "what is the optimal model?", can be reformulated more precisely as "what is the model that represents the best balance between fit and complexity?". In a Bayesian context, the latter question can be addressed by comparing the evidence, $p(y|m)$, of different models. According to Bayes theorem

$$p(\theta|y,m) = [p(y|\theta,m)p(\theta|m)]/p(y|m), \tag{7}$$

the model evidence can be considered as a normalization constant for the product of the likelihood of the data and the prior probability of the parameters; therefore,

$$p(y|m) = \int p(y|\theta,m)p(\theta|m)\, d\theta. \tag{8}$$

Here, the number of free parameters (as well as the functional form) are considered by the integration. Unfortunately, this integral cannot usually be solved analytically; therefore, an approximation to the model evidence is needed.

In the context of DCM, one potential solution could be to make use of the Laplace approximation, that is, to approximate the model evidence by a Gaussian that is centered on its mode. As shown by Penny *et al.*,[23] this yields the following expression for the natural logarithm (ln) of the model evidence ($\eta_{\theta|y}$ denotes the MAP estimate, $C_{\theta|y}$ is the posterior covariance of the parameters, C_ε is the error covariance, θ_p is the prior mean of the parameters, and C_p is the prior covariance):

$$\begin{aligned} \ln p(y|m) &= accuracy(m) - complexity(m) \\ &= \{(-1/2)\ln|C_\varepsilon| - (1/2)[y - h(u, \eta_{\theta|y})]^T C_\varepsilon^{-1}[y - h(u, \eta_{\theta|y})]\} - \\ &\quad [(1/2)\ln|C_p| - (1/2)\ln|C_{\theta|y}| + (1/2)(\eta_{\theta|y} - \theta_p)^T C_p^{-1}(\eta_{\theta|y} - \theta_p)]. \end{aligned} \tag{9}$$

This expression properly reflects the requirement, as discussed above, that the optimal model should represent the best compromise between model fit (accuracy) and model complexity. The complexity term depends on the prior density, for example, the prior covariance of the intrinsic connections (see eq. 9). This is problematic in the context of DCM for fMRI because this prior covariance is defined in a model-specific fashion to ensure that the probability of obtaining an unstable system is very small. (Specifically, this is achieved by choosing the prior covariance of the intrinsic coupling matrix A such that the probability of obtaining a positive Lyapunov exponent of A is $P < 0.001$; see Friston *et al.*[11] for details.) Consequently, one cannot easily compare models with different numbers of connections. Therefore, alternative approximations to the model evidence are useful for DCMs of this sort.

Suitable approximations, which do not depend on the prior density, are afforded by the Bayesian information criterion (BIC) and Akaike information criterion (AIC), respectively. As shown by Penny *et al.*,[23] these approximations for DCM are given by

$$\begin{aligned} \mathrm{BIC} &= accuracy(m) - (d_\theta/2)\ln N \\ \mathrm{AIC} &= accuracy(m) - d_\theta, \end{aligned} \qquad (10)$$

where d_θ is the number of parameters and N is the number of data points (scans). If one compares the complexity terms of BIC and AIC, it becomes obvious that BIC pays a heavier penalty than AIC as soon as one deals with 8 or more scans (which is virtually always the case for fMRI data):

$$(d_\theta/2)\ln N > d_\theta \Rightarrow N > e^2 \approx 7.39. \qquad (11)$$

Therefore, BIC will be biased towards simpler models, whereas AIC will be biased towards more complex models. This can lead to disagreement between the two approximations about which model should be favored. We have thus adopted the convention that, for any pairs of models m_i and m_j to be compared, a decision is only made if AIC and BIC concur; the decision is then based on that approximation that gives the smaller *Bayes factor* (BF):

$$\mathrm{BF}_{ij} = p(y|m_i)/p(y|m_j). \qquad (12)$$

This approach to BMS is a robust procedure to decide between competing hypotheses represented by different DCMs. These hypotheses can concern any part of the structure of the modeled system, for example, the pattern of intrinsic connections or which inputs affect the system and where they enter. Below, we will show a concrete example that demonstrates how the combination of DCM and BMS can be applied in practice to disclose previously unknown principles of interhemispheric integration.

Some Considerations on the Study of Interhemispheric Integration

Interhemispheric integration appears to be an ongoing process that is invoked by any cognitive task and cannot be abolished voluntarily. The question for the experimentalist is therefore not how to induce or prevent interhemispheric integration, but rather how to alter the form it takes. Two experimental manipulations are particularly effective for so doing. First, there are various ways of delivering sensory stimuli such that one hemisphere is initially or preferentially affected by these stimuli. For example, due to the topography of the anatomical connections from the retina to the visual cortex, presentation of visual stimuli in the periphery of one visual hemifield ensures that the contralateral visual cortex receives the stimulus information first. Therefore, one knows that any area that is in the hemisphere ipsilateral to stimulus presentation can only receive this information if it is transferred through the corpus callosum (or some alternative, e.g., subcortical, commissure). Second, if one uses a strongly lateralized task that draws on easily identified areas, one knows the target hemisphere and the target areas that require the stimulus information.

Combining both approaches allows one to predict precisely where the stimulus information initially enters the system and to which areas of the system it must be transferred. The challenge is then to characterize the potential paths of information flow and how these pathways are modulated by cognitive context. Rephrasing the question about the nature of interhemispheric integration in this way shows that we are dealing with a generic system identification problem that, in the context of neuroimaging, is best addressed using models of effective connectivity.[1,30] Unfortunately, very few neuroimaging studies have been conducted so far that directly tackle the question of interhemispheric integration on the basis of a precisely defined system model (see McIntosh *et al.*[31] for an exception).

In the next sections, we will provide an example for using DCM and BMS to address the mechanisms underlying interhemispheric integration. This example focuses on the ventral stream of the visual system and is motivated by a recent study that combined the two experimental manipulations described above, that is, using two strongly and inversely lateralized tasks operating on identical visual stimuli that were presented peripherally in the visual hemifields.[32]

Interhemispheric Integration in the Ventral Stream of the Visual System

In a previous fMRI study on the mechanisms underlying hemispheric specialization, we investigated whether lateralization of brain activity depends on the nature of the sensory stimuli or on the nature of the cognitive task performed.[32] For example, microstructural differences between homotopic areas in the left and right hemisphere have been reported, including visual[33] and language-related[34] areas. Within a given hemisphere, these differences could favor the processing of certain stimulus characteristics and disadvantage others and might thus support stimulus-dependent lateralization in a bottom-up fashion.[35] On the other hand, processing demands, mediated through cognitive control processes, might determine in a top-down fashion which hemisphere obtains precedence in a particular task context.[36,37] To decide between these two possibilities, we used a paradigm in which the stimuli were kept constant throughout the experiment, and subjects were alternately instructed to attend to certain stimulus features and ignore others.[32] The stimuli were concrete German nouns (each four letters in length) in which either the second or third letter was printed in red (the other letters were black). In a letter decision (LD) task, the subjects had to ignore the position of the red letter and indicate whether or not the word contained the target letter “A”. In a spatial decision (SD) task, they were required to ignore the language-related properties of the word and to judge whether the red letter was located left or right of the word center: 50% of the stimuli were presented in the non-foveal part of the right visual field (RVF) and the other 50% in the nonfoveal part of the left visual field (LVF).

The results of a conventional fMRI data analysis were clearly in favor of the top-down hypothesis: despite the use of identical word stimuli in all conditions, comparing spatial to letter decisions showed strongly right-lateralized activity in the parietal

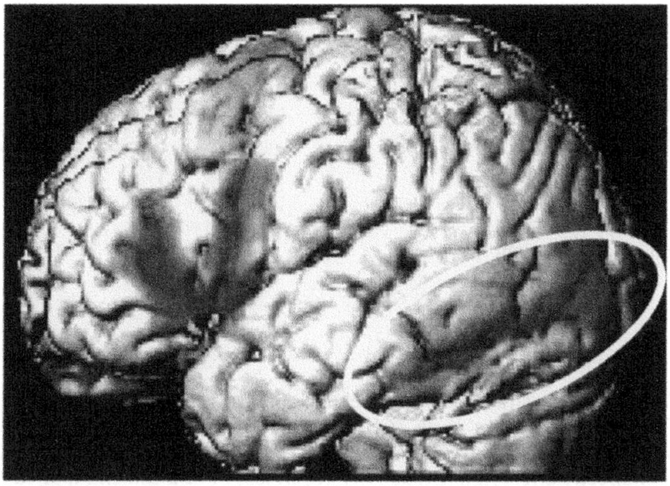

FIGURE 2. Results from a conventional analysis of the fMRI data by Stephan *et al.*[32] using SPM99. Comparing letter decisions to visuospatial decisions about identical stimuli showed strongly left-lateralized activity, including classical language areas in the left inferior frontal gyrus and visual areas in the left ventral visual stream (*white ellipse*), for example, in the fusiform gyrus, middle occipital gyrus, and lingual gyrus. Results are shown at $P < 0.05$, corrected at the cluster level for multiple comparisons across the whole brain. Adapted with permission from figure 1 in reference 32.

cortex, whereas comparing letter to visuospatial decisions showed strongly left-lateralized activity, including classical language areas in the left inferior frontal gyrus and visual areas in the left ventral visual stream, for example, in the fusiform gyrus (FG), middle occipital gyrus (MOG), and lingual gyrus (LG) (see FIG. 2).

We now want to demonstrate how one can use DCM to investigate interhemispheric interactions with this paradigm. We focus on the ventral stream of the visual system, which, as shown in FIGURE 2, is preferentially involved in letter decisions in this experiment. For simplicity, we initially omit MOG and concentrate on LG and FG. First, we need to define a model comprising these four areas (FIG. 3A). To start with the direct (driving) inputs to the system, we model the lateral stimulus presentation and the crossed course of the visual pathways by allowing all RVF stimuli to directly affect left LG activity and all LVF stimuli to directly affect right LG activity, regardless of task. Each stimulus lasted for 150 ms only; therefore, these inputs are represented as trains of short events (delta functions). The induced activity then spreads through the system according to the intrinsic connections of the model. For visual areas, it is biologically plausible to assume that both the intra- and the interhemispheric connections are reciprocal and that homotopic regions in both hemispheres are linked by interhemispheric connections.[14,21,22,38,39]

Note that up to this point there are few, if any, plausible alternatives for how a DCM of interhemispheric integration between LG and FG, respectively, should be constructed. The important question, however, is how transcallosal information

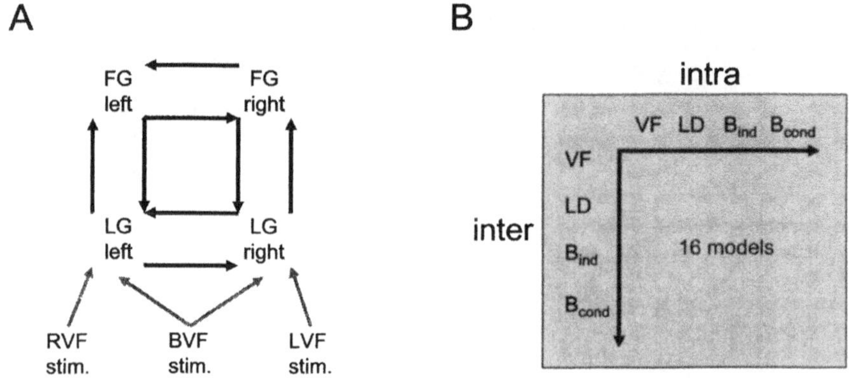

FIGURE 3. **(A)** Basic structure of a model that comprises the left and right lingual gyrus (LG) and left and right fusiform gyrus (FG). The areas are reciprocally connected (*black arrows*). Driving inputs are shown as *gray arrows*. RVF stimuli directly affect left LG activity and LVF stimuli directly affect right LG activity, regardless of task context. Individual stimuli lasted for 150 ms only; therefore, these inputs are represented as trains of events (delta functions). During the instruction periods, bilateral visual field input was provided for 6 s; this was modeled as a box-car input affecting LG in both hemispheres. **(B)** Schema showing how the 16 models tested were constructed by systematically combining four different types of modulatory inputs for inter- and intrahemispheric connections; these different inputs are described by FIGURE 4.

transfer is regulated by cognitive set. For example, one could assume that the strengths of interhemispheric interactions between visual areas are merely determined by the visual field of stimulus presentation, regardless of what the subject is instructed to do with the stimulus: whenever a stimulus is presented in the LVF, for example, and stimulus information is thus received initially by the right visual cortex, this information is transmitted transcallosally to the left visual cortex. Vice versa, whenever a stimulus is presented in the RVF, stimulus information is transmitted transcallosally from left to right visual cortex. In this scenario, the task performed is assumed to have no influence on callosal couplings. In contrast to this notion, results from previous analyses of these data by simple models of effective connectivity have indicated the importance of task demands on modulating functional couplings *within* hemispheres.[32] If this finding extends to interhemispheric interactions, one might expect that callosal connection strengths depend more on which task is performed than on which visual field the stimulus is presented in. That is, both right→left and left→right callosal connections could be enhanced during the LD task, enabling a tight cooperation of the two hemispheres during the task. As a third hypothesis, it is conceivable that both visual field and task exert an influence on callosal connection strengths, but *independently* of each other. As a fourth and final option, one might postulate that task demands modulate callosal connections, but are *conditional* on the visual field, that is, right→left connections are only modulated by LD during LVF stimulus presentation (LD|LVF), whereas left→right connections are only modulated by LD during RVF stimulus presentation (LD|RVF).

Each of these hypotheses of how cognitive set may modulate the callosal connections represents a different DCM, describing the mechanisms that caused the observed

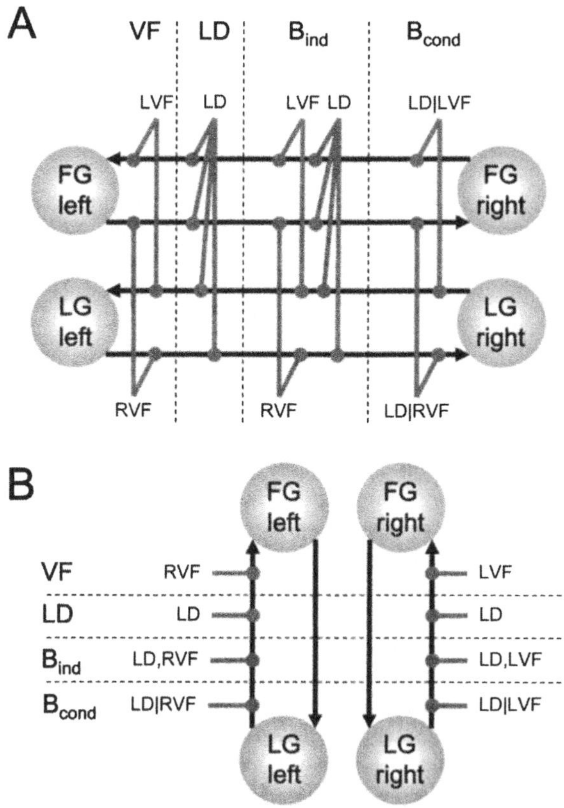

FIGURE 4. This figure describes four competing hypotheses about which experimental factors determine the strength of interhemispheric connections **(A)** and intrahemispheric connections **(B)**, respectively. These different types of modulatory inputs were combined to give 16 different models (compare FIG. 3B). In contrast to the driving inputs (FIG. 3A), which were modeled as streams of events, all modulatory inputs were modeled as extended processes, that is, box-car inputs of 24-s duration.

data. This model selection problem can be addressed by means of BMS. Briefly summarized, the four competing hypotheses are the following (see FIG. 4A):

- *(i)* Information transfer between hemispheres depends only on the visual field of stimulus presentation. This is referred to as the **VF model**.
- *(ii)* Information transfer between hemispheres depends only on whether the letter decision task is performed or not. This is the **LD model**.
- *(iii)* Information transfer between hemispheres depends on *both* the task and the visual field, but *independently* of each other (corresponding to a Boolean OR operation). This is the $\mathbf{B_{ind}}$ **model**.
- *(iv)* Information transfer between hemispheres depends on *both* the task and the visual field, but in a *conditional* fashion: modulation of connection strength by task is only present if the stimulus was presented in a

particular visual field (corresponding to a Boolean AND operation). This is the **B_{cond} model**.

Although less interesting in the present context, the same questions about the nature of modulatory inputs arise with respect to the intrahemispheric connections. Therefore, to perform a thorough model comparison, one needs to systematically compare all combinations of how inter- and intrahemispheric connections are changed in the four ways described above (note that, in the models presented here, we only allowed for modulation of the intrahemispheric connections from LG→FG, but not from FG→LG; see FIG. 4B). FIGURE 3B summarizes the combinatorial logic that resulted in 16 different models that were fitted to the same data. In the following, we refer to these 16 models by first listing the modulation of the inter- and then that of the intrahemispheric connections. For example, LD–VF is the model where the callosal connections are modulated by the letter decision task and the intrahemispheric connections are modulated by the visual field.

Once the best model and thus the factors that most strongly determine transcallosal information transfer in the context of the present task are identified, we can use the posterior density of the parameters of that model to characterize mechanisms of interhemispheric interactions. For example, we can attempt to clarify at what stage of the ventral stream contextual modulation of callosal connections is present. With the exception of some EEG studies,[40,41] which have rather low spatial resolution, this is a largely unexplored issue. Of even more interest, however, is whether the contextual modulation of callosal connections is asymmetric, that is, stronger for right→left connections than left→right connections or vice versa, and whether this asymmetry generalizes across the visual system or is specific to particular connections. This is a new dimension of hemispheric asymmetry that goes beyond the classical characterization of hemispheric specialization in terms of lateralized local activations and is directly related to the functional role of individual callosal "channels".

RESULTS

Here, we report the results from fitting the 16 DCMs described above to the fMRI data of a single subject from the study by Stephan *et al.*[32] We initially present a four-area model as shown in FIGURE 3, that is, comprising bilateral LG and FG, and subsequently extend the model to include the MOG in both hemispheres.

Starting with the four-area case, the BMS procedure indicated that, for the particular subject studied, the model that represented the best balance between model fit and model complexity was the B_{cond}–LD model, that is, modulation of interhemispheric connections by the letter decision task conditional on the visual field of stimulus presentation and modulation of intrahemispheric connection strengths by the task only. TABLE 1 shows the Bayes factors for the comparison of the B_{cond}–LD model with the other 15 models. The AIC and BIC approximations agreed for all comparisons. The second-best model was the LD–B_{cond} model (i.e., the "flipped" version of the B_{cond}–LD model). The Bayes factor of comparing the B_{cond}–LD with the LD–B_{cond} model was only 2.33, which according to the criteria summarized by Penny *et al.*[23] (see their table 1) could be interpreted as weak evidence in favor of the B_{cond}–LD model. All other comparisons gave Bayes factors larger than 3, representing positive, strong, or very strong evidence in favor of the B_{cond}–LD model (see TABLE 1).

TABLE 1. Bayes factors (BF) for comparison of the best model (B_{cond}–LD) with each of the other 15 models

B_{cond}–LD vs.	BF	Evidence in favor of B_{cond}–LD
B_{ind}–B_{ind}	477.31	Very strong
B_{ind}–B_{cond}	60.83	Strong
B_{ind}–LD	110.84	Strong
B_{ind}–VF	479.03	Very strong
B_{cond}–B_{ind}	3.92	Positive
B_{cond}–B_{cond}	4.48	Positive
B_{cond}–VF	46267.47	Very strong
LD–B_{ind}	19.96	Positive
LD–B_{cond}	2.33	Weak
LD–LD	3.43	Positive
LD–VF	29.74	Strong
VF–B_{ind}	16.85	Positive
VF–B_{cond}	4.81	Positive
VF–LD	5.59	Positive
VF–VF	1.35E+13	Very strong

NOTE: This table shows the Bayes factors (BF) (*middle column*) for the comparison of the best model (B_{cond}–LD) with each of the other 15 models (*left column*). The *right column* lists the interpretation of the evidence in favor of the B_{cond}–LD model according to the criteria summarized by Penny *et al.*[23] (see their table 1).

FIGURE 5 shows the MAP estimates of the modulatory parameters (± standard deviation, i.e., the square root of the posterior variances) for the B_{cond}–LD model. The numerical values of the modulatory parameter estimates indicated an obvious hemispheric asymmetry: at the levels of both LG and FG, the MAP estimates of the modulation of the right→left connections are much larger than those of the left→right connections. However, how secure is our inference about this asymmetry? This issue can be addressed by means of contrasts of the appropriate parameter estimates. The contrasts comparing modulation of the right→left connection by LD|LVF versus modulation of the left→right connection by LD|RVF are shown in FIGURE 6 (separately for connections at the levels of LG and FG, respectively). These plots indicate our certainty about asymmetrical modulation of callosal interactions through the probability that these contrasts exceed a value of zero (see eq. 6). For the particular subject shown here, we can be very certain (98.7%) that modulation of the right LG → left LG connection by LD|LVF (0.34 ± 0.14 Hz) is larger than the modulation of the left LG → right LG connection by LD|RVF (–0.08 ± 0.16 Hz) (compare FIGS. 4 and 5). Although there is also a clear difference in the MAP estimates of the modulatory parameters of the callosal connections at the level of the FG, the difference is smaller and the variance of these estimates is larger (0.13 ± 0.19 Hz vs. 0.01 ± 0.17 Hz). Consequently, for the specific subject studied, we have little confidence (68.0%) that the asymmetry observed for the LG connections also exists for callosal connections between right and left FG. Similarly, even though the MAP estimates indicated that the modulation of the intrahemispheric LG→FG connection by task demands was larger in the left hemisphere (0.44 ± 0.14 Hz) than in the right

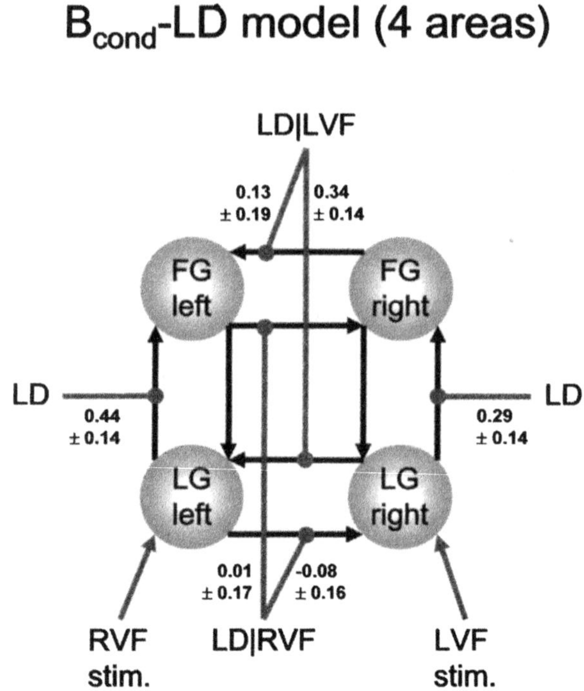

FIGURE 5. This figure shows the maximum a posteriori (MAP) estimates of the parameters (± square root of the posterior variances; units: 1/s = Hz) for the B_{cond}–LD model, which for the particular subject studied proved to be the best of all 16 models tested. For clarity, only the parameters of interest, that is, the modulatory parameters of inter- and intrahemispheric connections, are shown and the bilateral visual field input has been omitted.

hemisphere (0.29 ± 0.14 Hz), we only have a very modest certainty (75.2%) about the presence of this form of connectional asymmetry.

Since the difference between the best and the second-best model was not huge (see above), we also investigated the parameter estimates for the LD–B_{cond} model. In this model, task demands alone, independent of visual field, determined the strength of callosal connections. It was pleasing to find that this model gave compatible results in terms of the asymmetrical modulation of callosal connections. Here, there was a 95.6% confidence that modulation of the right LG → left LG connection by LD (0.43 ± 0.15 Hz) was larger than the equivalent modulation of the left LG → right LG connection by LD (0.03 ± 0.16 Hz). As with the B_{cond}–LD model, we were considerably less certain (67.7%) about this asymmetry for the callosal connections at the level of FG (0.31 ± 0.18 Hz vs. 0.19 ± 0.17 Hz).

Finally, we extended the B_{cond}–LD model to include left and right MOG. Fitting this six-area model to the data gave estimates that were only marginally different for the modulatory parameters of LG and FG connections discussed above (see FIG. 7). In contrast, the MAP estimates for modulation of the callosal connections between left and right MOG were very close to zero (modulation of the right MOG → left MOG

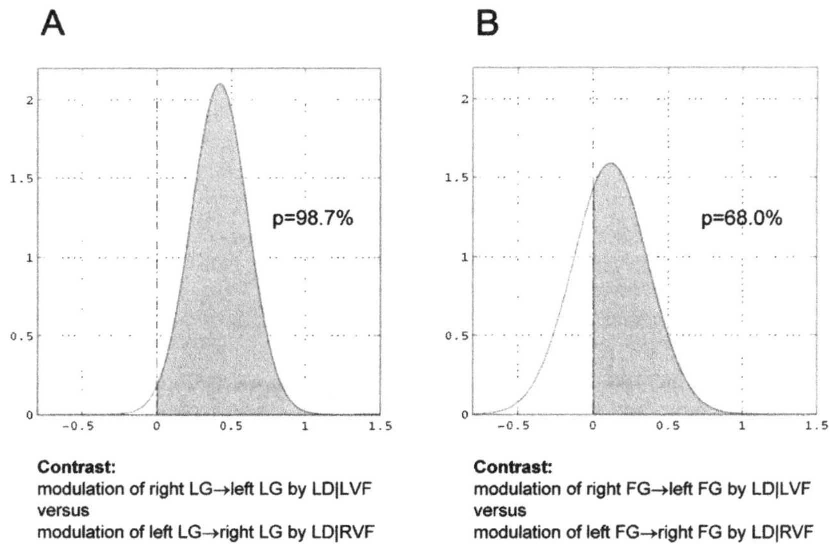

FIGURE 6. Asymmetry of callosal connections with regard to contextual modulation. The plots show the probability that the modulation of the right→left connection by task conditional on left visual field stimulation is stronger than the modulation of the left→right connection by task conditional on right visual field stimulation. **(A)** For the callosal connections between left and right LG, we can be very confident that this asymmetry exists: the probability of the contrast being larger than zero is $p(c^T\eta_{\theta|y} > 0) = 98.7\%$. **(B)** For the callosal connections between left and right FG, we are considerably less certain about the presence of this asymmetry: the probability of the contrast being larger than zero is $p(c^T\eta_{\theta|y} > 0) = 68.0\%$.

connection by LD|LVF: 0.01 ± 0.02 Hz; modulation of the left MOG → right MOG connection by LD|RVF: –0.01 ± 0.02 Hz). Given these very small effects, the contrast between these estimates unsurprisingly indicated that there was no support for the presence of an asymmetry with regard to contextual modulation of callosal connections between left and right MOG: the probability $p(c^T\eta_{\theta|y} > \gamma) = 54.2\%$ was very close to the 50% margin, indicating that it is equally likely that $c^T\eta_{\theta|y}$ is smaller or larger than zero.

DISCUSSION

In this article, we have summarized the theoretical foundations of DCM and BMS and have provided an example of the practical application of these techniques, using data from a single subject in the study by Stephan *et al.*[32] DCM enables the investigation of how neural systems (composed of large neural populations like cortical areas) that have relatively high connectional complexity, for example, multiple reciprocal connections and loops,[11,23] operate. Together with BMS, DCM is a powerful tool to clarify which of multiple experimental manipulations (e.g., stimulus type, induction

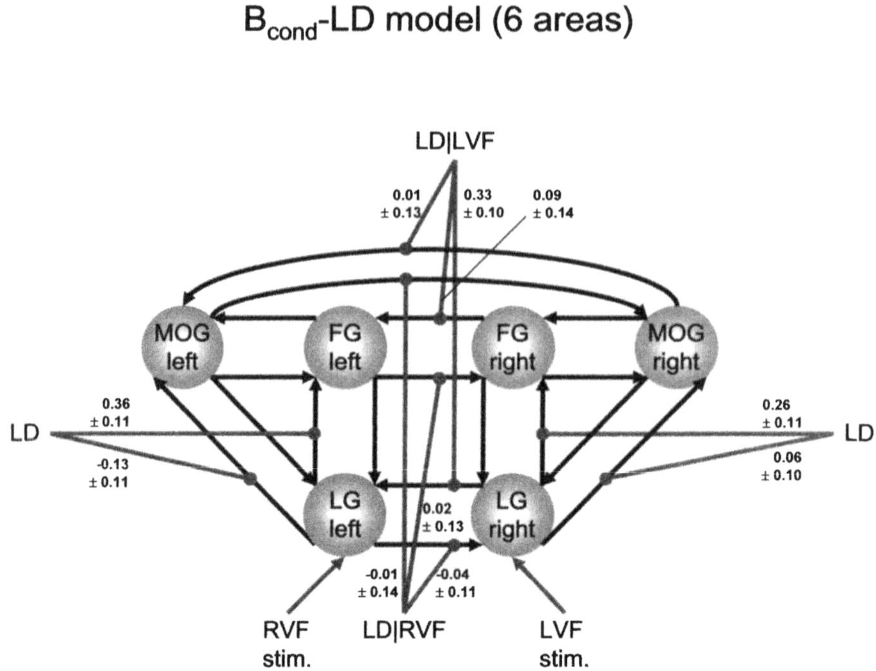

FIGURE 7. Parameter estimates for an extended version of the B_{cond}–LD model that includes the MOG in both hemispheres (see FIG. 5 for details). The estimates for callosal connections at the level of LG and FG were very similar to those from the four-area model (FIG. 5). In contrast to LG, there was no evidence for contextual modulation of callosal connections between right and left MOG.

of cognitive set, learning processes, etc.) have a significant impact on the dynamics of the network under investigation. By representing experimental factors as external inputs in the model, modeled effects can be interpreted fairly directly in neurobiological terms: any given DCM specifies precisely *where* inputs enter and whether they are *driving* (i.e., exert their effects through direct synaptic responses in the target area) or *modulatory* (i.e., exert their effects through changing synaptic responses in the target area to inputs from another area). This distinction, made at the level of neural populations, has a nice correspondence to empirical observations that single neurons can have either driving or modulatory effects on other neurons.[23]

There are several ways in which DCM and DWI techniques can profit from each other. We have given a simple example in this paper of how models of effective connectivity like DCM are important for our understanding of processes in neural systems, even if their anatomical connectivity is well understood. On the other hand, there are obviously many systems whose connectivity is less well known than that of the corpus callosum and visual areas. In this case, the specification of DCMs could be greatly facilitated by precise anatomical data on human brain connectivity obtained by DWI. Another potential opportunity is to use the likelihood of the existence of particular connections (as obtained from probabilistic tractography methods[7,8]) as priors on connection strengths in Bayesian models like DCM.

In our empirical example, we demonstrated two things: in the particular individual studied, there is good evidence that (i) the measured data are best explained by a model in which interhemispheric interactions depend on task demands, but conditional on the visual field of stimulus presentation, and (ii) there is a hemispheric asymmetry in context-dependent transcallosal interactions. Importantly, this asymmetry was not equally pronounced for all visual areas studied. It was particularly strong for the callosal connections between left and right LG: performance of the LD task specifically enhanced the strength of the influence of the right on the left LG, but only if the stimulus was presented in the left visual field, and thus the information was initially only available in the right hemisphere. The reversed conditional effect, that is, modulation of left LG $\rightarrow$ right LG by LD|RVF, was much weaker (and actually slightly negative; see FIGS. 5 and 7). This result means that, for the particular paradigm used, enhancement of callosal connections was only necessary if stimulus information was initially represented in the "suboptimal", that is, right, hemisphere. Interestingly, in the subject analyzed, this asymmetry was less pronounced at the level of the FG and virtually absent at the level of the MOG.

These results complement very nicely our previous results on local activations elicited by this paradigm.[32] While we had previously established, using identical stimuli in all conditions, that a change in task demands was sufficient to determine the lateralization of brain activity, our previous analyses were not suited to clarify the principles according to which the two hemispheres functionally interacted. In other words, despite empirical data on context-dependent activations of visual areas and good knowledge about the general anatomical connectivity between these areas, we could not infer the mechanism by which transcallosal interactions contributed to the observed activations. In this paper, we have shown how this can be achieved by comparing different models, which are fitted to the empirical data, and then making statistical inferences about the parameters of the optimal model. It should be emphasized, however, that in this paper we have applied DCM to data from a single subject only, and it remains to be seen whether or not the principles of interhemispheric interactions found for the particular individual studied here generalize to the population. To clarify this question, we are currently analyzing the data from all subjects of the study by Stephan *et al.*[32] following the strategy outlined in this article.

Group studies with DCM and BMS require somewhat different statistical procedures from those used in this paper. For example, when trying to find the optimal model for a group of individuals by BMS, it is likely that the optimal model will vary, at least to some degree, across subjects. An overall decision for n subjects can be made by computing an average Bayes factor, which corresponds to the n-th root of the product of the individual Bayes factors (note that multiplication is appropriate because model comparisons from different individuals are statistically independent). Moreover, in this paper, all inferences about coupling parameters were based on the magnitude of the effect of interest (i.e., the contrast of the appropriate MAP estimates) compared to the precision with which this estimate was obtained (i.e., the posterior variance). When dealing with a group of subjects, one may be interested in different types of inference. For example, one may want to establish that a certain effect (e.g., the modulation of a particular connection by some experimental condition) is consistently expressed across subjects. This second-level inference can take various forms.

For example, one can apply a classical statistical test to the parameters of interest or one can use a Bayesian approach, either treating the individual effects as fixed and combining the posterior variances or treating the individual parameters as random effects, thus enabling inference beyond the particular group of subjects studied.[42] These topics will be the subject of forthcoming methodological papers on DCM.

It should be noted that the idea underlying dynamic causal models is not restricted to fMRI data. David and Friston[43] have recently developed a neural mass model that, when combined with an appropriate forward model and estimation scheme, can be used as a DCM to derive neural coupling parameters from empirically measured electroencephalographic (EEG) or magnetoencephalographic (MEG) data (see David *et al.*[44] for a first application). This model has a much more sophisticated neural state equation than equation 2 in this paper, distinguishing between different cortical layers and different neural populations with specific time constants and connectivity. Furthermore, Harrison *et al.*[45] have developed a mean field model for event-related potentials that uses stochastic differential equations for the description of neural processes involving different types of transmitter receptors. The coming years will see a further refinement of dynamic causal models of this kind.

We conclude by emphasizing that one of the long-term goals of developments related to DCM is to obtain tools for clinical applications. Such tools are particularly important for the study of psychiatric diseases like schizophrenia whose phenotypes are often confusingly heterogeneous due to strong interactions between genotype and environmental influences. One aim is to determine disease-specific *endophenotypes*; these are biological markers at intermediate levels between genome and behavior, for example, particular neurophysiological or neurochemical indices.[46] For example, if the pathophysiological mechanism that underlies a specific disease is an abnormal functional coupling between two or more brain regions in a particular context, this would correspond to a disease-specific pattern of coupling parameters in an appropriate DCM. The challenge for the future will be to establish models of valid neural systems that can be fitted to measured imaging data and that are sensitive enough that their connectivity parameters can be used reliably for the diagnostic classification of individual patients. Ideally, such models should be used in conjunction with "implicit" paradigms that are minimally dependent on patient compliance, for example, mismatch negativity paradigms.[47] Given established validity and sufficient sensitivity of such a model, one could use it by analogy to a biochemical laboratory test, that is, to compare a particular model parameter (or combinations thereof) against reference distributions in order to obtain diagnostic classifications. If the model is sophisticated enough to distinguish between different transmitter receptors, it might also be possible to obtain predictions for the optimal pharmacological treatment of individual patients. DCM, as described in this paper, provides a generic framework and starting point for this long-term endeavor.

ACKNOWLEDGMENTS

This work was supported by the Wellcome Trust (to K. E. Stephan, W. D. Penny, and K. J. Friston), the Medical Research Council (to J. C. Marshall), and the Deutsche Forschungsgemeinschaft (to G. R. Fink).

REFERENCES

1. FRISTON, K.J. 2002. Beyond phrenology: what can neuroimaging tell us about distributed circuitry? Annu. Rev. Neurosci. **25:** 221–250.
2. MARSHALL, J.C. & G.R. FINK. 2003. Cerebral localization, then and now. NeuroImage **20**(suppl. 1): S2–S7.
3. KÖBBERT, C., R. APPS, I. BECHMANN *et al.* 2000. Current concepts in neuroanatomical tracing. Prog. Neurobiol. **62:** 327–351.
4. STEPHAN, K.E., L. KAMPER, A. BOZKURT *et al.* 2001. Advanced database methodology for the collation of connectivity data on the macaque brain (CoCoMac). Philos. Trans. R. Soc. Lond. B Biol. Sci. **356:** 1159–1186.
5. BÜRGEL, U., T. SCHORMANN, A. SCHLEICHER *et al.* 1999. Mapping of histologically identified long fiber tracts in human cerebral hemispheres to the MRI volume of a reference brain: position and spatial variability of the optic radiation. NeuroImage **10:** 489–499.
6. GALUSKE, R.A., W. SCHLOTE, H. BRATZKE *et al.* 2000. Interhemispheric asymmetries of the modular structure in human temporal cortex. Science **289:** 1946–1949.
7. BEHRENS, T.E.J., M.W. WOOLRICH, M. JENKINSON *et al.* 2003. Characterization and propagation of uncertainty in diffusion-weighted MR imaging. Magn. Reson. Med. **50:** 1077–1088.
8. PARKER, G.J.M., H.A. HAROON & C.A.M. WHEELER-KINGSHOTT. 2003. A framework for a streamline-based probabilistic index of connectivity (PICo) using a structural interpretation of MRI diffusion measurements. J. Magn. Reson. Imag. **18:** 242–254.
9. ALEXANDER, D.C., G.J. BARKER & S.R. ARRIDGE. 2002. Detection and modeling of non-Gaussian apparent diffusion coefficient profiles in human brain data. Magn. Reson. Med. **48:** 331–340.
10. STROGATZ, S.H. 2001. Exploring complex networks. Nature **410:** 268–276.
11. FRISTON, K.J., L. HARRISON & W. PENNY. 2003. Dynamic causal modelling. NeuroImage **19:** 1273–1302.
12. ROCKLAND, K.S. & D.N. PANDYA. 1979. Laminar origins and terminations of cortical connections of the occipital lobe in the rhesus monkey. Brain Res. **179:** 3–20.
13. RINGO, J.L., R.W. DOTY, S. DEMETER *et al.* 1994. Time is of the essence: a conjecture that hemispheric specialization arises from interhemispheric conduction delay. Cereb. Cortex **4:** 331–343.
14. CAVADA, C. & P.S. GOLDMAN-RAKIC. 1989. Posterior parietal cortex in rhesus monkey: I. Parcellation of areas based on distinctive limbic and sensory corticocortical connections. J. Comp. Neurol. **287:** 393–421.
15. MCGUIRE, P.K., J.F. BATES & P.S. GOLDMAN-RAKIC. 1991. Interhemispheric integration: I. Symmetry and convergence of the corticocortical connections of the left and the right principal sulcus (PS) and the left and the right supplementary motor area (SMA) in the rhesus monkey. Cereb. Cortex **1:** 390–407.
16. FUNNELL, M.G., P.M. CORBALLIS & M.S. GAZZANIGA. 2000. Insights into the functional specificity of the human corpus callosum. Brain **123:** 920–926.
17. WESTERHAUSEN, R., F. KREUDER, S. SEQUEIRA *et al.* 2004. Effects of handedness and gender on macro- and microstructure of the corpus callosum and its subregions: a combined high-resolution and diffusion-tensor MRI study. Cogn. Brain Res. **21:** 418–426.
18. HULSHOFF POL, H.E., H.G. SCHNACK, R.C. MANDL *et al.* 2003. Focal white matter density changes in schizophrenia: reduced inter-hemispheric connectivity. NeuroImage **21:** 27–35.
19. BANICH, M.T. 1998. The missing link: the role of interhemispheric interaction in attentional processing. Brain Cogn. **36:** 128–157.
20. BOLLEN, K.A. 1989. Structural Equations with Latent Variables. Wiley. New York.
21. KENNEDY, H., C. DEHAY & J. BULLIER. 1986. Organization of the callosal connections of visual areas V1 and V2 in the macaque monkey. J. Comp. Neurol. **247:** 398–415.
22. SEGRAVES, M.A. & A.C. ROSENQUIST. 1982. The afferent and efferent callosal connections of retinotopically defined areas in cat cortex. J. Neurosci. **2:** 1090–1107.
23. PENNY, W.D., K.E. STEPHAN, A. MECHELLI *et al.* 2004. Comparing dynamic causal models. NeuroImage **22:** 1157–1172.

24. SHERMAN, S.M. & R.W. GUILLERY. 1998. On the actions that one nerve cell can have on another: distinguishing "drivers" from "modulators". Proc. Natl. Acad. Sci. USA **95:** 7121–7126.
25. BUXTON, R.B., E.C. WONG & L.R. FRANK. 1998. Dynamics of blood flow and oxygenation changes during brain activation: the balloon model. Magn. Reson. Med. **39:** 855–864.
26. FRISTON, K.J., A. MECHELLI, R. TURNER *et al.* 2000. Nonlinear responses in fMRI: the balloon model, Volterra kernels, and other hemodynamics. NeuroImage **12:** 466–477.
27. FRISTON, K.J. 2002. Bayesian estimation of dynamical systems: an application to fMRI. NeuroImage **16:** 513–530.
28. PENNY, W.D., K.E. STEPHAN, A. MECHELLI *et al.* 2004. Modelling functional integration: a comparison of structural equation and dynamic causal models. NeuroImage **23:** S264–S274.
29. PITT, M.A. & I.J. MYUNG. 2002. When a good fit can be bad. Trends Cogn. Sci. **6:** 421–425.
30. STEPHAN, K.E. 2004. On the role of general systems theory for functional neuroimaging. J. Anat. **205:** 443–470.
31. MCINTOSH, A.R., C.L. GRADY, L.G. UNGERLEIDER *et al.* 1994. Network analysis of cortical visual pathways mapped with PET. J. Neurosci. **14:** 655–666.
32. STEPHAN, K.E., J.C. MARSHALL, K.J. FRISTON *et al.* 2003. Lateralized cognitive processes and lateralized task control in the human brain. Science **301:** 384–386.
33. JENNER, A.R., G.D. ROSEN & A.M. GALABURDA. 1999. Neuronal asymmetries in primary visual cortex of dyslexic and nondyslexic brains. Ann. Neurol. **46:** 189–196.
34. AMUNTS, K., A. SCHLEICHER, U. BÜRGEL *et al.* 1999. Broca's region revisited: cytoarchitecture and intersubject variability. J. Comp. Neurol. **412:** 319–341.
35. SERGENT, J. 1983. Role of the input in visual hemispheric asymmetries. Psychol. Bull. **93:** 481–512.
36. LEVY, J. & C. TREVARTHEN. 1976. Metacontrol of hemispheric function in human split-brain patients. J. Exp. Psychol. Hum. Percept. Perform. **2:** 299–312.
37. FINK, G.R., P.W. HALLIGAN, J.C. MARSHALL *et al.* 1996. Where in the brain does visual attention select the forest and the trees? Nature **382:** 626–628.
38. ABEL, P.L., B.J. O'BRIEN & J.F. OLAVARRIA. 2000. Organization of callosal linkages in visual area V2 of macaque monkey. J. Comp. Neurol. **428:** 278–293.
39. KÖTTER, R. & K.E. STEPHAN. 2003. Network participation indices: characterizing component roles for information processing in neural networks. Neural Networks **16:** 1261–1275.
40. SCHACK, B., S. WEISS & P. RAPPELSBERGER. 2003. Cerebral information transfer during word processing: where and when does it occur and how fast is it? Hum. Brain Map. **19:** 18–36.
41. NOWICKA, A., A. GRABOWSKA & E. FERSTEN. 1996. Interhemispheric transmission of information and functional asymmetry of the human brain. Neuropsychologia **34:** 147–151.
42. PENNY, W. & A.P. HOLMES. 2004. Random-effects analysis. *In* Human Brain Function, pp. 843–850. Elsevier. Amsterdam/New York/San Diego.
43. DAVID, O. & K.J. FRISTON. 2003. A neural mass model for MEG/EEG: coupling and neuronal dynamics. NeuroImage **20:** 1743–1755.
44. DAVID, O., L. HARRISON, J. KILNER *et al.* 2004. Studying effective connectivity with a neural mass model of evoked MEG/EEG responses. *In* Proceedings of the 14th International Conference on Biomagnetism (BIOMAG 2004), pp. 135–138. Boston.
45. HARRISON, L.M., O. DAVID & K.J. FRISTON. 2005. Stochastic models of neuronal dynamics. Philos. Trans. R. Soc. Lond. B Biol. Sci. **360:** 1075–1091.
46. GOTTESMAN, I.I. & T.D. GOULD. 2003. The endophenotype concept in psychiatry: etymology and strategic intentions. Am. J. Psychiatry **160:** 636–645.
47. BALDEWEG, T., A. KLUGMAN, J. GRUZELIER *et al.* 2004. Mismatch negativity potentials and cognitive impairment in schizophrenia. Schizophr. Res. **69:** 203–217.

Age-Related Changes in Prefrontal White Matter Measured by Diffusion Tensor Imaging

D. H. SALAT,[a] D. S. TUCH,[a] N. D. HEVELONE,[a] B. FISCHL,[a] S. CORKIN,[a,b] H. D. ROSAS,[a,c] AND A. M. DALE[d]

[a]*MGH/MIT/HMS Athinoula A. Martinos Center for Biomedical Imaging, Charlestown, Massachusetts, USA*

[b]*MIT Department of Brain and Cognitive Sciences, Cambridge, Massachusetts, USA*

[c]*MGH Department of Neurology, Boston, Massachusetts, USA*

[d]*UCSD Departments of Neurosciences and Radiology, La Jolla, California, USA*

ABSTRACT: Age-related degeneration of brain white matter (WM) has received a great deal of attention, with recent studies demonstrating that such changes are correlated with cognitive decline and increased risk for the development of age-related neurodegenerative disease. Past studies have used magnetic resonance imaging (MRI) to measure the volume of normal and abnormal tissue signal as an index of tissue pathology. More recently, diffusion tensor MRI (DTI) has been employed to obtain regional measures of tissue microstructure, such as fractional anisotropy (FA), providing better spatial resolution and potentially more sensitive metrics of tissue damage than traditional volumetric measures. We used DTI to examine the regional basis of age-related alterations in prefrontal WM. As expected from prior volumetric and DTI studies, prefrontal FA was reduced in older adults (OA) compared to young adults (YA). Although WM volume has been reported to be relatively preserved until late aging, FA was significantly reduced by middle age. Much of prefrontal WM showed reduced FA with increasing age. Ventromedial and deep prefrontal regions showed a somewhat greater reduction compared to other prefrontal areas. Prefrontal WM anisotropy correlated with prefrontal WM volume, but the correlation was significant only when the analysis was limited to participants over age 40. This evidence of widespread and regionally accelerated alterations in prefrontal WM with aging illustrates FA's potential as a microstructural index of volumetric measures.

KEYWORDS: aging; Alzheimer's disease; white matter; prefrontal; MRI; DTI

BRAIN AGING

Great effort has been put toward characterizing alterations that occur in the brain with healthy or "successful" aging. This issue is particularly important because the risk for a number of neurological diseases increases with increasing age, and

Address for correspondence: David H. Salat, Ph.D., MGH/MIT/HMS Athinoula A. Martinos Center for Biomedical Imaging, MGH Department of Radiology, Building 149, 13th Street, Mail Code 149(2301), Charlestown, MA 02129-2060. Voice: 617-726-4704; fax: 617-726-7422. salat@nmr.mgh.harvard.edu

Ann. N.Y. Acad. Sci. 1064: 37–49 (2005).
doi: 10.1196/annals.1340.009

knowledge of expected, nondisease-related changes would help to distinguish healthy individuals from those with impending disease. Thus, contemporary studies of brain aging extend across research domains and methods, from histological research in rats to *in vivo* neuroimaging of humans and nonhuman primates. Through these studies, it has become clear that the prefrontal region of the brain shows significant age-related alteration in various indices of tissue integrity, ranging from reduced brain volume[1] to layer-specific reduction in the size of dendritic fields.[2]

NEUROIMAGING STUDIES OF GRAY AND WHITE MATTER

Neuroimaging technologies have greatly enhanced the ability to perform studies on large numbers of well-characterized individuals *in vivo*, and magnetic resonance imaging (MRI) has been particularly important due to this technology's ability to perform high-resolution scanning noninvasively. Early MRI studies noted the obvious appearance of alterations in white matter (WM) tissue signal with aging (i.e., WM hyper/hypointensities),[3,4] a finding known from prior computed tomography studies. Importantly, changes were observed in cognitively healthy individuals. These changes were not apparent in individuals under the age of 45 or even in every older individual examined.[4] Thus, such changes are not an early or inevitable consequence of brain aging.

In addition to abnormal WM signal, gray matter (GM) and WM atrophy are apparent in MR scans of older adults (OA). Prior studies have reported disparate results, with some investigators emphasizing loss of GM[5–8] and others emphasizing alterations of WM[9,10] as the prominent feature of brain aging. For example, a recent study by Raz and colleagues measured regional GM and WM volumes throughout the brain in 148 adults aged 18 to 77 years.[1] Their results demonstrated that the most dramatic age-related loss was in prefrontal cortical volume, where the loss surpassed that in prefrontal WM volume and in all other WM and GM volumes. In contrast, Guttmann and colleagues examined 72 adults aged 18–81 and found that WM volume was significantly reduced in OA with minimal loss of GM.[11] This finding held when controlling for the presence WM abnormalities (hyperintensities), demonstrating that WM volume loss is independent of WM abnormalities. Procedural differences, including participant sample, scan type and resolution, volumetric method employed, and the type of statistical analysis, all contribute to discrepancies among prior reports. A growing consensus now exists that both GM and WM volumes decline with aging. Such loss may be regionally and temporally selective, with prefrontal GM volume declining early in the age span,[1] and WM volume loss most prominent sometime after age 40.[12] WM likely shows an age-related increase in volume prior to this middle-age (MA) period, that is, in adolescents and young adults (YA).[12–14]

PREFRONTAL TISSUE INTEGRITY MEASURED WITH VOLUMETRIC MRI

A number of studies have indirectly examined prefrontal integrity through measures of age-related changes in the volume of the corpus callosum (e.g., refs. 15–19). This structure is particularly informative in this regard because of its topographical

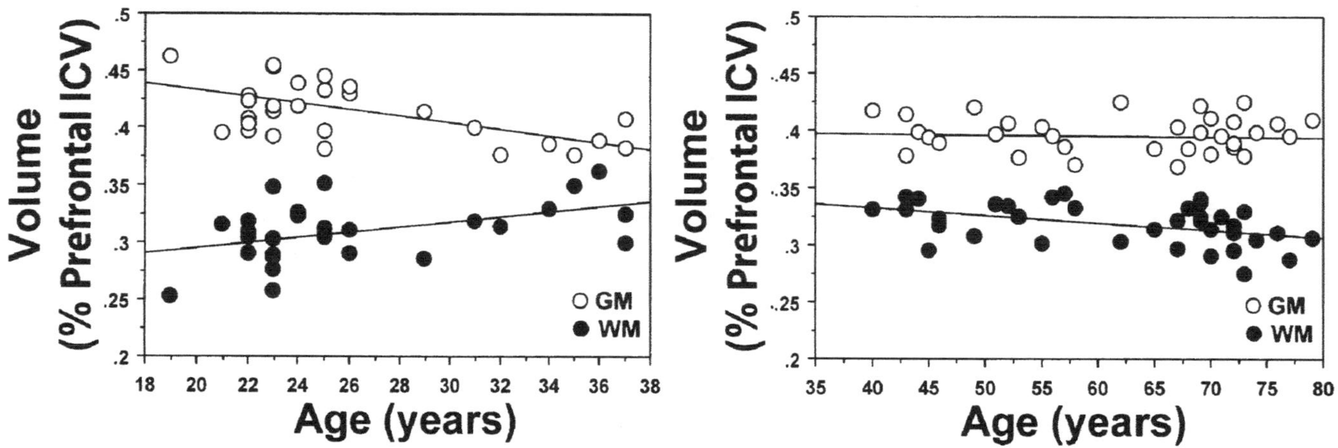

FIGURE 1. Prefrontal GM and WM volume across the adult age span. Adults younger than 40 years of age showed a reduction in prefrontal GM volume with an increase in prefrontal WM volume (*left panel*). In contrast, adults older than 40 showed a reduction in prefrontal WM volume with only modest reductions in prefrontal GM volume (*right panel*).

organization, with frontal commissural fibers projecting through the anterior portion, the genu, and occipital fibers projecting through the posterior portion, the splenium.[20] Thus, regional division of the corpus callosum provides an index of tissue integrity on a functional and/or lobar level. Examination of age-related changes in the corpus callosum shows that the genu and anterior portions decrease in volume with increasing age, whereas the splenium and posterior sections of the callosum are relatively preserved.[15–19] These findings support the view that anterior tissue changes are more prominent than posterior changes with aging.

Age-related reduction in prefrontal GM and WM volume has also been directly measured with MRI (e.g., refs. 1 and 21). As reviewed above, discrepancies as to which tissue is most affected could be explained by the population sample: cortical loss appears to be more prominent in the adolescent to middle age range, with WM loss occurring later. This pattern of early reduction in cortical volume and later loss of WM volume is apparent in prefrontal tissue. Investigators have demonstrated that orbitofrontal cortical volume[1] and superior frontal GM density[13] show a strong reduction in early adulthood, with minimal decline in middle to late aging. Similarly, we find a decrease in total prefrontal GM and an increase in prefrontal WM in participants less than 40 years of age, and a decline in WM volume with only modest reductions in cortical volume in participants older than 40 years of age (FIG. 1). Other studies show a trend for volume loss in more posterior frontal GM with later aging (e.g., ref. 12), and we see similar reductions in posterior frontal cortical thickness that continue later in the age span.[22] The finding of increased WM volume in early adulthood is supported by recent studies demonstrating a curvilinear pattern of WM volume across the adult age span (e.g., ref. 12); the finding of early cortical volume reduction is supported by recent studies in adolescents[14,23] and young adults.[22]

Our prior work specifically examined how prefrontal tissue volume is altered with very late aging.[21] A younger group comprised adults 60–75 years of age, while an older group consisted of individuals 85 years and older. In addition, the study included a group of patients with Alzheimer's disease (AD) who were age and demographically matched to the "younger" group. This investigation addressed two important questions: (i) is late aging accompanied by a preferential decline in GM or WM volume?; (ii) does the pathology of AD look like "accelerated aging" (i.e., do AD patients show the same pattern of volume loss in prefrontal tissue as the >85-year-old adults)? Contrary to expectations at the time, we found minimal prefrontal cortical volume loss with late aging, whereas a profound decline in WM volume emerged.[21] Additionally, although AD patients showed a reduction in cortical volume, the loss of prefrontal WM was highly variable, suggesting that certain patients have a propensity towards WM loss, while other patients are relatively spared from these changes. Thus, prefrontal WM changes are more prominent than GM loss with late aging, and prefrontal WM loss could be an important contributor to clinical heterogeneity in AD patients.

Importantly, volumetric reductions in WM occurred in the absence of significant abnormal prefrontal WM signal in these participants (abnormalities occupied less than 1% of the total prefrontal WM volume). Thus, similar to the results of Guttmann and colleagues, this finding clarifies that volumetric loss in brain WM is not simply due to a change in the amount of tissue being classified as "abnormal". Instead, volumetric loss likely represents a process that is related to microstructural alterations in tissue, ultimately resulting in gross volume reduction. A method for measuring such microstructural changes *in vivo* would facilitate sorting out the various components contributing to the diminution in WM integrity.

DTI AS A TOOL TO MEASURE TISSUE PATHOLOGY

DTI is one technique that could, in the least, quantify alterations in tissue structure regardless of pathophysiological mechanism and, much more ambitiously, differentiate among pathologies and specify affected fiber pathways. Diffusion imaging in humans using MRI has existed for over 15 years (e.g., ref. 24), but recently has gained significant popularity due to new acquisition and analysis techniques. Specifically, a strong interest has been developed in the use of diffusion data to describe normal and abnormal brain anatomy through anisotropy, tensormap, and tractography techniques that provide information about the anatomical orientation and microstructure of major brain fasciculi. Although these new technologies have been applied in a number of recent clinical investigations, the precise biophysical basis of anisotropy measured by DTI techniques is still being explored (e.g., ref. 25) and the histological bases of alterations in this signal due to age or disease are yet to be defined.

REGIONAL CHANGES IN PREFRONTAL WHITE MATTER MEASURED WITH DTI

DTI has allowed an indirect assessment of prefrontal tissue through measurement of "fractional anisotropy" (FA)[26] in callosal tissue (e.g., ref. 27). The FA metric is a summary measure of the restrictional microenvironment of brain water diffusion, and this measure changes regionally in the brain with age and disease. Similar to volumetric results, the anterior regions of the callosum are most susceptible to alterations in tissue microstructure (e.g., refs. 27–30). These findings support the view that anterior tissue is preferentially affected by age-related alterations. Further, they give credence to the theory that DTI measures represent a microstructural index of tissue volume. Still, it is unclear from studies of the callosum alone whether FA changes are due to altered prefrontal WM or a result of alterations in cortex. Recent DTI experiments in our lab and others have more directly uncovered preferential alterations in frontal WM relative to other areas of the brain[29–33] (FIG. 2).

DTI may also be useful for characterizing the regional nature of changes within prefrontal WM because FA is calculated on a voxel by voxel basis throughout the volume. Our prior work examined age-related changes in FA across all brain WM and in a number of regions of interest (ROIs) across the adult age span (adults aged 21–76).[30] Alterations within prefrontal WM were heterogeneous, with ventromedial and "deep" prefrontal WM particularly affected, and WM in the inferior frontal gyrus not statistically affected. More recently, we have used tensormapping techniques to define regional boundaries for ROI measurements. We placed ROIs in the WM of each of the major frontal gyri and, using the tensor information, attempted to sample from within a homogeneous region of WM and from homologous regions across participants. This procedure showed a statistical reduction in FA in all regional measures, except for a trend in middle frontal WM (FIG. 3). In certain regions, such as ventromedial prefrontal WM, an accelerated decline in FA with age was apparent

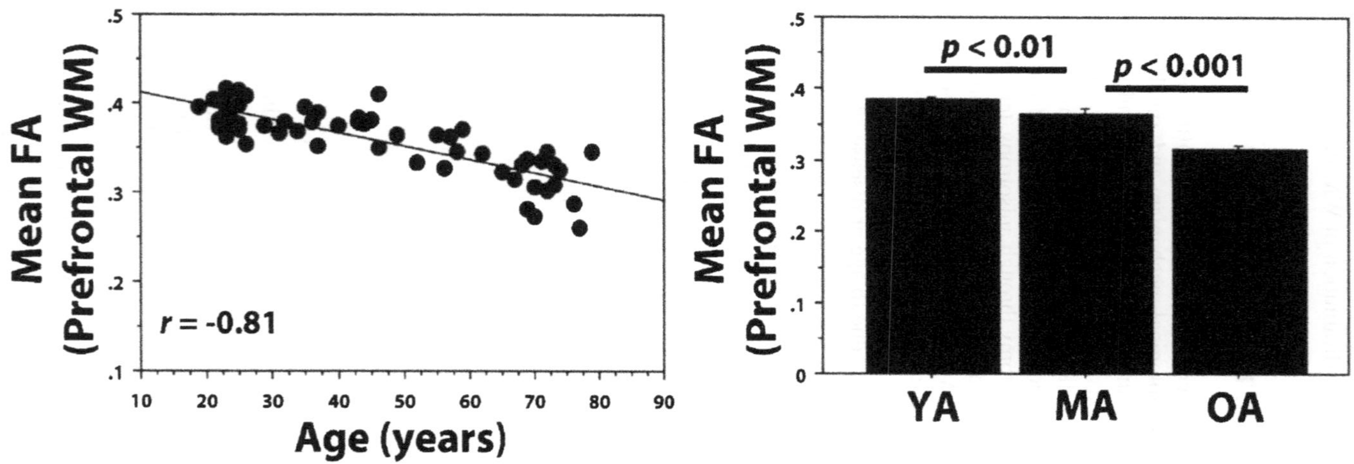

FIGURE 2. Age-related reduction in prefrontal WM FA. We found a strong reduction in the FA of prefrontal WM with increasing age (*left panel*), with significant reductions in MA (40- to 59-year-old adults) compared to YA (18- to 39-year-old adults), and in OA (60- to 80-year-old adults) compared to MA (*right panel*).

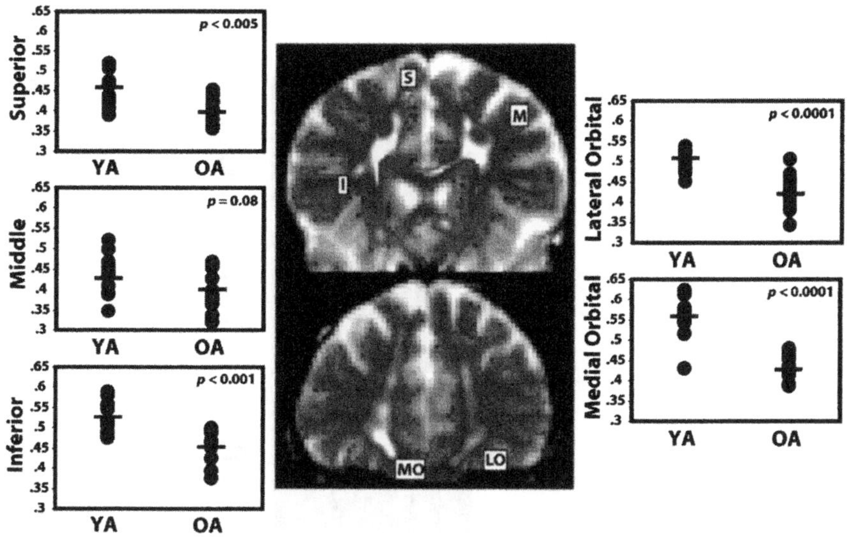

FIGURE 3. Regional measures of prefrontal FA. ROIs were placed in superior (S), middle (M), inferior (I) (*left panel*), lateral orbital (LO), and medial orbital (MO) (*right panel*) WM in YA (18- to 39-year-old adults) and OA (60- to 80-year-old adults). FA was statistically reduced in OA in all regions, except for a trend in the middle frontal ROI. ROI measures and voxel-based statistical maps (*center panel*) suggested that ventromedial prefrontal WM showed accelerated reduction in FA with aging.

(FIG. 4). We find similar results with whole-brain voxel-based comparisons of FA in YA and OA. Thus, although alterations in prefrontal WM are widespread, certain regions, such as ventromedial WM, appear to show an accelerated decline in FA.

To further test whether prefrontal fibers generally show greater alteration than more posterior fiber systems, we also placed ROIs within the anterior and posterior limb of the internal capsule, expecting alteration of the frontally projecting fibers of the anterior limb and preservation of the posterior limb, which is associated with primary motor function. In fact, we found the opposite result. The posterior limb showed a significant age-related reduction in FA, while the anterior limb was relatively preserved (FIG. 5). This finding was particularly interesting given recent data demonstrating a reduction in thickness of the precentral gyrus,[22] where fibers of the posterior limb are believed to originate. Additionally, although prefrontal WM shows an appreciable decline in FA, these results suggest that the loss is selective and does not generalize to all or entire fiber systems projecting to and from prefrontal cortex.

IS FA A MICROSTRUCTURAL INDEX OF WM VOLUME?

An important question in the application of these innovative technologies is how they relate to traditionally employed imaging measures. For example, although FA is believed to be an index of tissue microstructure, it is currently unknown how FA

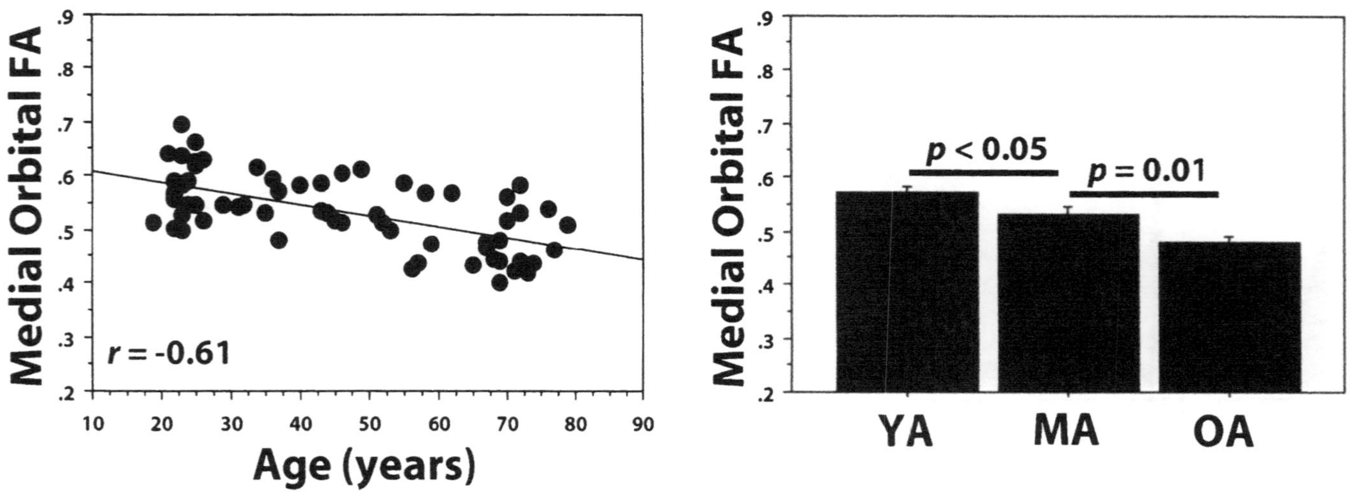

FIGURE 4. Age-related reductions in medial orbital (ventromedial) prefrontal FA. We found a strong statistical reduction in FA in the ventromedial prefrontal ROI. This effect was apparent when examining participants across the age span (*left panel*), and was significant by middle age when examining participants grouped into YA (18- to 39-year-old adults), MA (40- to 59-year-old adults), and OA (60- to 80-year-old adults) (*right panel*).

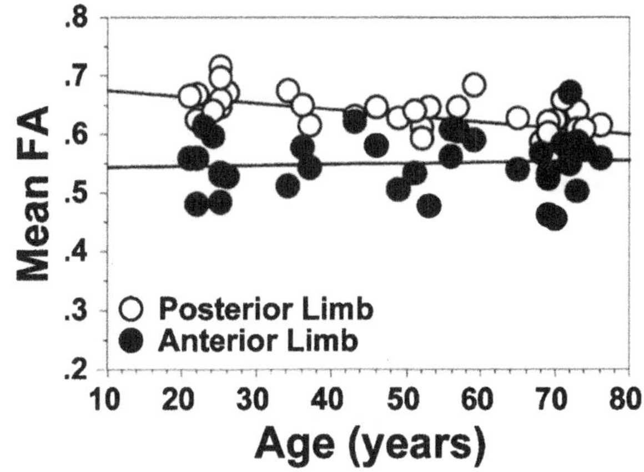

FIGURE 5. Regional measurements of FA in the anterior and posterior limbs of the internal capsule. We obtained regional measurements by sampling a line of voxels through the center of each limb in a single axial slice. The results showed a significant age-related reduction in FA in the posterior limb, with relative preservation of the anterior limb.

relates to WM volume, a classic imaging metric of tissue integrity. We addressed this issue by examining the relation between prefrontal WM FA and prefrontal WM volume. Mean prefrontal WM FA was correlated with total prefrontal volume, but only when the analyses were limited to adults over age 40 (FIG. 6). This result could be expected given the findings that FA is reduced during the YA to MA period, a time when WM volume in this region is still increasing. Thus, FA could provide a meaningful metric that is potentially a microstructural index of volumetric measures. One would need to interpret this index with caution, however, because FA and volume could be temporally shifted such that FA begins to decline significantly earlier than gross effects measured through volume. Our prefrontal GM and FA measures were not correlated within any segment of the age span. We suggest that this finding is due to the fact that FA is measuring a meaningful property of WM as opposed to more generic, global brain changes. A clear description of the full relation between WM FA and volume awaits future research.

WHAT CONTRIBUTES TO AGE-RELATED CHANGES IN PREFRONTAL FA?

Evidence suggests that changes in FA with aging are due to microstructural alterations that are antecedent to WM hyperintensities or volume loss commonly measured in MR studies. Regionally, we find reduced FA in brain areas that have high hyperintensity distribution.[34] Prior studies have shown that these patches are heterogeneously pathologic, resulting from infarction, gliosis, demyelination, or cysts.[35] Many OA, however, are relatively free from WM abnormalities, yet still show marked decline in WM FA and volume. Additionally, changes in FA occur in regions such as the corpus callosum, where hyperintensities are not typically found. These

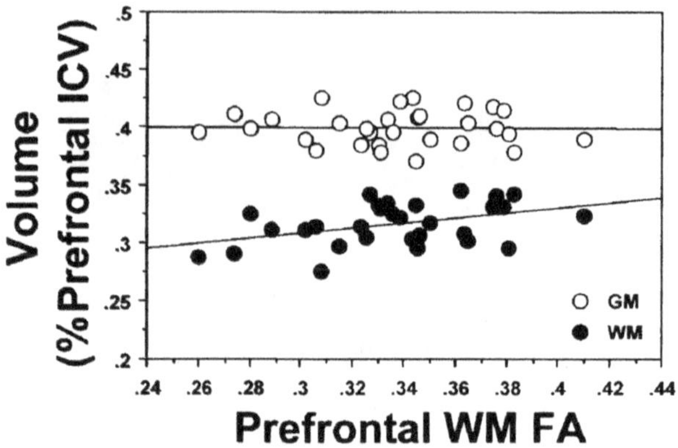

FIGURE 6. Correlation between prefrontal GM and WM volume and prefrontal WM FA. Greater prefrontal WM volume was related to higher prefrontal WM FA, but only when limiting the analysis to adults over age 40. In contrast, prefrontal GM volume was not related to prefrontal WM FA.

facts suggest that large signal abnormalities are not a hallmark of aging, but stem from medical conditions with increased incidence in the aging population, such as hypertension and cerebrovascular risk. Thus, FA provides information about additional alterations in WM independent of obvious abnormalities. It is possible that these reductions in FA could even be related to subclinical cerebrovascular risk. Recent research demonstrates that untreated as well as treated hypertension can result in negative changes in prefrontal brain structure,[36] supporting a role for subclinical cerebrovascular risk in accelerating brain pathology. Alternatively, FA reduction may represent an independent and distinct process from hyperintensity formation or cerebrovascular risk. Decreases in normal WM volume are found in well-characterized, healthy adults. Thus, whatever the basis of age-related reduction in FA, it is not necessarily associated with current or impending disease.

ARE FA MEASURES OF WM INTEGRITY CLINICALLY SIGNIFICANT?

DTI as described in this chapter will be useful as a clinical technique only if one can demonstrate that the measures are related to cognitive and other symptomology. The clinical significance of WM changes has been difficult to establish because of the pathologic, regional, and quantitative variability associated with WM damage. Studies appearing in the literature suggest that DTI can contribute to clinical diagnosis. One impressive study demonstrated a strong relation between DTI measures in the temporal stem WM and disease severity in patients with AD.[37] Preliminary data from our laboratory and published reports from others suggest that DTI measures predict functional capacity in patients with Huntington's disease[38] and cognitive abilities in OA.[39] In fact, our data even suggest that regional anisotropy and cognitive performance are related within young healthy participants.[40] Functional MRI (fMRI) studies (e.g., refs. 41–44) demonstrated age-related alterations in frontal lobe neural activity, and a promising future endeavor will be the combination of DTI with fMRI measures to determine the significance of WM alterations on brain function. Thus, although requiring further study and evaluation, recent data indicate that DTI measures are an important clinical index of regional pathology.

CAVEATS AND FUTURE DIRECTIONS

A number of caveats exist in the use of diffusion imaging to address questions about age-related tissue degeneration. The most common current methods do not allow for whole brain acquisition with much higher resolution than 2 mm^3. Thus, sampling from appreciable regions of WM without partial volume contamination from GM or cerebrospinal fluid is difficult, particularly when examining participant populations where WM atrophy is expected. Although partial voluming affects measures of FA, recent studies demonstrated that it does not account for the entire effect found in studies of aging.[30,45] A confounding factor to comparing regional measurements arises because some regions can be measured and spatially normalized more reliably than others. Similarly, FA is strongly influenced by the orientational homogeneity of the fiber system being measured. It is likely that measurements from regions comprising less homogeneous fiber systems have inherently reduced statistical power to detect significant alterations. Further, ROI and voxel-based studies will be affected by image distortions common to fast imaging. Thus, DTI must overcome these significant hurdles to advance as a research tool with clinical applicability. Novel methods in data acquisition and analysis will be critical towards attaining this goal.[46,47] In spite of these limitations, DTI has been, and will continue to be, an important research method for examining brain aging and age-related disease.

REFERENCES

1. Raz, N. *et al.* 1997. Selective aging of the human cerebral cortex observed *in vivo*: differential vulnerability of the prefrontal gray matter. Cereb. Cortex **7:** 268–282.
2. Uylings, H.B. & J.M. de Brabander. 2002. Neuronal changes in normal human aging and Alzheimer's disease. Brain Cogn. **49:** 268–276.
3. Brant-Zawadzki, M. *et al.* 1985. MR imaging of the aging brain: patchy white-matter lesions and dementia. AJNR Am. J. Neuroradiol. **6:** 675–682.
4. George, A.E. *et al.* 1986. Leukoencephalopathy in normal and pathologic aging: 2. MRI of brain lucencies. AJNR Am. J. Neuroradiol. **7:** 567–570.
5. Coffey, C.E. *et al.* 1992. Quantitative cerebral anatomy of the aging human brain: a cross-sectional study using magnetic resonance imaging. Neurology **42:** 527–536.
6. Lim, K.O. *et al.* 1992. Decreased gray matter in normal aging: an *in vivo* magnetic resonance study. J. Gerontol. **47:** B26–B30.
7. Murphy, D.G. *et al.* 1992. Age-related differences in volumes of subcortical nuclei, brain matter, and cerebrospinal fluid in healthy men as measured with magnetic resonance imaging. Arch. Neurol. **49:** 839–845.
8. Pfefferbaum, A. *et al.* 1994. A quantitative magnetic resonance imaging study of changes in brain morphology from infancy to late adulthood. Arch. Neurol. **51:** 874–887.
9. Christiansen, P. *et al.* 1994. Age dependent white matter lesions and brain volume changes in healthy volunteers. Acta Radiol. **35:** 117–122.
10. Jernigan, T.L. *et al.* 1991. Cerebral structure on MRI. Part I: Localization of age-related changes. Biol. Psychiatry **29:** 55–67.
11. Guttmann, C.R. *et al.* 1998. White matter changes with normal aging. Neurology **50:** 972–978.

12. BARTZOKIS, G. *et al.* 2001. Age-related changes in frontal and temporal lobe volumes in men: a magnetic resonance imaging study. Arch. Gen. Psychiatry **58:** 461–465.
13. SOWELL, E.R. *et al.* 2003. Mapping cortical change across the human life span. Nat. Neurosci. **6:** 309–315.
14. GIEDD, J.N. 2004. Structural magnetic resonance imaging of the adolescent brain. Ann. N.Y. Acad. Sci. **1021:** 77–85.
15. COWELL, P.E. *et al.* 1992. A developmental study of sex and age interactions in the human corpus callosum. Brain Res. Dev. Brain Res. **66:** 187–192.
16. DORAISWAMY, P.M. *et al.* 1991. Aging of the human corpus callosum: magnetic resonance imaging in normal volunteers. J. Neuropsychiatry Clin. Neurosci. **3:** 392–397.
17. JANOWSKY, J.S., J.A. KAYE & R.A. CARPER. 1996. Atrophy of the corpus callosum in Alzheimer's disease versus healthy aging. J. Am. Geriatr. Soc. **44:** 798–803.
18. SALAT, D. *et al.* 1997. Sex differences in the corpus callosum with aging. Neurobiol. Aging **18:** 191–197.
19. WEIS, S., K. JELLINGER & E. WENGER. 1991. Morphometry of the corpus callosum in normal aging and Alzheimer's disease. J. Neural Transm. Suppl. **33:** 35–38.
20. LOMBER, S.G., B.R. PAYNE & A.C. ROSENQUIST. 1994. The spatial relationship between the cerebral cortex and fiber trajectory through the corpus callosum of the cat. Behav. Brain Res. **64:** 25–35.
21. SALAT, D.H., J.A. KAYE & J.S. JANOWSKY. 1999. Prefrontal gray and white matter volumes in healthy aging and Alzheimer disease. Arch. Neurol. **56:** 338–344.
22. SALAT, D.H. *et al.* 2004. Thinning of the cerebral cortex in aging. Cereb. Cortex **14:** 721–730.
23. GIEDD, J.N. *et al.* 1999. Brain development during childhood and adolescence: a longitudinal MRI study. Nat. Neurosci. **2:** 861–863.
24. THOMSEN, C., O. HENRIKSEN & P. RING. 1987. *In vivo* measurement of water self diffusion in the human brain by magnetic resonance imaging. Acta Radiol. **28:** 353–361.
25. BEAULIEU, C. 2002. The basis of anisotropic water diffusion in the nervous system—a technical review. NMR Biomed. **15:** 435–455.
26. PIERPAOLI, C. & P.J. BASSER. 1996. Toward a quantitative assessment of diffusion anisotropy. Magn. Reson. Med. **36:** 893–906.
27. PFEFFERBAUM, A. *et al.* 2000. Age-related decline in brain white matter anisotropy measured with spatially corrected echo-planar diffusion tensor imaging. Magn. Reson. Med. **44:** 259–268.
28. ABE, O. *et al.* 2002. Normal aging in the central nervous system: quantitative MR diffusion-tensor analysis. Neurobiol. Aging **23:** 433–441.
29. HEAD, D. *et al.* 2004. Differential vulnerability of anterior white matter in nondemented aging with minimal acceleration in dementia of the Alzheimer type: evidence from diffusion tensor imaging. Cereb. Cortex **14:** 410–423.
30. SALAT, D.H. *et al.* 2005. Age-related alterations in white matter microstructure measured by diffusion tensor imaging. Neurobiol. Aging **26:** 1215–1227.
31. NUSBAUM, A.O. *et al.* 2001. Regional and global changes in cerebral diffusion with normal aging. AJNR Am. J. Neuroradiol. **22:** 136–142.
32. SULLIVAN, E.V. *et al.* 2001. Equivalent disruption of regional white matter microstructure in ageing healthy men and women. Neuroreport **12:** 99–104.
33. O'SULLIVAN, M. *et al.* 2001. Evidence for cortical "disconnection" as a mechanism of age-related cognitive decline. Neurology **57:** 632–638.
34. DECARLI, C. *et al.* 2005. Anatomical mapping of white matter hyperintensities (WMH): exploring the relationships between periventricular WMH, deep WMH, and total WMH burden. Stroke **36:** 50–55.
35. BRAFFMAN, B.H. *et al.* 1988. Brain MR: pathologic correlation with gross and histopathology. 1. Lacunar infarction and Virchow-Robin spaces. AJR Am. J. Roentgenol. **151:** 551–558.
36. RAZ, N., K.M. RODRIGUE & J.D. ACKER. 2003. Hypertension and the brain: vulnerability of the prefrontal regions and executive functions. Behav. Neurosci. **117:** 1169–1180.
37. HANYU, H. *et al.* 1998. Diffusion-weighted MR imaging of the hippocampus and temporal white matter in Alzheimer's disease. J. Neurol. Sci. **156:** 195–200.

38. Rosas, H.D. *et al.* 2005. Early microstructural changes in Huntington's disease: a diffusion tensor imaging study. Submitted.
39. Madden, D.J. *et al.* 2004. Diffusion tensor imaging of adult age differences in cerebral white matter: relation to response time. Neuroimage **21:** 1174–1181.
40. Tuch, D.S. *et al.* 2005. Choice reaction time performance correlates with diffusion anisotropy in white matter pathways supporting visuospatial attention. Proc. Natl. Acad. Sci. USA **102:** 12212–12217.
41. Rypma, B. & M. D'Esposito. 2000. Isolating the neural mechanisms of age-related changes in human working memory. Nat. Neurosci. **3:** 509–515.
42. Logan, J.M. *et al.* 2002. Under-recruitment and nonselective recruitment: dissociable neural mechanisms associated with aging. Neuron **33:** 827–840.
43. Stebbins, G.T. *et al.* 2002. Aging effects on memory encoding in the frontal lobes. Psychol. Aging **17:** 44–55.
44. Kensinger, E.A. & S. Corkin. 2004. Two routes to emotional memory: distinct neural processes for valence and arousal. Proc. Natl. Acad. Sci. USA **101:** 3310–3315.
45. Pfefferbaum, A. & E.V. Sullivan. 2003. Increased brain white matter diffusivity in normal adult aging: relationship to anisotropy and partial voluming. Magn. Reson. Med. **49:** 953–961.
46. Liu, C. *et al.* 2004. Self-navigated interleaved spiral (SNAILS): application to high-resolution diffusion tensor imaging. Magn. Reson. Med. **52:** 1388–1396.
47. Tuch, D.S. 2004. Q-ball imaging. Magn. Reson. Med. **52:** 1358–1372.

Diffusion Tensor Imaging of the Spinal Cord

STEPHAN E. MAIER AND HATSUHO MAMATA

Department of Radiology, Brigham and Women's Hospital, Harvard Medical School, Boston, Massachusetts 02115, USA

ABSTRACT: The spinal cord is an important part of the nervous system and provides the connection of the brain with the periphery. It consists not only of a large number of longitudinal fibers, but also contains collateral fibers and a central gray matter structure, which are part of autonomous circuits. Magnetic resonance diffusion tensor imaging can reveal this complex fiber architecture in great detail. This report summarizes the normal findings for ADC, diffusion anisotropy, and diffusion eigenvector directions in the spinal cord. Sagittal and axial diffusion-weighted images of the spinal cord were obtained with line scan diffusion imaging (LSDI) in adults, children, infants, and a spinal cord specimen.

KEYWORDS: spinal cord; collateral nerve fibers; white matter; gray matter; second eigenvector; LSDI

INTRODUCTION

Diffusion tensor imaging of the brain is well established. The existing magnetic resonance (MR) diffusion imaging methods permit reliable assessment of the three-dimensional nerve fiber anatomy in the normal brain, during normal development, and in disease. While the basic correlation between the observed diffusion tensor geometry and nerve fiber direction[1,2] also applies to the spinal cord, diffusion tensor imaging is not as readily and reliably performed in the spinal cord as it is in the brain. The reasons for this difficulty are diverse:

- *(a)* The presence of nearby bony structures causes susceptibility variations.[3] Although smaller than the susceptibility variations at air-tissue interfaces, they nevertheless may lead to serious image distortions.
- *(b)* Chemical-shift artifacts arise from lipids in the vertebral bodies and other nearby structures.
- *(c)* The surrounding cerebrospinal fluid (CSF) is likely to confound the measurements due to partial volume effects.
- *(d)* Motion of the CSF and organs in the thorax and abdomen may lead to ghosting artifacts, unless single-shot diffusion imaging techniques are employed.

Address for correspondence: Stephan E. Maier, M.D., Ph.D., Radiology (MRI), Brigham and Women's Hospital, Harvard Medical School, 75 Francis Street, Boston, MA 02115. Voice: 617-732-5065; fax: 617-264-5275.
stephan@bwh.harvard.edu

**Ann. N.Y. Acad. Sci. 1064: 50–60 (2005). © 2005 New York Academy of Sciences.
doi: 10.1196/annals.1340.011**

(e) Finally, the small cross section requires high-resolution imaging in order to obtain any meaningful anatomical information. Since the spinal cord runs deep inside the body, it is difficult to achieve adequate signal-to-noise ratio (SNR) in conjunction with high spatial resolution.

The first reported *in vivo* spinal cord diffusion images in humans[4] were obtained with a navigated pulsed-gradient spin-echo sequence. This acquisition method yields images largely free of susceptibility-related distortions and chemical-shift artifacts, but requires cardiac gating and a scan time of 30 min or more for the measurement of the diffusion tensor. Dietrich *et al.*[5] used an improved method with self-navigated radial *k*-space trajectories to obtain images of the spinal cord. All more recent studies with navigator correction used interleaved echo-planar diffusion imaging[6–10] to achieve a considerably shorter scan time. Fast spin-echo diffusion-weighted sequences[11,12] have also proven useful to obtain diffusion-weighted images of the spinal cord void of severe image artifacts. However, the technique is considered inferior to the navigated methods,[13] particularly due to a low SNR. Single-shot echo-planar techniques, despite the tendency to produce images with the most severe distortion artifacts, have likewise been employed for spinal cord diffusion-weighted imaging.[14] These artifacts have been successfully diminished by combining single-shot echo-planar imaging with reduced field-of-view acquisition[15] or parallel imaging methods.[16,17]

For the work presented here, we used line scan diffusion imaging (LSDI),[18,19] a spin-echo technique that relies on the sequential single-shot acquisition of columns. LSDI is relatively insensitive to magnetic field inhomogeneities, eddy currents, and bulk motion, and can provide rectangular images, diffusion-weighted along 6 encoding axes in as little as 25 s per section.[20] These properties make LSDI well suited to spinal cord imaging, and application of this technique at 0.2 and 1.5 tesla in normal subjects and patients (infants, children, and adults) has been demonstrated.[20–25]

The spinal cord consists of a peripheral zone with white matter and a central zone with gray matter. The white matter contains a large number of longitudinally running fibers, which interconnect the brain with the periphery. The gray matter, unlike cortical gray matter, includes a large number of myelinated fibers, which arise in the white matter at an angle with the longitudinal tracts. These collateral nerve fibers interconnect different areas and levels of the spinal cord and form part of many functional connections within the spinal cord, such as the spinothalamic tract and the deep spinal reflex circuit. The reflex circuit enables rapid muscle contraction, without any brain intervention, after an appropriate sensory stimulus, for example, heat. Here, we show how diffusion tensor imaging can differentiate spinal cord white and gray matter as well as detect longitudinal and collateral fiber architecture.

MATERIALS AND METHODS

Normal Subjects, Patients, and Spinal Cord Specimen

This report summarizes our experience with LSDI tensor imaging in the human spinal cord. In particular, we have obtained LSDI data in the spinal cord of 15 normal subjects[23,25] at ages between 32 and 48 years, 11 premature born infants[21] at post-menstrual ages (PMA) between 28 and 40 weeks, 10 children[20] with different pathologies at ages between 2 months and 18 years, and 72 spondylosis patients[25] with ages ranging between 26 and 82 years. Volunteer MRI localizer scans revealed no abnormalities of the spinal cord.

Details of the human spinal cord fiber architecture were also studied in a spinal cord specimen,[26] which was excised at the C1–C2 level of a female, 34-year-old deceased patient, who had no known abnormality of the spinal cord. The specimen was suspended in physiological saline solution at approximately 22°C and scans were performed within 24 h after patient death. Since diffusion in the nonvital tissue sample at this temperature was much lower than under normal vital conditions, a very high *b*-factor of 2000 s/mm^2 was used. After axial MR diffusion tensor scans, the specimen was fixed in formaldehyde, and histology sections, from locations congruent with the slices scanned, were stained according to Bielschowsky's method of silver impregnation. Subimages of the histology sample were digitally captured with a DP11 microscope camera (Olympus America, Melville, NY) in a loss-less compressed 24-bit color TIFF format. Each of the subimages contained 1712×1368 pixels at a resolution of 0.53 μm per pixel. The collection of subimages was then combined into a mosaic to permit a composite view of the tissue sample. Contraction of the specimen due to the histology preparation process was adjusted. From the composite image, subimages, congruent to the area of individual pixels within the MR scan matrix, were extracted. For each extracted subimage, the course of collateral fibers was determined by visual inspection and compared with the findings of the MR diffusion tensor measurements.

All studies were conducted within the guidelines of the institutional internal review boards (IRBs) at Brigham and Women's Hospital and Children's Hospital, Boston.

Magnetic Resonance Imaging (MRI)

Diffusion tensor MRI scans were performed with LSDI on 1.5-tesla, whole-body LX Horizon Echospeed and Highspeed MRI systems (General Electric Medical Systems, Milwaukee, WI) with 40 and 22 mT/m magnetic field gradients, respectively. The imaging protocol included T1-weighted spin-echo localizing scans and LSDI tensor scans with the imaging parameters listed in TABLE 1. The parameters listed should be considered guidelines, and particularly parameters such as field of view (FOV), *b*-factor, and echo time TE may be slightly different from what was used in the earlier studies.[20,21,23,25] These protocols can easily be adjusted for imaging at different field strengths: that is, at 3 tesla, a higher receiver bandwidth of ±8 kHz or fat signal suppression is advisable to avoid severe chemical-shift artifacts; in contrast, at lower field strength, a lower receiver bandwidth and more signal averaging is recommended. Diffusion weighting was applied along six non-collinear and non-coplanar directions [(Gx, Gy, Gz) = (1, 1, 0), (0, 1, 1), (1, 0, 1), (0, 1, −1), (1, −1, 0), (−1, 0, 1)]. The T2-weighted images that were acquired to compute the apparent diffusion coefficients (ADCs) included crusher gradients that produced a slight diffusion weighting of 5 s/mm^2. At this low diffusion weighting, only two gradient directions were collected since the directional dependence of the MR signal was considered negligible. Image processing, which included the interpolation to a 256×256 square image matrix and calculation of the diffusion eigenvectors and eigenvalues for each image voxel, was performed off-line. Average values of ADC

TABLE 1. LSDI tensor scan protocols suitable for spinal cord diffusion imaging at 1.5 tesla

Protocol	Fast spinal cord scan	High resolution spinal cord scan	High resolution/ high SNR spinal cord scan	Spinal cord specimen scan
Section	Sagittal	Sagittal	Axial	Axial
Field of view (mm)	220×110	220×110	160×80	80×40
Slice thickness (mm)	3	3	4	3
Scan matrix size	128×64	128×128	128×64	128×64
Receiver coil	Phased array	Phased array	5-inch surface	3-inch surface
Bandwidth (kHz)	±3.91	±3.91	±3.91	±3.91
TR/TE (ms)	1650/59	3311/59	1694/63	2442/76
b-factor (s/mm^2)	750	750	750	2000
Excitations	1	2	16	16
Scan time per slice	40 s	2.5 min	13 min	16 min

and anisotropy were then determined for manually drawn regions of interest with the aid of the XPhase program, a software written by one of the authors. The same software was used to display the eigenvector directions.

RESULTS

Image data from an LSDI spinal cord tensor scan is presented in FIGURE 1. There is no evidence of severe image distortion, ghosting, or motion-related signal loss. On the diffusion-weighted image and the ADC map, the spinal cord appears largely uniform; in contrast, on the fractional anisotropy (FA) map, a narrow longitudinal strip along the ventral side of the spinal cord exhibits a lower anisotropy value than the rest of the spinal cord. FIGURE 2 shows a plot of average spinal cord ADC values of individuals versus age. The highest ADC values are observed in the spinal cord of infants and the lowest in adults with an age range of 30 to 50 years. In subjects older than 50 years, the spinal cord ADC appears to be higher, but on average not as high as in children. The mean ADC was 1.25 ± 0.10 μm^2/ms in infants; 0.96 ± 0.05 μm^2/ms in children; 0.81 ± 0.03 μm^2/ms in normal adults at the relatively wide C2–C3 level; and 0.86 ± 0.08 μm^2/ms in spondylosis patients also at the relatively wide C2–C3 level, presumably unaffected by spondylosis. Average spinal cord FA values of individual adults versus age are presented in the plot of FIGURE 3. Contrary to the ADC values, there appears to be a slight negative correlation with age. The mean FA was 0.70 ± 0.05 in normal adults at the C2–C3 level and 0.69 ± 0.06 in spondylosis patients at the same level. For the diffusion tensor study in premature born infants,[21] the relative anisotropy (RA)[27] was measured and the average of the 11 infants scanned equaled 0.243 ± 0.049. The mean RA value in the spinal cord of 2 adults was 0.425. For the diffusion tensor study in children,[20] anisotropy was assessed as the ratio of the maximum to minimum of the ADC values measured along the 3 or 6

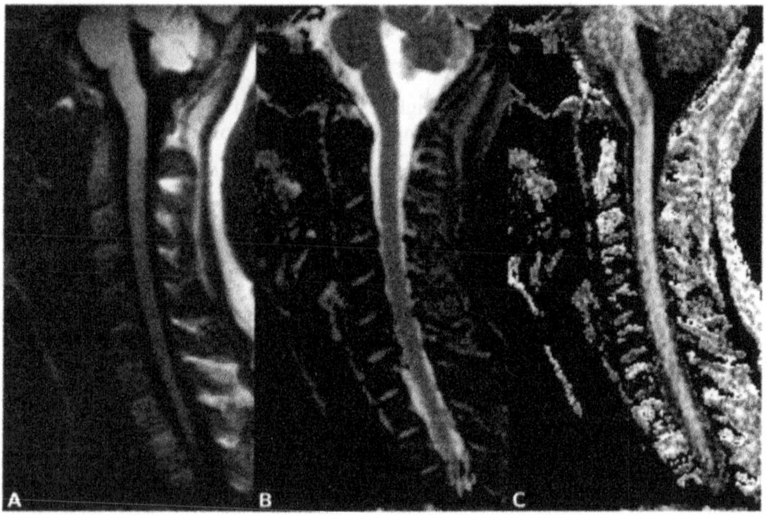

FIGURE 1. Sagittal line scan diffusion tensor image data of the cervical spinal cord in a healthy 40-year-old female volunteer. Images were acquired according to the *High Resolution Spinal Cord Scan* protocol of TABLE 1. **(A)** Diffusion-weighted image reconstructed from image data along 6 diffusion encoding directions and extrapolated to $b = 1000$ s/mm^2. **(B)** Trace ADC map. **(C)** Fractional anisotropy (FA) map. Note the band of slightly lower FA values in the ventral portion of the spinal cord.

diffusion encoding directions. The average spinal cord ADC ratio of the 10 children studied amounted to 1.70 ± 0.42.

Axial scans of the cervical spinal cord were performed in 4 normal volunteers.[23] An FA map is presented in FIGURE 4. On the axial map, it can be clearly seen that spinal cord white and gray matter exhibit different FA values. The mean FA values in white and gray matter of the 4 subjects scanned were 0.65 ± 0.04 and 0.34 ± 0.05, respectively. Without signal averaging, the FA values were higher, that is, 0.72 ± 0.03 in white matter and 0.56 ± 0.04 in gray matter. The mean ADC was 0.87 ± 0.04 μm^2/ms in white matter and 0.81 ± 0.02 μm^2/ms in gray matter. In agreement with the underlying fiber anatomy, the first eigenvectors point in the direction of the longitudinal fibers and spine nerve roots. The second eigenvectors are arranged in a centrifugal pattern, which was also observed in the spinal cord specimen shown in FIGURE 5. Visual inspection (see FIG. 5B) revealed that the second eigenvector follows the direction of the collateral fibers. For a more quantitative validation of this observation, the in-plane direction of the second eigenvector of each pixel of the MR scan matrix was compared with the estimated aggregate direction of the collateral fibers within the corresponding area of the histology section shown in FIGURE 5. For the 146 pixels present in the entire spinal cord cross section, the mean angle difference between the directions determined with different modalities was $0.5 \pm 8.5°$.

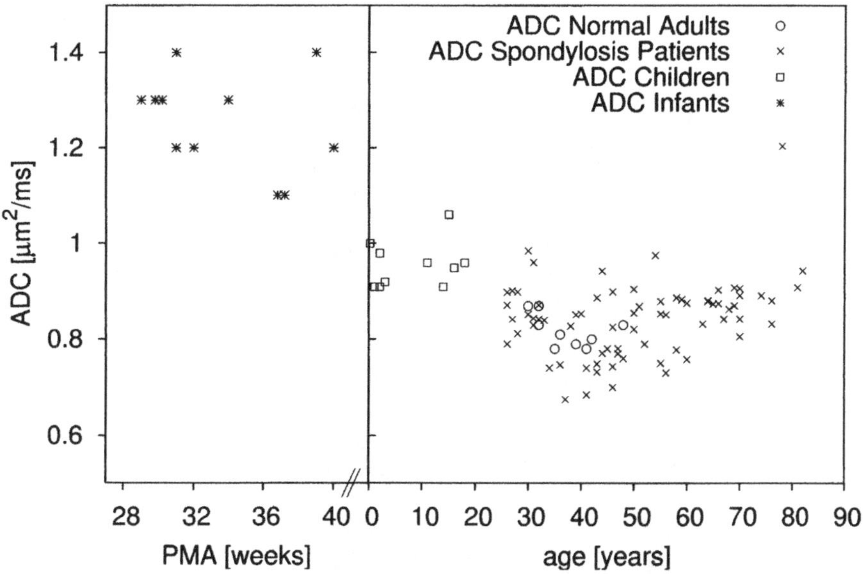

FIGURE 2. Trace ADC values of the spinal cord in 11 premature born infants[21] at post-menstrual ages (PMA) between 28 and 40 weeks, 10 children[20] with different pathologies at ages between 2 months and 18 years, 11 normal adults at ages between 32 and 48 years, and 72 spondylosis patients[25] with ages ranging between 26 and 82 years. ADC data were determined in the cervical spinal cord, with the exception of the ADC data obtained in 3 children, where the ADC was measured in the thoracal spinal cord.

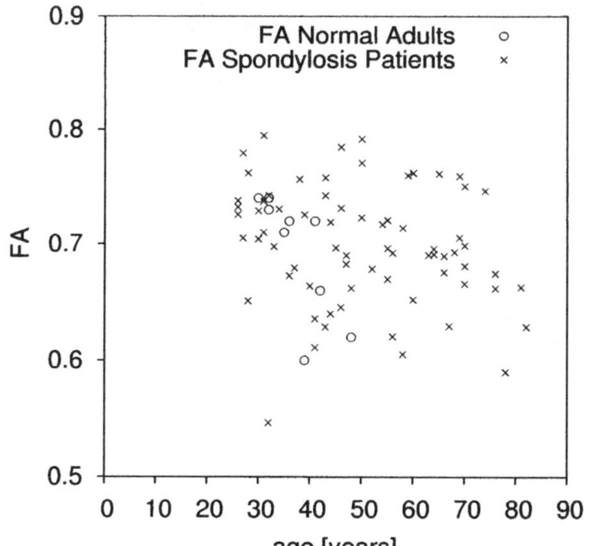

FIGURE 3. FA values of the cervical spinal cord in 11 normal adults between 32 and 48 years of age, and in 72 spondylosis patients[25] with ages ranging between 26 and 82 years.

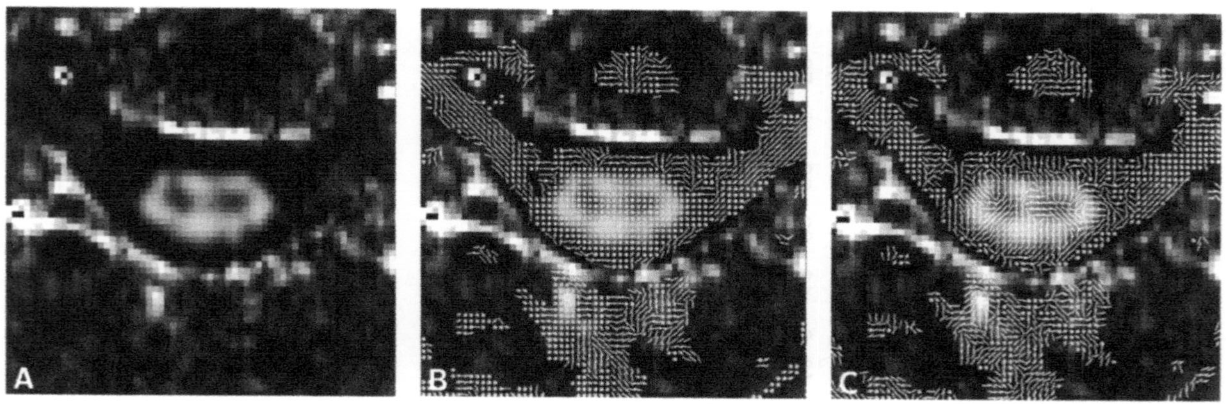

FIGURE 4. Axial line scan diffusion tensor image data obtained at the C6 level of the cervical spinal cord in a healthy 32-year-old female volunteer. Images were acquired according to the *High Resolution/High SNR Spinal Cord Scan* protocol of TABLE 1. **(A)** FA map showing an area of 40×40 mm. In the center of the image, the spinal cord, surrounded by CSF, which appears dark due to low anisotropy, can be seen. Within the spinal cord cross section, the dark central zone of low anisotropy corresponds to the central gray matter, whereas the surrounding bright zone of high anisotropy corresponds to the peripheral white matter. **(B)** The same FA map as in part A overlaid with the first eigenvector directions. The out-of-plane component is represented by dots of different sizes and the in-plane component by lines. The eigenvector map has been thresholded based on the intensity of the diffusion-weighted image. In agreement with known anatomy, within the spinal cord the first eigenvector is largely orthogonal to the axial plane. Right and left to the spinal cord, the eigenvectors clearly follow the direction of the spine nerve roots. **(C)** The same FA map as in part A overlaid with the second eigenvector directions. The second eigenvector directions exhibit a striking arrangement, which defies the notion of pure cylindrical diffusion[31] with random second eigenvector directions.

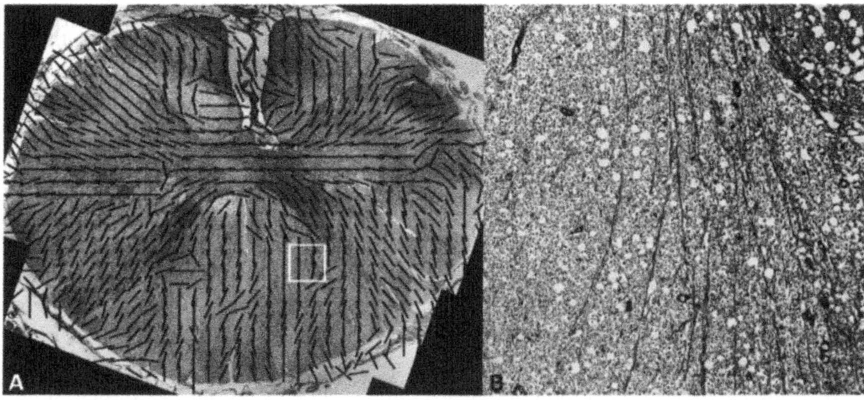

FIGURE 5. Human cervical spinal cord histology stained according to Bielschowsky's method of silver impregnation. **(A)** Complete axial cross section with overlaid second diffusion eigenvector directions, which were measured in the specimen with line scan diffusion imaging prior to obtaining the histology. **(B)** Detail of the histology shown in part A. The position of the area shown is indicated by a white square in part A. The predominant caudiocranial running fibers can be seen as small dots. Gray matter is present in the upper right corner. Collateral fibers running in-plane, which evidently share the same direction as the second eigenvectors shown in part A, are clearly visualized.

DISCUSSION

A large number of successful spinal cord scans are evidence of the reliability of LSDI. The measured ADC and FA values agree with the results of other studies, which used navigated pulsed-gradient multishot spin-echo diffusion imaging or single-shot fast spin-echo diffusion imaging. The age dependence of ADC (i.e., highest ADC during infancy, lowest ADC during early adulthood, and a slight ADC increase with higher age) and FA (i.e., lowest FA during infancy, highest FA during early adulthood, and a slight FA decrease with higher age) has also been observed in the brain.[28,29] However, not necessarily all of these changes may really be present in the spinal cord tissue, but may rather be a manifestation of the limited spatial resolution together with CSF partial volume effects and age-related changes in spinal cord geometry. Unfortunately, the FA values for the scans obtained in infants and children were not available. Nevertheless, the infant RA values were substantially lower than the RA values in 2 adults. Under the assumption of cylindrical diffusion, an FA value of 0.32 can be derived from the ADC ratio that was measured in children. However, since the ratio was not derived from the eigenvalues, but from the maximum and minimum ADC along 3 or 6 directions measured, the true FA value certainly would be higher. Independent of age, CSF contamination may lead to an overestimation of ADC values and an underestimation of FA values. With the anisotropic voxel geometry that was used, the contamination should be less severe for axial than sagittal scans. However, based on the very similar ADC and FA values observed in the sagittal spinal cord sections and the white matter area of the axial sections, the CSF contribution seems to be negligible.

Diffusion in spinal cord white and gray matter is clearly different. The narrow longitudinal strip of reduced FA along the ventral side of the spinal cord that is evident on sagittal high-resolution scans can most likely be attributed to the gray matter core of the spinal cord. In agreement with the absence of such an enhancement on the sagittal ADC map, the gray and white matter ADC values measured on the axial ADC map were similar.

A factor that strongly determines the accuracy of the anisotropy measurements is the low SNR of single-shot diffusion-weighted scans. Without signal averaging, FA may be overestimated in areas of low anisotropy.[23,30] Indeed, the gray and white matter FA values were significantly higher when only a single measurement was considered instead of all 16 available. Thus, the sagittal FA data, which agree perfectly with the axial FA white matter data without signal averaging, most likely are slightly overestimated.

Without the presence of secondary structures, such as collateral fibers, spinal cord diffusion would be purely cylindrical[31] and noise would randomize the second eigenvector directions. Alignment of the second eigenvector in the presence of crossing fibers has also been observed in the brain.[32] The comparison between MR diffusion tensor and histology confirms the alignment between second eigenvector direction and collateral fiber direction. The identical finding of an orderly centrifugal second eigenvector alignment in normal subjects confirms that it is not merely an artifact present in nonvital tissue. The fiber architecture in spinal cord is special insofar as the secondary collateral nerve fibers run at a right angle to the primary longitudinal fibers and, thus, the simplified diffusion tensor model with 3 orthogonal eigenvector directions is perfectly adequate. In the presence of nonorthogonal fiber crossings, fiber architecture can be assessed by measuring tissue diffusion properties along a large number of directions.[33]

In summary, MR diffusion tensor imaging reveals the spinal cord fiber architecture in great detail, although improvements in SNR and spatial resolution are highly desirable. Detailed knowledge of spinal cord ADC and FA values and their changes during normal development and aging form an important basis for the interpretation of spinal cord diffusion tensor data under pathological conditions.

ACKNOWLEDGMENTS

This research was supported by a grant from the National Institute of Neurological Disorders and Stroke (NIH R01 NS39335 to S. E. Maier).

REFERENCES

1. Hsu, E.W., A.L. Muzikant, S.A. Matulevicius *et al.* 1998. Magnetic resonance myocardial fiber-orientation mapping with direct histological correlation. Am. J. Physiol. **274**(5, part 2): H1627–H1634.
2. Lin, C.P., W.Y. Tseng, H.C. Cheng & J.H. Chen. 2001. Validation of diffusion tensor magnetic resonance axonal fiber imaging with registered manganese-enhanced optic tracts. NeuroImage **14:** 1035–1047.
3. Hopkins, J.A. & F.W. Wehrli. 1997. Magnetic susceptibility measurement of insoluble solids by NMR: magnetic susceptibility of bone. Magn. Reson. Med. **37:** 494–500.

4. CLARK, C.A., G.J. BARKER & P.S. TOFTS. 1999. Magnetic resonance diffusion imaging of the human cervical spinal cord *in vivo*. Magn. Reson. Med. **41**(6): 1269–1273.
5. DIETRICH, O., A. HERLIHY, W. DANNELS *et al.* 2001. Diffusion-weighted imaging of the spine using radial k-space trajectories. MAGMA **12**(1): 23–31.
6. BAMMER, R., F. FAZEKAS, M. AUGUSTIN *et al.* 2000. Diffusion-weighted MR imaging of the spinal cord. Am. J. Neuroradiol. **21**(3): 587–591.
7. RIES, M., R.A. JONES, V. DOUSSET & C.T. MOONEN. 2000. Diffusion tensor MRI of the spinal cord. Magn. Reson. Med. **44**(6): 884–892.
8. HOLDER, C.A., R. MUTHUPILLAI, S. MUKUNDAN *et al.* 2000. Diffusion-weighted MR imaging of the normal human spinal cord *in vivo*. Am. J. Neuroradiol. **21**(10): 1799–1806.
9. DEMIR, A., M. RIES, C.T. MOONEN *et al.* 2003. Diffusion-weighted MR imaging with apparent diffusion coefficient and apparent diffusion tensor maps in cervical spondylotic myelopathy. Radiology **229**(1): 37–43.
10. ZHANG, J., Y. HUAN, Y. QIAN *et al.* 2005. Multishot diffusion-weighted imaging features in spinal cord infarction. J. Spinal Disord. Tech. **18**(3): 277–282.
11. LE ROUX, P., A. DARQUIE, P.G. CARLIER & C.A. CLARK. 2002. Feasibility study of non Carr Purcell Meiboom Gill single shot fast spin echo in spinal cord diffusion imaging. MAGMA **14**(3): 243–247.
12. TSUCHIYA, K., S. KATASE, A. FUJIKAWA *et al.* 2003. Diffusion-weighted MRI of the cervical spinal cord using a single-shot fast spin-echo technique: findings in normal subjects and in myelomalacia. Neuroradiology **45**(2): 90–94.
13. BAMMER, R., M. AUGUSTIN, R. PROKESCH *et al.* 2002. Diffusion-weighted imaging of the spinal cord: interleaved echo-planar imaging is superior to fast spin-echo. J. Magn. Reson. Imaging **15**(4): 364–373.
14. NAGAYOSHI, K., S. KIMURA, M. OCHI *et al.* 2000. Diffusion-weighted echo planar imaging of the normal human cervical spinal cord. J. Comput. Assisted Tomogr. **24**(3): 482–485.
15. WHEELER-KINGSHOTT, C.A., S.J. HICKMAN, G.J. PARKER *et al.* 2002. Investigating cervical spinal cord structure using axial diffusion tensor imaging. NeuroImage **16**(1): 93–102.
16. CERCIGNANI, M., M.A. HORSFIELD, F. AGOSTA & M. FILIPPI. 2003. Sensitivity-encoded diffusion tensor MR imaging of the cervical cord. Am. J. Neuroradiol. **24**(6): 1254–1256.
17. TSUCHIYA, K., A. FUJIKAWA & Y. SUZUKI. 2005. Diffusion tractography of the cervical spinal cord by using parallel imaging. Am. J. Neuroradiol. **26**(2): 398–400.
18. GUDBJARTSSON, H., S.E. MAIER, R.V. MULKERN *et al.* 1996. Line scan diffusion imaging. Magn. Reson. Med. **36**(4): 509–519.
19. MAIER, S.E., H. GUDBJARTSSON, S. PATZ *et al.* 1998. Line scan diffusion imaging: characterization in healthy subjects and stroke patients. Am. J. Roentgenol. **17**(1): 85–93.
20. ROBERTSON, R.L., S.E. MAIER, R.V. MULKERN *et al.* 2000. MR line-scan diffusion imaging of the spinal cord in children. Am. J. Neuroradiol. **21:** 1344–1348.
21. MURPHY, B.P., G.P. ZIENTARA, P.S. HUPPI *et al.* 2001. Line scan diffusion tensor MRI of the cervical spinal cord in preterm infants. J. Magn. Reson. Imaging **13:** 949–953.
22. HORI, M., T. OKUBO, S. AOKI *et al.* 2002. Line scan diffusion weighted imaging (LSDI) on 0.2 tesla MRI of the normal cervical cord *in vivo*: preliminary study. Nippon Igaku Hoshasen Gakkai Zasshi **62**(5): 221–223.
23. MAMATA, H., F.A. JOLESZ & S.E. MAIER. 2004. Characterization of central nervous system structures by magnetic resonance diffusion anisotropy. Neurochem. Int. **45**(4): 553–560.
24. SHINOYAMA, M., T. TAKAHASHI, H. SHIMIZU *et al.* 2005. Spinal cord infarction demonstrated by diffusion-weighted magnetic resonance imaging. J. Clin. Neurosci. **12**(4): 466–468.
25. MAMATA, H., F.A. JOLESZ & S.E. MAIER. 2005. Apparent diffusion coefficient and fractional anisotropy in spinal cord: age and cervical-spondylosis related changes. J. Magn. Reson. Imaging **22**(1): 38–43.
26. MAMATA, H., C.F. WESTIN, U. DE GIROLAMI *et al.* 2002. Visualization of collateral nerve fibers of human cervical spinal cord: direct histologic validation of diffusion tensor imaging. *In* Book of Abstracts: Tenth Annual Meeting, Hawaii. Society of Magnetic Resonance. Berkeley, CA.
27. PAPADAKIS, N.G., D. XING, G.C. HOUSTON *et al.* 1999. A study of rotationally invariant and symmetric indices of diffusion anisotropy. Magn. Reson. Imaging **17**(6): 881–892.

28. HUPPI, P.S., S.E. MAIER, S.S. PELED *et al.* 1998. Microstructural development of the human newborn cerebral white matter assessed *in vivo* by diffusion tensor MRI. Pediatr. Res. **44**(4): 584–590.
29. PFEFFERBAUM, A., E.V. SULLIVAN, M. HEDEHUS *et al.* 2000. Age-related decline in brain white matter anisotropy measured with spatially corrected echo-planar diffusion tensor imaging. Magn. Reson. Med. **44**(2): 259–268.
30. BASTIN, M., P. ARMITAGE & I. MARSHALL. 1998. A theoretical study of the effect of experimental noise on the measurement of anisotropy in diffusion imaging. Magn. Reson. Imaging **16**(7): 773–785.
31. WESTIN, C.F., S.E. MAIER, H. MAMATA *et al.* 2002. Processing and visualization for diffusion tensor MRI. Med. Image Anal. **6**(2): 93–108.
32. WIEGELL, M.R., H.B. LARSSON & V.J. WEDEEN. 2000. Fiber crossing in human brain depicted with diffusion tensor MR imaging. Radiology **217**(3): 897–903.
33. TUCH, D.S., T.G. REESE, M.R. WIEGELL *et al.* 2002. High angular resolution diffusion imaging reveals intravoxel white matter fiber heterogeneity. Magn. Reson. Med. **48**(4): 577–582.

Amyotrophic Lateral Sclerosis and Primary Lateral Sclerosis

The Role of Diffusion Tensor Imaging and Other Advanced MR-Based Techniques as Objective Upper Motor Neuron Markers

SUMEI WANG AND ELIAS R. MELHEM

Department of Radiology, Division of Neuroradiology, University of Pennsylvania, Philadelphia, Pennsylvania, USA

ABSTRACT: Amyotrophic lateral sclerosis (ALS), also called Lou Gehrig's disease, is a motor neuron disease characterized by progressive degeneration of upper motor neuron (UMN) and lower motor neuron (LMN), while primary lateral sclerosis (PLS) is defined by pure UMN involvement. A reliable objective marker of UMN involvement is critical for the early diagnosis and monitoring of disease progression in patients with ALS and PLS. Diffusion tensor imaging (DTI), magnetization transfer imaging (MTI), and magnetic resonance spectroscopy (MRS), which provide insight into the pathophysiological process of ALS and PLS, show great promise in this regard. Further investigation is needed to determine and to compare the utility of various neuroimaging markers.

KEYWORDS: amyotrophic lateral sclerosis (ALS); primary lateral sclerosis (PLS); diffusion tensor imaging (DTI); magnetization transfer imaging (MTI); magnetic resonance spectroscopy (MRS); upper motor neuron (UMN)

INTRODUCTION

Amyotrophic lateral sclerosis (ALS), also called Lou Gehrig's disease, is a motor neuron disease characterized by progressive degeneration of upper motor neuron (UMN) and lower motor neuron (LMN),[1,2] while primary lateral sclerosis (PLS) is defined by pure UMN involvement.[3] The incidence and prevalence of ALS are about 2 per 100,000 annually and 8 per 100,000, respectively. In the United States, approximately 30,000 people are living with ALS. There are about 5000 new cases per year, or 15 per day.[4]

There are two known types of ALS: sporadic and familial. The sporadic type is the most common one and accounts for 90% of ALS cases. Familial ALS is hereditary, which is passed on by a dominant gene and makes up the remaining 10% of cases.[4,5]

Address for correspondence: Elias R. Melhem, M.D., Department of Radiology, Division of Neuroradiology, Hospital of the University of Pennsylvania, 3400 Spruce Street, Second Floor Dulles, Philadelphia, PA 19104. Voice: 215-662-6865; fax: 215-662-3283.
emelhem@rad.upenn.edu

**Ann. N.Y. Acad. Sci. 1064: 61–77 (2005). © 2005 New York Academy of Sciences.
doi: 10.1196/annals.1340.013**

Patients with ALS experience a relentlessly progressive paralysis of the skeletal muscles, resulting in loss of mobility, loss of the ability to speak and eat, and eventual loss of respiratory function, making this disease one of the most devastating diseases.[2] PLS patients typically manifest as prominent spasticity, with only slight weakness in the lower limbs, eventually leading to pseudobulbar symptoms (dysarthria and compulsive laughing or crying).[3,6] The cause of ALS and PLS is unknown. Approximately 2% of cases are due to mutations in the superoxide dismutase (SOD1) gene.[5]

No effective medical or surgical cure exists for ALS and PLS. Only one drug, riluzole, has been found to be somewhat effective in prolonging life. Fifty percent of ALS patients die within 18 months after diagnosis. Only 20% survive 5 years, and 10% live longer than 10 years.[5] The mean survival of PLS patients after the onset of symptoms (~18–19 years) is substantially longer than that of ALS.[3,6]

Effective and objective disease markers are needed to diagnose and monitor disease progression in ALS and PLS. Early diagnosis is critical because a delay may result in loss of irreversible motor function that may not be corrected by therapeutic interventions. Moreover, longitudinal studies will be required to monitor the disease progression and treatment efficacy.

PATHOLOGY

The classical neuropathological features of ALS include loss and degeneration of the large motor neurons in the gray matter of the spinal cord, brain stem, and cortex, as well as degeneration of the corticospinal tracts (CSTs) that contain axons of the cortical UMNs. Other extramotor systems are also involved to various degrees.[7–9]

Histological studies of ALS patients demonstrate loss of the giant pyramidal Betz cells in cortical layer 5 accompanied by reactive gliosis, which is demonstrable as astrocytosis and as a diffuse microgliosis in the motor cortex.[7] Myelin pallor, secondary to CST degeneration, is the most conspicuous change along the pathway. This change is most evident in the internal capsule.[9] Involvement of spinal cord can include loss of anterior horn cells (AHCs), shrinkage of remaining AHCs, and gliosis. Lipofuscin deposits also accumulate in AHCs, although some postulate that this is an artifact of cell atrophy.[10]

Pathological changes in PLS include neuron loss and CST degeneration, which is similar to ALS. However, the AHCs of the spinal cord are well preserved.[6] Some reported that PLS can also be distinguished from ALS by its extensive atrophy in the precentral gyrus.[11,12]

DIAGNOSIS OF UMN INVOLVEMENT

The diagnosis of ALS and PLS is currently based on clinical features, electromyography (EMG), and exclusion of other diseases with similar symptoms.[2] LMN signs include weakness, muscle wasting, muscle cramps, fasciculations, and eventually hyporeflexia. UMN signs include hyperreflexia, extensor plantar response, increased muscle tone, and weakness in a topographical representation.[2] ALS is diagnosed when UMN and LMN signs are present, while PLS is diagnosed when UMN signs are found in the absence of LMN signs. EMG and muscle biopsy can

TABLE 1. Revised El Escorial criteria for ALS

- **Suspected ALS:** Condition is considered a pure LMN syndrome.
- **Possible ALS:** UMN and LMN signs are present together in one region, or UMN signs alone in two or more regions, or UMN signs caudal to LMN signs.
- **Clinically probable laboratory-supported ALS:** Clinical signs of UMN and LMN dysfunction are present in only one region, or UMN signs alone are present in one region, and LMN signs defined by EMG criteria are present in at least two limbs, with proper application of neuroimaging and clinical laboratory protocols to exclude other causes.
- **Probable ALS:** UMN and LMN signs are present in more than two regions, but some UMN signs must be rostral to LMN signs.
- **Definite ALS:** UMN and LMN signs are present in more than three regions.

identify LMN involvement, whereas UMN dysfunction has been assessed solely by neurological examination. The revised El Escorial criteria classify ALS as suspected, possible, probable, or definite according to the UMN and LMN signs in different body regions (bulbar, cervical, thoracic, and lumbosacral) (TABLE 1).[13]

Objective UMN markers are needed because clinical evaluation alone is insufficient in distinguishing among certain ALS-related syndromes. Purely LMN signs can be caused by ALS, progressive muscular atrophy (a purely LMN disease), or motor neuropathy. At autopsy, CST degeneration is found in 50–75% of patients clinically diagnosed with progressive muscular atrophy.[1,14–16] These autopsy findings suggest that clinical examination is probably an unreliable or insensitive method for detecting UMN pathology when LMN signs are pronounced. Better *in vivo* UMN markers may improve diagnostic certainty and allow patients with suspected ALS more timely access to treatment and clinical trials. Reliable UMN markers may also improve our understanding of the disease process and our ability to provide the prognosis.

In the search for the diagnostic test, several imaging modalities have been used with varying success: specifically, diffusion tensor imaging (DTI),[17–23] magnetization transfer imaging (MTI),[24–27] and proton magnetic resonance spectroscopy (^{1}H-MRS).[23,28–31] In the following sections, we review the current recognized roles of conventional magnetic resonance imaging (MRI), DTI, MTI, and ^{1}H-MRS in the diagnosis of ALS and PLS.

MAGNETIC RESONANCE IMAGING

Conventional MRI in ALS and PLS is currently used to exclude other pathologies rather than confirming the diagnosis. For example, MRI of the neck easily establishes the presence of cervical spondylosis, a degenerative disease of the spinal column causing spinal cord and nerve root compression that can produce the cardinal signs of ALS—namely, the combination of UMN and LMN dysfunction. Once diseases mimicking ALS are ruled out, one can look for the subtle neuroimaging signs that are supportive of ALS and PLS. Several studies have investigated the abnormal findings on T1-, T2-, and proton density–weighted images and fluid-attenuated inversion recovery (FLAIR) images in patients with ALS and PLS.[32–35]

Important features that are consistent with (but not required for) the diagnosis of ALS and PLS include hyperintensity along the CST, hypointensity in the motor cortex, and atrophy of the precentral gyrus. There is a general lack of agreement regarding how sensitive and specific the presence of these abnormalities is for the detection of UMN involvement. This is probably due to differences in the pulse sequence parameters used in the published studies and due to the subjectivity of the imaging findings. Generally, positive MR findings correlate with average or rapid progression of the disease.[36]

Hyperintensity along the CST

Hyperintense signals along the CST on T2WI and FLAIR images are most readily identified in the posterior limb of the internal capsule (PLIC), and are also seen in the centrum semiovale and ventral brain stem. There is, however, some overlap with healthy subjects. Mirowitz *et al.*[37] proposed that the high signal foci in the PLIC in normal controls are due to reduced myelination. Therefore, the hyperintensity may reflect the current level of demyelination and gliosis rather than the absolute degree of CST axonal degeneration.

Hypointensity in the Motor Cortex

Hypointensity in the motor cortex (motor dark line) on T2-weighted and FLAIR images has been observed in a proportion of patients with ALS (FIG. 1). It is reported that the frequency of detection of the hypointensity in the precentral gyrus is associated with the severity of clinical UMN signs, from 50% seen in patients with mild or moderate disease to 100% seen in patients with severe disease.[29] The mechanisms of this phenomenon may be a T2 shortening effect[38] due to excessive iron deposition,

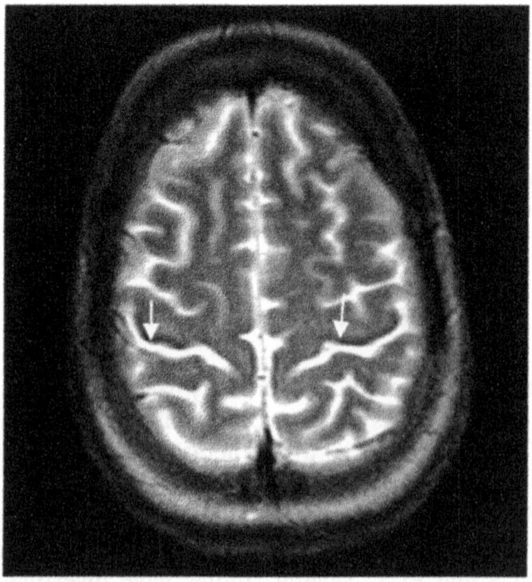

FIGURE 1. Hypointensity in the motor cortex in a patient with ALS. Axial T2-weighted image demonstrates linear hypointensity along the anterior rim of central sulcus (*arrows*).

fibrillary gliosis, and macrophage infiltration.[32,39] Theoretically, paramagnetic or ferromagnetic materials will cause visually apparent susceptibility effects when they are high enough in concentration, and a low water content tissue will also lead to hypointensity. However, this MR abnormality can also be identified in up to 38% of healthy subjects.[34] Such inconsistency may be explained by the natural accumulation of iron in the motor cortex with advancing age.[40]

Atrophy of the Precentral Gyrus

Atrophy of precentral gyrus may be identified indirectly as enlargement of adjacent central sulcus. Pringle *et al.*[3] described this abnormality in cases of PLS; however, later studies also suggested similar, but less severe, findings among some patients with ALS.[30] The prominent atrophy in PLS patients may be explained by the long duration of the disease process in PLS as a result of longer survival.[6,11,12] Although visualization of a clearly enlarged central sulcus may be helpful for the presence of motor cortex atrophy, reliable determination requires quantitative evaluation.

DIFFUSION TENSOR IMAGING

DTI provides microscopic structural information about tissue *in vivo.* Diffusion is the molecular movement of bulk water. When unimpeded, water molecules move in a random manner (isotropic diffusion); however, the presence of obstacles to free motion, such as axonal membranes and myelin sheaths in white matter fiber tracts, restricts molecular motion in a particular direction, resulting in anisotropic diffusion.[41] Diffusivity is generally higher in directions along fiber tracts than perpendicular to them.[42] This can be described mathematically by a tensor, which is characterized by its three eigenvectors and the corresponding eigenvalues. The eigenvector associated with the largest eigenvalue indicates the predominant orientation of fibers in the given voxel.

Two DTI-based scalar indices have often been used to characterize microstructure of the local brain tissue—apparent diffusion coefficient (ADC) and fractional anisotropy (FA), which can be calculated according to equations 1 and 2, respectively:

$$\mathrm{ADC} = (\lambda_1 + \lambda_2 + \lambda_3)/3 \qquad \textbf{(1)}$$

$$\mathrm{FA} = \sqrt{\frac{3}{2}}\sqrt{\frac{(\lambda_1 - \bar{\lambda})^2 + (\lambda_2 - \bar{\lambda})^2 + (\lambda_3 - \bar{\lambda})^2}{\lambda_1^2 + \lambda_2^2 + \lambda_3^2}} \qquad \textbf{(2)}$$

where λ_1, λ_2, and λ_3 are the three eigenvalues of the diffusion tensor and $\bar{\lambda}$ denotes the mean of the three eigenvalues, a measure of directionally averaged diffusivity.

ADC is a measure of the directionally averaged magnitude of diffusion and is related to the integrity of the local brain tissue. FA represents the degree of diffusion anisotropy and reflects the degree of alignment of cellular structures.[43]

The CST is a major white matter fiber tract in the brain conveying information from the precentral gyrus (motor strip) to the spinal cord. There is strong restriction to water diffusion across this highly coherent fiber pathway. The pathologic hallmark of ALS and PLS is CST degeneration. Generally, there are three ways to evaluate the CST degeneration using DTI.

Measurements of FA and ADC

Changes in tissue structure, such as CST degeneration in ALS, can lead to a modification of diffusion characteristics, which can be detected by the measurements of FA and ADC values.

DTI studies in patients with ALS and PLS have generally focused on the measurements of FA and ADC values along the CST in the PLIC,[17–22] cerebral peduncle,[17,21] corona radiata,[20,23] and pons and pyramids.[20,21] These studies reported decreased FA and elevated ADC values only in the PLIC. Difference in diffusion characteristics observed at various anatomical levels of the CST may be related to its architecture and/or to the unequal distribution of pathologic damage in ALS.[21,44] Diffusion anisotropy is high in the internal capsule, which contains very coherent and tightly packed CST fibers, whereas it is quite low in the pons and medulla, where the CST fibers are less coherent due to the presence of transverse pontine fibers as well as nuclei and roots of cranial nerves.[45] Several postmortem studies have described uneven involvement of CST and variable patterns of degeneration.[9,15] Despite widespread loss of myelin throughout the CST in ALS patients, the loss is most consistently present in the PLIC.[9,15] Therefore, we believe that the PLIC is the optimal site for the quantification of CST degeneration.

Quantitative image analyses in ALS and PLS patients have been performed using a region-of-interest (ROI) approach. One major limitation of this approach is the ability to accurately determine the actual boundaries of the CST and avoid partial volume contamination from non-CST fibers. This may lead to site selection bias, resulting in additional interobserver variability in the measurements. A better method

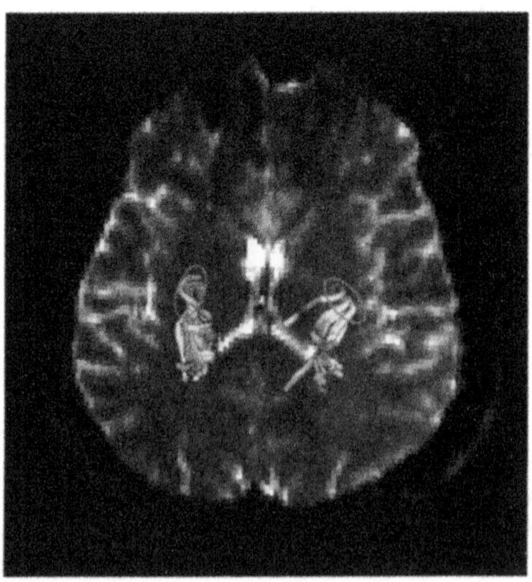

FIGURE 2. The corticospinal tracts (CSTs, green) are reconstructed and overlaid on b0 images in a control subject. ROIs (red) are placed manually in the left and right side of the posterior limb of internal capsule on the axial slice based on the location of the CSTs.

may be to employ DTI-based tractography, which will be discussed later, to map the CST location and use this as an unbiased guide for ROI placement (FIG. 2).

Reduced FA values in the PLIC of ALS and PLS patients have been widely reported, indicating a breakdown of the barriers that restrict free water movement, probably reflecting axonal degeneration.[17–22] An increase of ADC values may represent an increase in the extracellular water volume secondary to axonal loss.[17–22] It has been reported that FA and ADC assess different pathologic effects.[18] FA correlates with the severity of the disease,[17–19] whereas ADC correlates with the duration of the disease.[18] Hence, DTI may be useful to document UMN involvement objectively and quantitatively in ALS and PLS. These studies support the potential role of DTI as an objective marker for detecting and monitoring UMN involvement in ALS and PLS patients.

DTI-Based Color Map

DTI also provides information about the direction of the principal eigenvector, which denotes the direction of maximum diffusivity. The principal eigenvector represents the major orientation of the interrogated white matter tract. Hence, DTI allows mapping of the white matter tracts in the brain, where the orientation is coded using red, green, and blue (RGB) color channels, and the brightness of the assigned color is modulated by the degree of anisotropy, which is often represented by FA. This display technique results in a convenient orientation-based color map in which both the degree of anisotropy and the local fiber orientation can be determined. Application of this technique to the brain has been demonstrated to be useful in showing white matter architecture.[46–49] The CST, therefore, can be readily identified in color on every cross-sectional slice along its course, making it feasible to segment the CST and quantify the volume in ALS and PLS patients (FIG. 3). Based on our experience, ALS and PLS patients exhibit a decrease in CST volume compared to healthy subjects. Thus, the CST volume may be used as an objective UMN marker for the diagnosis and disease progression.

Diffusion Tensor Tractography

DTI techniques also allow interregional fiber tracking, known as diffusion tensor tractography, which allows tracking of major white matter tracts.[44,48,50–52] Fiber tracking is generally performed using a line propagation technique based on continuous number fields[50,53] and a multiple ROI approach.[51,54] Tracking is launched from a "seed" voxel from which a line is propagated in both retrograde and antegrade directions according to the major eigenvector at each voxel. Tracking propagates on the basis of the orientation of the eigenvector that is associated with the largest eigenvalue. The propagation terminates when it reaches a voxel with FA lower than a specified FA threshold or when the angle between two principal eigenvectors is greater than an angle threshold. Using this technique, several authors have reported depiction of the CST.[51,52,54] ALS and PLS patients with severe clinical deficits demonstrate decreased number of CST fibers compared to normal subjects (FIG. 4).

Diffusion tensor tractography provides a unique opportunity for localizing white matter tracts in the brain; however, there are important questions still unanswered. Fiber tracking is a user-dependent process, the result of which varies significantly

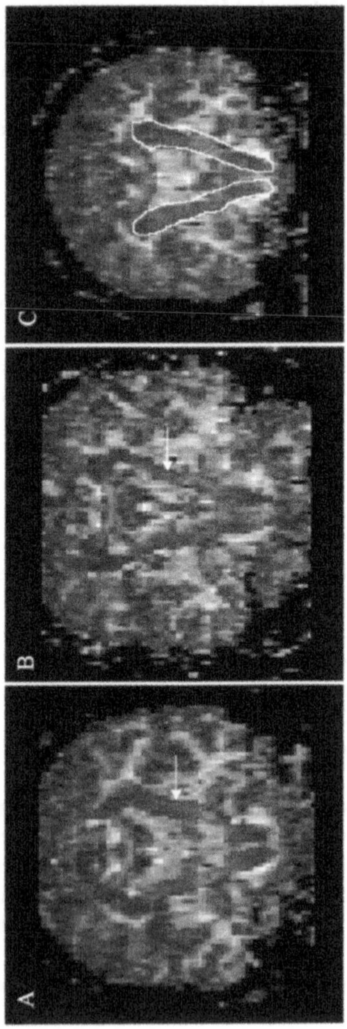

FIGURE 3. DTI-based color maps of a control subject (**A**) and ALS patient (**B**). The left CST is shown by *arrows*, which appears thinner in the ALS patient (**B**). The *white line* delineates manually segmented CST (**C**) for the purpose of volume calculation of the CST.

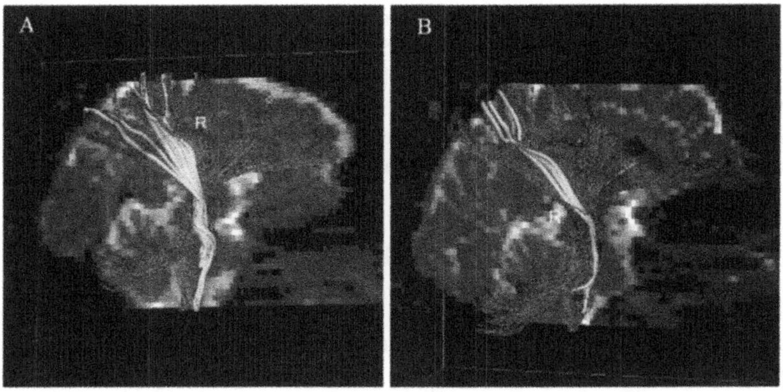

FIGURE 4. Fiber tracking images of a control subject **(A)** and ALS patient **(B)**. Descending fibers connecting the cortex and brain stem are shown in purple. CSTs are in green. The CST fiber density is diminished in ALS patients **(B)**.

depending on the FA and angular threshold, vector step lengths, and seed ROIs. The tracking result is also affected by image resolution and signal-to-noise ratio. The validity of the technique could be addressed by defining a gold standard volume for the structure of interest based on anatomical atlases. Threshold parameters, step length, and seed ROIs could then be optimized by determining the appropriate parameter choices that provide the most faithful reconstruction of the gold standard volume.[55] Due to the above-mentioned limitation, we believe that a quantitative estimation of CST volume using fiber tracking may not be accurate; however, this technique does help in mapping the location of the CST.

MAGNETIZATION TRANSFER IMAGING

MTI is based on the exchange of magnetization in biologic tissues between a pool of protons in water and a pool of protons that is bound to macromolecules. The magnitude of the effect depends on the ratio of water and macromolecules and on the surface chemistry and biophysical dynamics of macromolecules and may be quantified by MT ratio (MTR), which can be calculated according to the following equation:

$$MTR = [(M_0 - Ms)/M_0] \times 100,$$

where M_0 is the measured signal intensity on the reference image and Ms is the measured signal intensity on the MT image.[56] In the brain, macromolecules that contribute to the MT effect are the cholesterol component of myelin, cerebrosides, and phospholipids. MT measurements can be obtained to determine the local tissue status by assessing MTR values in ROIs or performing MTR histogram analysis of the whole brain. Low MTR indicates damage to myelin or to the axonal membrane.

MTI also has been able to detect CST pathology in ALS and PLS. Two studies have reported a reduction of MTR in the internal capsule among a group of patients with ALS when compared with a control group,[26,27] implying the presence of tissue damage. Da Rocha *et al.*,[25] in a series of 25 patients, described hyperintensity along the CST in 80% of patients with ALS on T1-weighted spin-echo (SE) magnetization transfer contrast (MTC) images. These studies suggest that MTI is more sensitive and accurate than conventional MRI for the detection of CST degeneration and may be a useful, objective marker of UMN pathology in ALS and PLS.

MR SPECTROSCOPY

^{1}H-MR spectroscopy (MRS) of the brain exhibits three major metabolites: namely, *N*-acetylaspartate (NAA), choline-containing compounds (Cho), and creatine/phosphocreatine (Cr).[57,58] Hitherto published ^{1}H-MRS studies in ALS and PLS have demonstrated that NAA concentrations[10,11] or ratios of NAA/Cr,[12–14] NAA/Cho,[11,15] and NAA/(Cr + Cho)[16] are reduced in the motor cortex. Because NAA is present primarily in neurons, these metabolite changes reflect a loss or dysfunction of motor neurons. Therefore, reduced NAA might be used as a spectroscopic marker for UMN involvement.

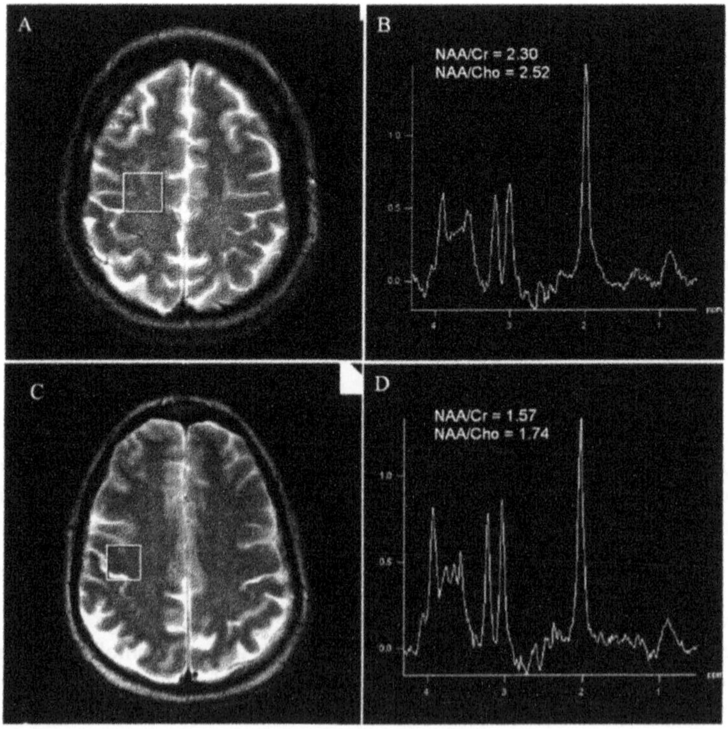

FIGURE 5. Single-voxel MR spectroscopy of the motor cortex in a control subject **(A, B)** and ALS patient **(C, D)**. Axial T2-weighted image shows 2 × 2 × 2 cm voxel placement in a control subject **(A)** and ALS patient **(C)**. Spectra of the motor cortex from the ALS patient **(D)** demonstrate reduced NAA/Cr and NAA/Cho ratios compared to the control subject **(B)**.

Technical Issues

The technical methods used in evaluating NAA vary widely among published reports. These differences mainly include the following:

(1) Single-voxel versus multivoxel technique (also known as chemical shift imaging, CSI): The single-voxel method is easier and faster. However, the volume of interest is relatively large, usually encompassing more than just the motor cortex (FIG. 5). Compared to the single-voxel method, multivoxel provides spectra from an array of voxels, which are much smaller than single voxels. Thus, the measurements of metabolites can be obtained more precisely from the motor cortex (FIGS. 6 and 7). Furthermore, multivoxel can sample different brain regions simultaneously, which is important since previous studies have found spectroscopic abnormalities in regions outside the motor cortex, including the postcentral gyrus, medulla,[59] and internal capsule.[60,61] Since most MRS studies of ALS that indicated significant intergroup differences were based on single-voxel techniques, these techniques will probably continue to be in use. Nevertheless, multivoxel will become more prevalent because of the above-stated advantages.

(2) Short echo time versus long echo time: Many of the MR spectroscopy studies of ALS used long echo time, which limits analysis to NAA, creatine, and choline (FIG. 6). Preliminary work with short echo time techniques in ALS patients has revealed the presence of elevated myoinositol in the motor cortex region,[61] and glutamate and glutamine in the affected brain stem of patients.[59] Future improvements to reduce lipid contamination and improve spectral resolution should facilitate the use of short echo time MRS and allow the study of additional metabolites (FIG. 7).

(3) Metabolite ratios vs. absolute metabolite concentration: Accurate measurement of absolute concentrations of metabolites is very difficult. Metabolite ratios can be more valid because standardization to Cr and Cho minimizes the interindividual differences that may occur in computing absolute concentrations. Calculating metabolic ratios is more straightforward than absolute quantification and has been used extensively in the literature.

Decreased NAA as a Spectroscopic Marker for UMN Involvement

Both single-voxel and multivoxel spectroscopic studies have shown a reduction of NAA concentrations,[62,63] NAA/Cho,[28,63] or NAA/Cr[29,30,61] ratios in the motor cortex of ALS and PLS patients. The degree of NAA reduction seems to vary quantitatively owing to differences in MR spectroscopic technique, the severity of disease, definition of patient and control populations, and the method used to calculate NAA concentration (relative vs. absolute). There is no consensus on which metabolite ratios can accurately reflect UMN involvement.[29,30,61] While some studies have reported NAA/Cho ratios correlated with disease severity, others have concluded NAA/Cr ratios as the sensitive indicator in the diagnosis of ALS and PLS.[29,30,61] Ideally, future studies should increasingly use absolute quantification of metabolite concentration.

Finally, metabolite ratios may also be affected by exogenous treatments in ALS patients. Riluzole has been reported to increase the NAA/Cr ratio in ALS patients after 3 weeks of treatment,[64] a finding that has not been confirmed so far by other

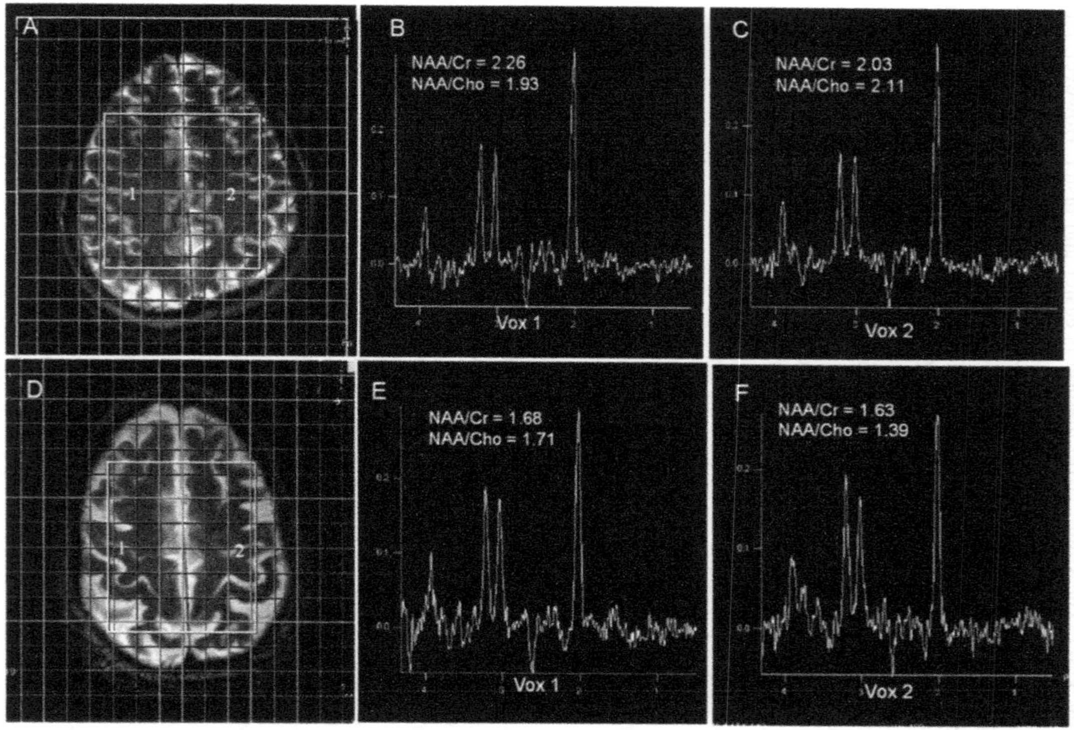

FIGURE 6. Two-dimensional CSI spectroscopy of the motor cortex with long echo time (135 ms) in a control subject (**A, B, C**) and ALS patient (**D, E, F**). Axial T2-weighted image shows the grid and the volume of interest (VOI, *solid white rectangle*) in a control subject (**A**) and ALS patient (**D**). Representative spectra of the motor cortex from the ALS patient (**E, F**) demonstrate reduced NAA/Cr and NAA/Cho ratios compared to the control subject (**B, C**).

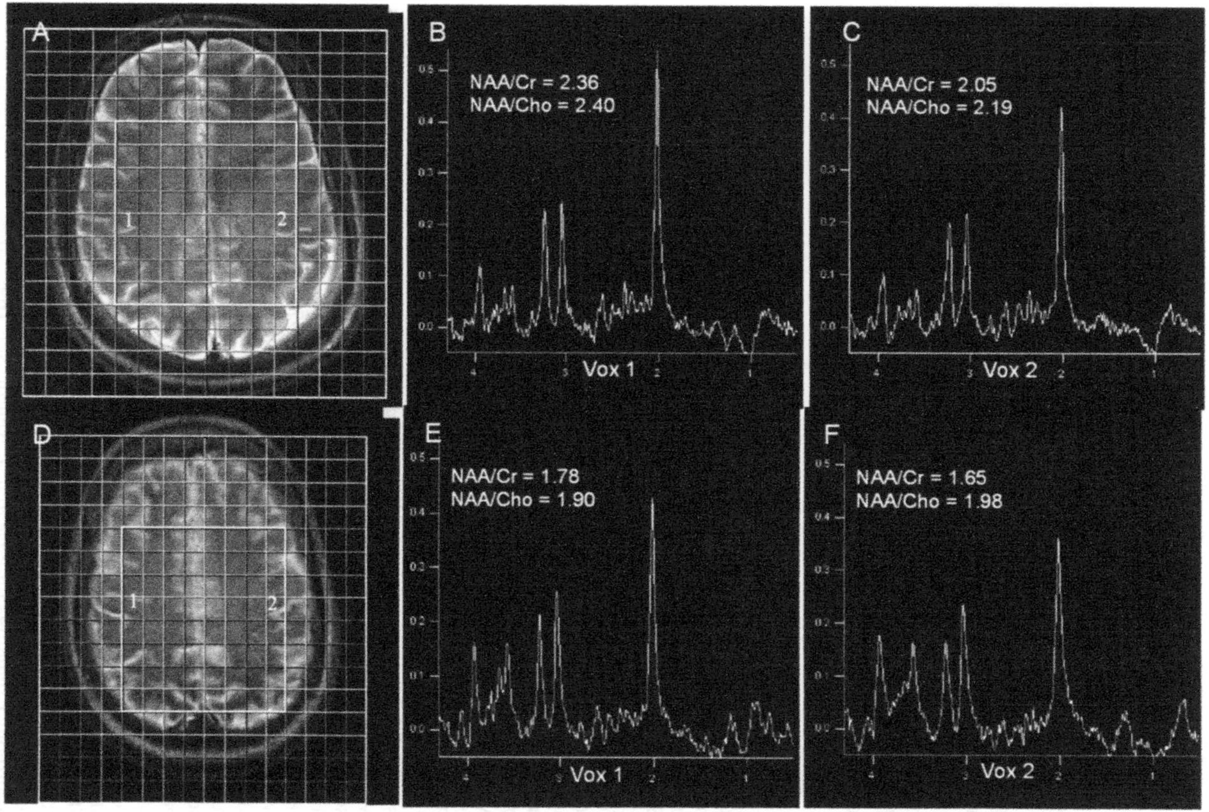

FIGURE 7. Two-dimensional CSI spectroscopy of the motor cortex with short echo time (30 ms) in a control subject (**A, B, C**) and ALS patient (**D, E, F**). Axial T2-weighted image shows the grid and the volume of interest (VOI, *solid white rectangle*) in a control subject (**A**) and ALS patient (**D**). Representative spectra of the motor cortex from the ALS patient (**E, F**) demonstrate reduced NAA/Cr and NAA/Cho ratios compared to the control subject (**B, C**).

investigators. High-dose Cr therapy (20 g/day for 4 weeks) increased brain Cr MRS estimates in normal volunteers.[65] However, a later study of brain metabolites found no effect of 1-month treatment with 15 g/day of Cr (in patients with ALS).[66]

SUMMARY

A reliable objective marker of UMN involvement is critical for the early diagnosis and monitoring of disease progression in patients with ALS and PLS. DTI, MTI, and MRS, which provide insight into the pathophysiological process of ALS and PLS, show great promise in this regard. Further investigation is needed to determine and to compare the utility of various neuroimaging markers. The goal of finding an objective marker will require more uniform study design with respect to cohort composition, clinical rating scales for severity of disease, and medication history. Also, DTI, MTI, and MRS acquisition methods, analysis programs, and reporting of metabolite results in terms of absolute or relative concentration must be more uniform.

REFERENCES

1. Brownell, B., D.R. Oppenheimer & J.T. Hughes. 1970. The central nervous system in motor neurone disease. J. Neurol. Neurosurg. Psychiatry **33:** 338–357.
2. Rowland, L.P. 1998. Diagnosis of amyotrophic lateral sclerosis. J. Neurol. Sci. **160:** S6–S24.
3. Pringle, C.E., A.J. Hudson, D.G. Munoz *et al.* 1992. Primary lateral sclerosis: clinical features, neuropathology, and diagnostic criteria. Brain **115:** 495–520.
4. Chan, S., P. Kaufmann, D.C. Shungu & H. Mitsumoto. 2003. Amyotrophic lateral sclerosis and primary lateral sclerosis: evidence-based diagnostic evaluation of the upper motor neuron. Neuroimaging Clin. N. Am. **13:** 307–326.
5. Rowland, L.P. & N.A. Shneider. 2001. Amyotrophic lateral sclerosis. N. Engl. J. Med. **344:** 1688–1700.
6. Kuipers-Upmeijer, J., A.E. de Jager, J.M. Hew Snoek & T.W. van Weerden. 2001. Primary lateral sclerosis: clinical, neurophysiological, and magnetic resonance findings. J. Neurol. Neurosurg. Psychiatry **71:** 615–620.
7. Eisen, A. & M. Weber. 2001. The motor cortex and amyotrophic lateral sclerosis. Muscle Nerve **24:** 564–573.
8. Fischer, L.R., D.G. Culver, P. Tennant *et al.* 2004. Amyotrophic lateral sclerosis is a distal axonopathy: evidence in mice and man. Exp. Neurol. **185:** 232–240.
9. Takahashi, T., S. Yagishita, N. Amano *et al.* 1997. Amyotrophic lateral sclerosis with numerous axonal spheroids in the corticospinal tract and massive degeneration of the cortex. Acta Neuropathol. (Berl.) **94:** 294–299.
10. Tsuchiya, K., S. Shintani, M. Kikuchi *et al.* 1999. Sporadic amyotrophic lateral sclerosis of long duration mimicking spinal progressive muscular atrophy: a clinicopathological study. J. Neurol. Sci. **162:** 174–178.
11. Beal, M.F. & E.P. Richardson, Jr. 1981. Primary lateral sclerosis: a case report. Arch. Neurol. **38:** 630–633.
12. Smith, C.D. 2002. Serial MRI findings in a case of primary lateral sclerosis. Neurology **58:** 647–649.
13. Miller, R.G., T.L. Munsat, M. Swash & B.R. Brooks. 1999. Consensus guidelines for the design and implementation of clinical trials in ALS: World Federation of Neurology Committee on Research. J. Neurol. Sci. **169:** 2–12.
14. Ince, P.G., J. Evans, M. Knopp *et al.* 2003. Corticospinal tract degeneration in the progressive muscular atrophy variant of ALS. Neurology **60:** 1252–1258.
15. Iwanaga, K., S. Hayashi, M. Oyake *et al.* 1997. Neuropathology of sporadic amyotrophic lateral sclerosis of long duration. J. Neurol. Sci. **146:** 139–143.

16. LAWYER, T., JR. & M.G. NETSKY. 1953. Amyotrophic lateral sclerosis. AMA Arch. Neurol. Psychiatry **69:** 171–192.
17. HONG, Y.H., K.W. LEE, J.J. SUNG *et al.* 2004. Diffusion tensor MRI as a diagnostic tool of upper motor neuron involvement in amyotrophic lateral sclerosis. J. Neurol. Sci. **227:** 73–78.
18. ELLIS, C.M., A. SIMMONS, D.K. JONES *et al.* 1999. Diffusion tensor MRI assesses corticospinal tract damage in ALS. Neurology **53:** 1051–1058.
19. GRAHAM, J.M., N. PAPADAKIS, J. EVANS *et al.* 2004. Diffusion tensor imaging for the assessment of upper motor neuron integrity in ALS. Neurology **63:** 2111–2119.
20. SACH, M., G. WINKLER, V. GLAUCHE *et al.* 2004. Diffusion tensor MRI of early upper motor neuron involvement in amyotrophic lateral sclerosis. Brain **127:** 340–350.
21. TOOSY, A.T., D.J. WERRING, R.W. ORRELL *et al.* 2003. Diffusion tensor imaging detects corticospinal tract involvement at multiple levels in amyotrophic lateral sclerosis. J. Neurol. Neurosurg. Psychiatry **74:** 1250–1257.
22. ULUG, A.M., T. GRUNEWALD, M.T. LIN *et al.* 2004. Diffusion tensor imaging in the diagnosis of primary lateral sclerosis. J. Magn. Reson. Imaging **19:** 34–39.
23. YIN, H., C.C. LIM, L. MA *et al.* 2004. Combined MR spectroscopic imaging and diffusion tensor MRI visualizes corticospinal tract degeneration in amyotrophic lateral sclerosis. J. Neurol. **251:** 1249–1254.
24. DA ROCHA, A.J., A.C. MAIA, JR., R.G. NOGUEIRA & H.M. LEDERMAN. 1999. Magnetic resonance findings in amyotrophic lateral sclerosis using a spin echo magnetization transfer sequence: preliminary report. Arq. Neuropsiquiatr. **57:** 912–915.
25. DA ROCHA, A.J., A.S. OLIVEIRA, R.B. FONSECA *et al.* 2004. Detection of corticospinal tract compromise in amyotrophic lateral sclerosis with brain MR imaging: relevance of the T1-weighted spin-echo magnetization transfer contrast sequence. AJNR Am. J. Neuroradiol. **25:** 1509–1515.
26. KATO, Y., K. MATSUMURA, Y. KINOSADA *et al.* 1997. Detection of pyramidal tract lesions in amyotrophic lateral sclerosis with magnetization-transfer measurements. AJNR Am. J. Neuroradiol. **18:** 1541–1547.
27. TANABE, J.L., M. VERMATHEN, R. MILLER *et al.* 1998. Reduced MTR in the corticospinal tract and normal T2 in amyotrophic lateral sclerosis. Magn. Reson. Imaging **16:** 1163–1169.
28. BLOCK, W., J. KARITZKY, F. TRABER *et al.* 1998. Proton magnetic resonance spectroscopy of the primary motor cortex in patients with motor neuron disease: subgroup analysis and follow-up measurements. Arch. Neurol. **55:** 931–936.
29. BOWEN, B.C., P.M. PATTANY, W.G. BRADLEY *et al.* 2000. MR imaging and localized proton spectroscopy of the precentral gyrus in amyotrophic lateral sclerosis. AJNR Am. J. Neuroradiol. **21:** 647–658.
30. CHAN, S., D.C. SHUNGU, A. DOUGLAS-AKINWANDE *et al.* 1999. Motor neuron diseases: comparison of single-voxel proton MR spectroscopy of the motor cortex with MR imaging of the brain. Radiology **212:** 763–769.
31. KAUFMANN, P., S.L. PULLMAN, D.C. SHUNGU *et al.* 2004. Objective tests for upper motor neuron involvement in amyotrophic lateral sclerosis (ALS). Neurology **62:** 1753–1757.
32. HECHT, M.J., F. FELLNER, C. FELLNER *et al.* 2001. MRI-FLAIR images of the head show corticospinal tract alterations in ALS patients more frequently than T2-, T1-, and proton-density-weighted images. J. Neurol. Sci. **186:** 37–44.
33. HECHT, M.J., F. FELLNER, C. FELLNER *et al.* 2002. Hyperintense and hypointense MRI signals of the precentral gyrus and corticospinal tract in ALS: a follow-up examination including FLAIR images. J. Neurol. Sci. **199:** 59–65.
34. WARAGAI, M. 1997. MRI and clinical features in amyotrophic lateral sclerosis. Neuroradiology **39:** 847–851.
35. ZHANG, L., A.M. ULUG, R.D. ZIMMERMAN *et al.* 2003. The diagnostic utility of FLAIR imaging in clinically verified amyotrophic lateral sclerosis. J. Magn. Reson. Imaging **17:** 521–527.
36. CHEUNG, G., M.J. GAWEL, P.W. COOPER *et al.* 1995. Amyotrophic lateral sclerosis: correlation of clinical and MR imaging findings. Radiology **194:** 263–270.

37. MIROWITZ, S., K. SARTOR, M. GADO & R. TORACK. 1989. Focal signal-intensity variations in the posterior internal capsule: normal MR findings and distinction from pathologic findings. Radiology **172:** 535–539.
38. KARANTANAS, A.H. 1998. Amyotrophic lateral sclerosis: unilateral T2-shortening on MRI. Comput. Med. Imaging Graphics **22:** 353–355.
39. OBA, H., T. ARAKI, K. OHTOMO *et al.* 1993. Amyotrophic lateral sclerosis: T2 shortening in motor cortex at MR imaging. Radiology **189:** 843–846.
40. HALLGREN, B. & P. SOURANDER. 1958. The effect of age on the non-haemin iron in the human brain. J. Neurochem. **3:** 41–51.
41. BEAULIEU, C. 2002. The basis of anisotropic water diffusion in the nervous system—a technical review. NMR Biomed. **15:** 435–455.
42. CHENEVERT, T.L., J.A. BRUNBERG & J.G. PIPE. 1990. Anisotropic diffusion in human white matter: demonstration with MR techniques *in vivo*. Radiology **177:** 401–405.
43. BASSER, P.J. & C. PIERPAOLI. 1996. Microstructural and physiological features of tissues elucidated by quantitative-diffusion-tensor MRI. J. Magn. Reson. **B111:** 209–219.
44. VIRTA, A., A. BARNETT & C. PIERPAOLI. 1999. Visualizing and characterizing white matter fiber structure and architecture in the human pyramidal tract using diffusion tensor MRI. Magn. Reson. Imaging **17:** 1121–1133.
45. SHIMONY, J.S., R.C. MCKINSTRY, E. AKBUDAK *et al.* 1999. Quantitative diffusion-tensor anisotropy brain MR imaging: normative human data and anatomic analysis. Radiology **212:** 770–784.
46. ALBAYRAM, S., E.R. MELHEM, S. MORI *et al.* 2002. Holoprosencephaly in children: diffusion tensor MR imaging of white matter tracts of the brainstem—initial experience. Radiology **223:** 645–651.
47. HOON, A.H., JR., W.T. LAWRIE, JR., E.R. MELHEM *et al.* 2002. Diffusion tensor imaging of periventricular leukomalacia shows affected sensory cortex white matter pathways. Neurology **59:** 752–756.
48. MORI, S., W.E. KAUFMANN, C. DAVATZIKOS *et al.* 2002. Imaging cortical association tracts in the human brain using diffusion-tensor-based axonal tracking. Magn. Reson. Med. **47:** 215–223.
49. PAJEVIC, S. & C. PIERPAOLI. 1999. Color schemes to represent the orientation of anisotropic tissues from diffusion tensor data: application to white matter fiber tract mapping in the human brain. Magn. Reson. Med. **42:** 526–540.
50. MORI, S., B.J. CRAIN, V.P. CHACKO & P.C. VAN ZIJL. 1999. Three-dimensional tracking of axonal projections in the brain by magnetic resonance imaging. Ann. Neurol. **45:** 265–269.
51. MORI, S. & P.C. VAN ZIJL. 2002. Fiber tracking: principles and strategies—a technical review. NMR Biomed. **15:** 468–480.
52. STIELTJES, B., W.E. KAUFMANN, P.C. VAN ZIJL *et al.* 2001. Diffusion tensor imaging and axonal tracking in the human brainstem. Neuroimage **14:** 723–735.
53. XUE, R., P.C. VAN ZIJL, B.J. CRAIN *et al.* 1999. *In vivo* three-dimensional reconstruction of rat brain axonal projections by diffusion tensor imaging. Magn. Reson. Med. **42:** 1123–1127.
54. MELHEM, E.R., S. MORI, G. MUKUNDAN *et al.* 2002. Diffusion tensor MR imaging of the brain and white matter tractography. AJR Am. J. Roentgenol. **178:** 3–16.
55. CLARK, C.A., T.R. BARRICK, M.M. MURPHY & B.A. BELL. 2003. White matter fiber tracking in patients with space-occupying lesions of the brain: a new technique for neurosurgical planning? Neuroimage **20:** 1601–1608.
56. WOLFF, S.D. & R.S. BALABAN. 1994. Magnetization transfer imaging: practical aspects and clinical applications. Radiology **192:** 593–599.
57. BLOCK, W., F. TRABER, S. FLACKE *et al.* 2002. *In-vivo* proton MR-spectroscopy of the human brain: assessment of *N*-acetylaspartate (NAA) reduction as a marker for neurodegeneration. Amino Acids **23:** 317–323.
58. CASTILLO, M., L. KWOCK & S.K. MUKHERJI. 1996. Clinical applications of proton MR spectroscopy. AJNR Am. J. Neuroradiol. **17:** 1–15.
59. PIORO, E.P., A.W. MAJORS, H. MITSUMOTO *et al.* 1999. ^{1}H-MRS evidence of neurodegeneration and excess glutamate + glutamine in ALS medulla. Neurology **53:** 71–79.

60. ROONEY, W.D., R.G. MILLER, D. GELINAS *et al.* 1998. Decreased *N*-acetylaspartate in motor cortex and corticospinal tract in ALS. Neurology **50:** 1800–1805.
61. SCHUFF, N., W.D. ROONEY, R. MILLER *et al.* 2001. Reanalysis of multislice (1)H MRSI in amyotrophic lateral sclerosis. Magn. Reson. Med. **45:** 513–516.
62. BRADLEY, W.G., B.C. BOWEN, P.M. PATTANY & F. ROTTA. 1999. ^{1}H-magnetic resonance spectroscopy in amyotrophic lateral sclerosis. J. Neurol. Sci. **169:** 84–86.
63. POHL, C., W. BLOCK, J. KARITZKY *et al.* 2001. Proton magnetic resonance spectroscopy of the motor cortex in 70 patients with amyotrophic lateral sclerosis. Arch. Neurol. **58:** 729–735.
64. KALRA, S., N.R. CASHMAN, A. GENGE & D.L. ARNOLD. 1998. Recovery of *N*-acetylaspartate in corticomotor neurons of patients with ALS after riluzole therapy. Neuroreport **9:** 1757–1761.
65. DECHENT, P., P.J. POUWELS, B. WILKEN *et al.* 1999. Increase of total creatine in human brain after oral supplementation of creatine-monohydrate. Am. J. Physiol. **277:** R698–R704.
66. VIELHABER, S., J. KAUFMANN, M. KANOWSKI *et al.* 2001. Effect of creatine supplementation on metabolite levels in ALS motor cortices. Exp. Neurol. **172:** 377–382.

White Matter Tractography by Means of Turboprop Diffusion Tensor Imaging

KONSTANTINOS ARFANAKIS,[a,b] MINZHI GUI,[a] AND MARIANA LAZAR[c]

[a]*Department of Biomedical Engineering, Illinois Institute of Technology, Chicago, Illinois, USA*

[b]*Brain Research Imaging Center, University of Chicago, Chicago, Illinois, USA*

[c]*Keck Laboratory for Functional Brain Imaging and Behavior, University of Wisconsin–Madison, Madison, Wisconsin, USA*

ABSTRACT: White matter fiber-tractography by means of diffusion tensor imaging (DTI) is a noninvasive technique that provides estimates of the structural connectivity of the brain. However, conventional fiber-tracking methods using DTI are based on echo-planar image acquisitions (EPI), which suffer from image distortions and artifacts due to magnetic susceptibility variations and eddy currents. Thus, a large percentage of white matter fiber bundles that are mapped using EPI-based DTI data are distorted, and/or terminated early, while others are completely undetected. This severely limits the potential of fiber-tracking techniques. In contrast, Turboprop imaging is a multiple-shot gradient and spin-echo (GRASE) technique that provides images with significantly fewer susceptibility and eddy current–related artifacts than EPI. The purpose of this work was to evaluate the performance of fiber-tractography techniques when using data obtained with Turboprop-DTI. All fiber pathways that were mapped were found to be in agreement with the anatomy. There were no visible distortions in any of the traced fiber bundles, even when these were located in the vicinity of significant magnetic field inhomogeneities. Additionally, the Turboprop-DTI data used in this research were acquired in less than 19 min of scan time. Thus, Turboprop appears to be a promising DTI data acquisition technique for tracing white matter fibers.

KEYWORDS: tractography; DTI; PROPELLER; Turboprop; distortion

INTRODUCTION

Estimates of white matter fiber orientation obtained with diffusion tensor imaging[1–6] (DTI) can be used to construct paths that connect different brain regions and are interpreted as representations of the underlying white matter fiber system. This technique is called tractography[7] and is currently the only noninvasive method that can provide estimates of brain structural connectivity. Fiber-tractography has

Address for correspondence: Konstantinos Arfanakis, Ph.D., Department of Biomedical Engineering, Illinois Institute of Technology, 10 West 32nd Street, E1-116, Chicago, IL 60616. Voice: 312-567-3864; fax: 312-567-5707.
arfanakis@iit.edu

**Ann. N.Y. Acad. Sci. 1064: 78–87 (2005). © 2005 New York Academy of Sciences.
doi: 10.1196/annals.1340.014**

been used in studies of white matter structure and brain function,[8,9] and is gradually becoming an important tool for clinical applications such as presurgical planning.[10–12] Although several algorithms[7,13–25] have been proposed to produce white matter connectivity estimates, based on fiber orientation information obtained with DTI, the accuracy and reproducibility of fiber-tracking results depend heavily on the noise level and amount of artifacts of the DTI data.

Conventional tractography is performed based on DTI data acquired using spin-echo echo-planar DTI[22] (SE-EPI-DTI) sequences. These acquisitions suffer from susceptibility-related image distortions, signal loss and pileup,[26] and image warping due to eddy currents.[27] Thus, a large percentage of white matter fiber bundles that are mapped using EPI-based DTI data are distorted, and/or terminated early, while others are completely undetected.[28] This severely limits the potential of fiber-tractography. In contrast, PROPELLER imaging is an MRI data acquisition and reconstruction technique with greatly reduced sensitivity to various sources of image artifacts.[29,30] PROPELLER acquisitions follow a multiple-shot fast spin-echo (FSE) approach in which several k-space lines are acquired in each TR, forming a blade that is then rotated around its center, and acquisition is repeated to cover k-space.[29,30] Due to the FSE nature of PROPELLER MRI, the images produced are relatively immune to B_0-related artifacts and image warping due to eddy currents.[30] Also, in each blade, the same central disk of k-space is acquired that can be used as a 2D navigator to correct data between shots without requiring additional echoes.[31] Comparison of this central k-space disk between blades allows correction of the subject's rotation and translation, as well as identification of blades with corrupted data and exclusion of such blades from the final reconstruction. Hence, PROPELLER MRI is less sensitive to motion than conventional multiple-shot FSE. Furthermore, PROPELLER acquisitions are radial in nature and thus uncorrected errors are expressed in a benign fashion, similar to projection reconstruction methods.[32] For these reasons, PROPELLER MRI was originally used to compensate for motion.[29,33] Recently, diffusion-weighting (DW) gradients were incorporated in the PROPELLER pulse sequence and it was shown, in studies of the brain,[28,30,34] spinal cord,[35] prostate gland,[36] and cartilage,[37] that the results in regions with significant magnetic susceptibility differences and in objects undergoing motion were superior when using diffusion imaging with PROPELLER acquisition compared to conventional diffusion imaging methods. However, the imaging time in PROPELLER MRI is considerably longer than in EPI since PROPELLER is based on multiple-shot FSE. This issue is addressed in Turboprop,[38] where data acquisition is accelerated by reading out multiple lines of k-space during the spin-echo produced after each 180° pulse, similar to the gradient and spin-echo (GRASE) sequence.[39] In addition to the shorter imaging time, Turboprop also allows acquisition of more lines per blade, which leads to more robust motion correction.

The purpose of this work was to evaluate the performance of white matter tractography techniques when using Turboprop-DTI data acquisitions. Our hypothesis was that fiber bundles mapped from data obtained with Turboprop-DTI would be in agreement with the anatomy, even in regions near significant magnetic field inhomogeneities, and without any additional compensation for susceptibility effects or eddy currents. If our hypothesis is valid, then the use of Turboprop as a DTI data acquisition technique might significantly enhance the role of white matter tractography in neuroscience and clinical applications.

METHODS

Scans were performed on a clinical 3T GE Signa MRI scanner (Waukesha, WI) equipped with high-speed gradients (40 mT/m maximum amplitude, 150 mT/m/ms slew rate). All subjects signed a consent in accordance with institutional policy. Turboprop-DTI and high-resolution anatomical data were acquired in each subject. The subjects were asked to maintain the same head position throughout the examination. The imaging parameters for Turboprop-DTI were TR = 5000 ms, 8 spin-echoes per TR (ETL = 8), 5 k-space lines per spin-echo (turbo-factor = 5) (thus each blade contained $8 \times 5 = 40$ lines), 16 k-space blades per image, field of view = 24 cm × 24 cm, 36 contiguous axial slices, and slice thickness = 3 mm. The 36 axial slices covered most of the brain. DW images with $b = 900$ s/mm^2 were acquired for 12 non-collinear diffusion directions uniformly distributed in 3D space. Two $b = 0$ s/mm^2 images were also acquired. The duration of the Turboprop-DTI scan was 18 min and 55 s. The diffusion tensors were estimated in each voxel. Maps of the trace of the diffusion tensor and fractional anisotropy[5] (FA) were produced for all slices. Image distortions due to eddy currents were assessed by superimposing the outline of the $b = 0$ s/mm^2 image of each slice on all DW images of the same slice.

High-resolution anatomical data were also acquired in each subject and compared to the results from Turboprop-DTI in order to assess the amount of distortions in the DTI data. The high-resolution anatomical images were acquired using the 3D magnetization-prepared rapid acquisition gradient echo (MP-RAGE) sequence with the following parameters: TE = 3.2 ms, TR = 8 ms, preparation time = 725 ms, flip angle = 6°, field of view = 24 cm × 24 cm, 124 slices, slice thickness = 1.5 mm, and 192×256 image matrix reconstructed to 256×256. The scan time for this sequence was 9 min and 58 s. An edge detection algorithm was used to produce maps of the edges in the high-resolution anatomical images. These edges were then superimposed on the FA maps in order to assess the amount of distortions present in the Turboprop-DTI data. In addition, vector maps were produced, which displayed fiber orientation information overlaid on the high-resolution anatomical data.

Seed regions for fiber tracking were selected in all subjects. They enclosed fibers from the following pathways: cingulum, fornix, corpus callosum (CC), U-fibers of the frontal lobe, corticospinal tract (CST) and corona radiata, inferior and superior longitudinal fasciculi (ILF, SLF), anterior commissure (AC), inferior and superior occipitofrontal fasciculi (IOF, SOF), and uncinate fasciculus (UF). The streamline tracking (ST) algorithm with constant step-size was used in this study.[7,14,19] Fibers were displayed using as background the FA maps, as well as the high-resolution anatomical images.

RESULTS

No visible distortions, signal dropout or pileup, were present in the Turboprop-DTI images with $b = 0$ s/mm^2, even in regions with significant magnetic field inhomogeneities, such as the frontal and temporal lobes, and the midbrain (FIG. 1). The same was true for all DW images. Additionally, for all slices, the borders of the brain, as shown in the DW images, matched well with the outline of the brain obtained from the $b = 0$ s/mm^2 images (FIG. 2). Overlaying edge-maps produced

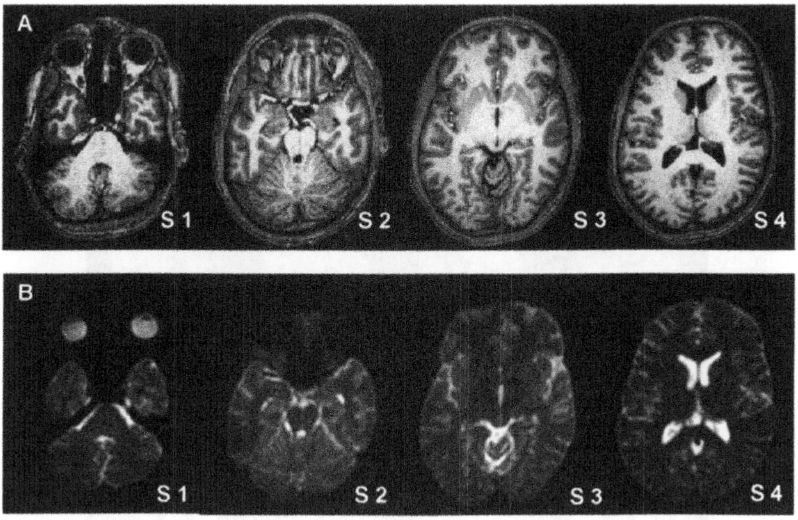

FIGURE 1. This figure shows MP-RAGE high-resolution anatomical images (**A**) and Turboprop-DTI images with $b = 0$ s/mm^2 (**B**) from 4 slices of the same healthy volunteer. There are no visible distortions, signal dropout or pileup, in the Turboprop-DTI images, even in regions with significant magnetic field inhomogeneities, such as the frontal and temporal lobes, and the midbrain.

from the MP-RAGE images, onto FA maps, showed no visible distortions nor misregistration of the Turboprop-DTI data with the high-resolution anatomical images (FIG. 3). White matter tissue shown in FA maps (high FA values) was in agreement with the anatomical information obtained from the MP-RAGE images. Finally, vector maps demonstrated that the fiber orientation information obtained with Turboprop-DTI registered well with the high-resolution anatomical images (FIG. 4).

Fibers of the cingulum, fornix, CC, U-fibers of the frontal lobe, CST and corona radiata, ILF, SLF, AC, IOF, SOF, and UF were successfully mapped (FIG. 5). All fibers traced from Turboprop-DTI data were determined to be anatomically accurate representations of the corresponding fiber pathways. Similar to the FA maps that were not distorted nor misregistered to the MP-RAGE data, the fiber bundles mapped were also not distorted nor misregistered to the high-resolution anatomical data. Excluding fiber endings, which cannot be located with certainty using any existing tractography procedure, there was no major part of the aforementioned pathways that was not traced. Robust fiber tracts were produced even in the frontal and temporal lobes, and the brain stem.

DISCUSSION

Fiber tractography by means of DTI offers the unique opportunity to localize white matter pathways of the brain *in vivo*. However, conventional tractography is

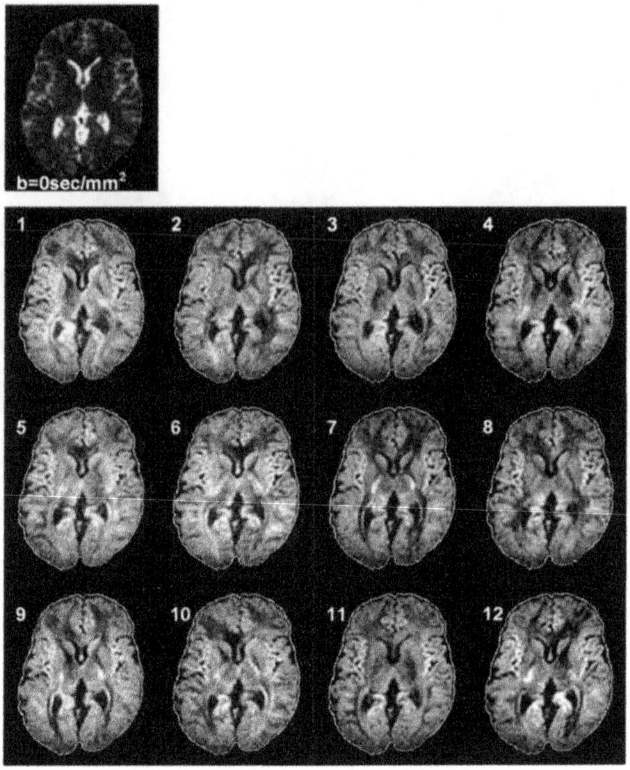

FIGURE 2. This figure shows the $b = 0$ s/mm^2 image for a slice of a healthy volunteer, and overlays of the outline of this image on the DW images obtained for the 12 diffusion directions. This figure demonstrates that the DW images acquired with Turboprop-DTI are immune to image distortions due to eddy currents.

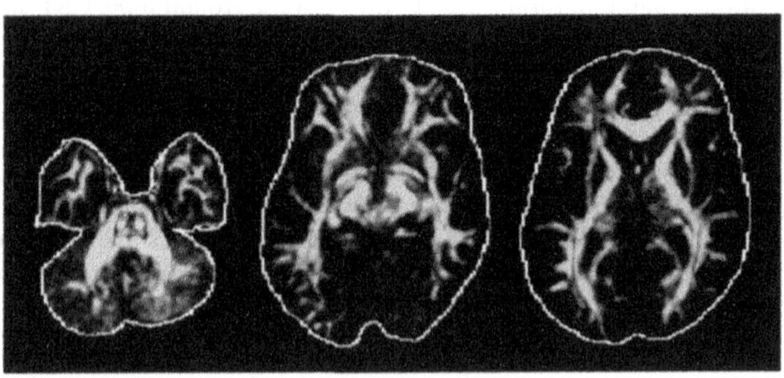

FIGURE 3. The result of edge detection on MP-RAGE images from 3 slices of a healthy volunteer, superimposed onto the corresponding FA maps from Turboprop-DTI. White matter tissue shown in the FA maps is in agreement with the anatomical information obtained from the MP-RAGE images.

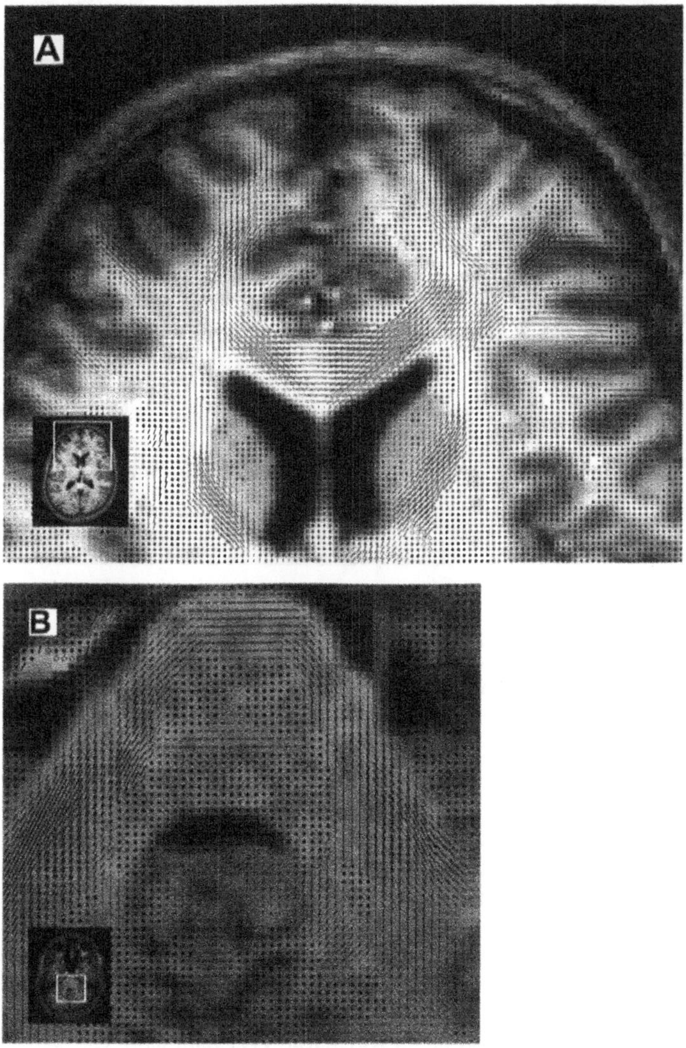

FIGURE 4. Vector maps showing estimates of fiber orientation (based on Turboprop-DTI data), superimposed on high-resolution MP-RAGE images. The orientation of fibers in the frontal lobe **(A)** and the brain stem **(B)** is in agreement with the anatomical information. Vectors correspond to fiber orientations that form less than 30° angles with the image plane. The length of the vectors is proportional to the FA value for the specific voxel. Out-of-plane fiber orientations are shown with dots. Information about the exact direction of out-of-plane vectors is not displayed.

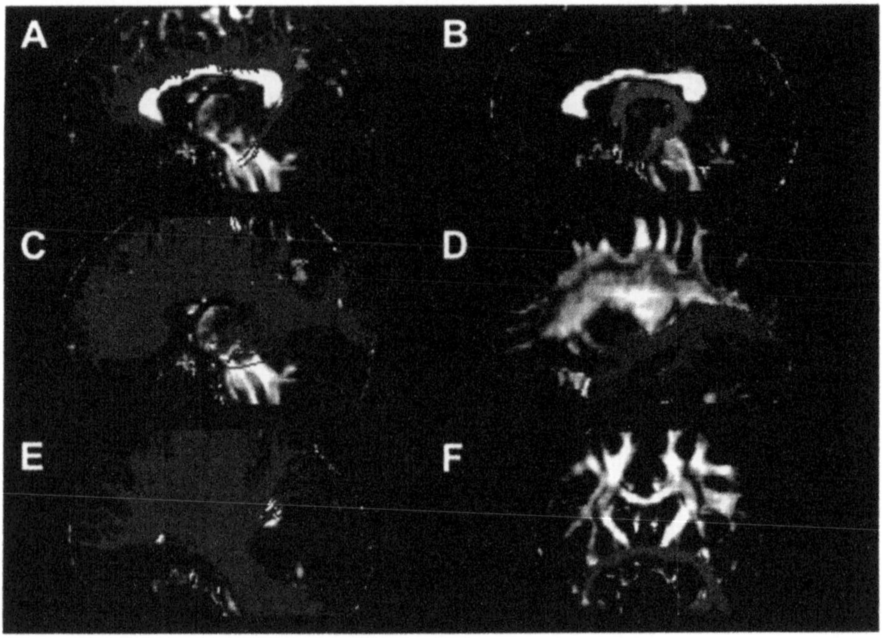

FIGURE 5. Fibers of the cingulum **(A)**, fornix **(B)**, corpus callosum **(C)**, inferior longitudinal fasciculus **(D)**, corticospinal tract and corona radiata **(E)**, and anterior commissure **(F)** reconstructed from Turboprop-DTI data from a healthy subject. Fibers traced from Turboprop-DTI data appear to be anatomically accurate representations of the corresponding fiber pathways.

performed based on DTI sequences with EPI readouts, which suffer from image distortions and artifacts due to susceptibility differences and eddy currents. Thus, a large percentage of white matter fiber bundles that are mapped using EPI-based DTI data are distorted, and/or terminated early, while others are completely undetected.[28] This severely limits the potential of fiber-tractography. In this work, we tested the performance of white matter tractography techniques when using Turboprop-DTI data acquisitions.

Fiber bundles often mapped using EPI-based DTI data were successfully traced using Turboprop-DTI data. All fibers that were mapped based on Turboprop-DTI were not characterized by any significant distortions, were good representations of known anatomy, and were in agreement with the corresponding high-resolution anatomical images of each subject. These properties are particularly important for clinical applications such as presurgical planning, where knowledge of the exact location of the lesion with respect to eloquent white matter pathways is of great value since the principal goal in neurosurgery is preservation of vital brain tissue while maximizing resection of diseased tissue.

Several variations of the SE-EPI-DTI pulse sequence, as well as postprocessing algorithms, have been developed in order to reduce the effects of magnetic field inhomogeneities[26,40–44] and eddy currents[27,45–48] when using EPI-based DTI sequences. However, all of these techniques increase the total scan time or are computationally intensive, or both, and none of them completely corrects susceptibility and eddy current–related artifacts. Residual distortions and artifacts in DTI data obtained with SE-EPI-DTI acquisitions may introduce fiber-tracking errors. In contrast, Turboprop-DTI was shown to be immune to image distortions and artifacts from susceptibility differences and eddy currents. Therefore, data acquired with Turboprop-DTI can be used for tractography directly, without requiring additional calibration scans or preprocessing corrections.

Recently, in addition to PROPELLER-DTI,[30] other novel DTI acquisition methods (line-scan-DTI,[49] STEAM-DTI[50]) have been proposed that provide distortion-free DTI data. However, if a certain number of DW images from the whole brain is acquired in about 6 min with SE-EPI-DTI, the same number of images requires about 2 h of scan time with the new techniques,[30,49,50] which renders these methods inappropriate for clinical use. Turboprop-DTI is a significantly accelerated version of PROPELLER-DTI, which however maintains the advantages of PROPELLER imaging. In this work, Turboprop-DTI acquisitions were completed in less than 19 min. Further acceleration may be achieved by combining Turboprop-DTI with parallel imaging[51,52] or by undersampling k-space.[53]

In conlusion, Turboprop-DTI can provide whole-brain, undistorted DTI data, with satisfactory signal-to-noise levels, in a clinically acceptable time. Therefore, Turboprop-DTI is an attractive DTI data acquisition and reconstruction technique for tractography applications.

REFERENCES

1. Basser, P.J., J. Mattiello & D. Le Bihan. 1994. MR diffusion tensor spectroscopy and imaging. Biophys. J. **66:** 259–267.
2. Basser, P.J., J. Mattiello & D. Le Bihan. 1994. Estimation of the effective self-diffusion tensor from the NMR spin echo. J. Magn. Reson. **B103:** 247–254.
3. Pierpaoli, C., P. Jezzard, P.J. Basser *et al.* 1996. Diffusion tensor MR imaging of the human brain. Radiology **201:** 637–648.
4. Basser, P.J. 1995. Inferring microstructural features and the physiological state of tissues from diffusion-weighted images. NMR Biomed. **8:** 333–344.
5. Basser, P.J. & C. Pierpaoli. 1996. Microstructural and physiological features of tissues elucidated by quantitative-diffusion-tensor MRI. J. Magn. Reson. **B111:** 209–219.
6. Pierpaoli, C. & P.J. Basser. 1996. Toward a quantitative assessment of diffusion anisotropy. Magn. Reson. Med. **36:** 893–906.
7. Basser, P.J., S. Pajevic, C. Pierpaoli *et al.* 2000. *In vivo* tractography using DT-MRI data. Magn. Reson. Med. **44:** 625–632.
8. Toosy, A.T., O. Ciccarelli, G.J. Parker *et al.* 2004. Characterizing function-structure relationships in the human visual system with functional MRI and diffusion tensor imaging. Neuroimage **21:** 1452–1463.
9. Guye, M., G.J. Parker, M. Symms *et al.* 2003. Combined functional MRI and tractography to demonstrate the connectivity of the human primary motor cortex *in vivo.* Neuroimage **19:** 1349–1360.
10. Berman, J.I., M.S. Berger, P. Mukherjee & R.G. Henry. 2004. Diffusion-tensor imaging-guided tracking of fibers of the pyramidal tract combined with intraoperative cortical stimulation mapping in patients with gliomas. J. Neurosurg. **101:** 66–72.
11. Clark, C.A., T.R. Barrick, M.M. Murphy & B.A. Bell. 2003. White matter fiber tracking in patients with space-occupying lesions of the brain: a new technique for neurosurgical planning? Neuroimage **20:** 1601–1608.

12. Henry, R.G., J.I. Berman, S.S. Nagarajan *et al.* 2004. Subcortical pathways serving cortical language sites: initial experience with diffusion tensor imaging fiber tracking combined with intraoperative language mapping. Neuroimage **21:** 616–622.
13. Le Bihan, D., J.F. Mangin, C. Poupon *et al.* 2001. Diffusion tensor imaging: concepts and applications. J. Magn. Reson. Imaging **13:** 534–546.
14. Conturo, T.E., N.F. Lori, T.S. Cull *et al.* 1999. Tracking neuronal fiber pathways in the living human brain. Proc. Natl. Acad. Sci. USA **96:** 10422–10427.
15. Jones, D.K., A. Simmons, S.C.R. Williams & M.A. Horsfield. 1999. Noninvasive assessment of axonal fiber connectivity in the human brain via diffusion tensor MRI. Magn. Reson. Med. **42:** 37–41.
16. Poupon, C., C.A. Clark, V. Frouin *et al.* 2000. Regularization of diffusion-based direction maps for the tracking of brain white matter fascicles. Neuroimage **12:** 184–195.
17. Stieltjes, B., W.E. Kaufmann, P.C.M. van Zijl *et al.* 2001. Diffusion tensor imaging and axonal tracking in the human brain. Neuroimage **14:** 723–735.
18. Tench, C.R., P.S. Morgan, M. Wilson & L.D. Blumhardt. 2002. White matter mapping using diffusion tensor MRI. Magn. Reson. Med. **47:** 967–972.
19. Mori, S., B. Crain, V.P. Chacko & P.C.M. van Zijl. 1999. Three-dimensional tracking of axonal projections in the brain by magnetic resonance imaging. Ann. Neurol. **45:** 265–269.
20. Mori, S., W.E. Kaufmann, C. Davatzikos *et al.* 2002. Imaging cortical association tracts in the human brain using diffusion-tensor-based axonal tracking. Magn. Reson. Med. **47:** 215–223.
21. Lori, N.F., E. Akbudak, J.S. Shimony *et al.* 2002. Diffusion tensor fiber tracking of human brain connectivity: acquisition methods, reliability analysis, and biological results. NMR Biomed. **15:** 494–515.
22. Lazar, M., D.M. Weinstein, J.S. Tsuruda *et al.* 2003. White matter tractography using diffusion tensor deflection. Hum. Brain Map. **18:** 306–321.
23. Koch, M.A., D.G. Norris & M. Hund-Georgiadis. 2002. An investigation of functional and anatomical connectivity using magnetic resonance imaging. Neuroimage **16:** 241–250.
24. Behrens, T.E., M.W. Woolrich, M. Jenkinson *et al.* 2003. Characterization and propagation of uncertainty in diffusion-weighted MR imaging. Magn. Reson. Med. **50:** 1077–1088.
25. Parker, G.J.M., H.A. Haroon & C.A.M. Wheeler-Kingshott. 2003. A framework for a streamline-based probabilistic index of connectivity (PICo) using a structural interpretation of MRI diffusion measurements. J. Magn. Reson. Imaging **18:** 242–254.
26. Zeng, H. & R.T. Constable. 2002. Image distortion correction in EPI: comparison of field mapping with point spread function mapping. Magn. Reson. Med. **48:** 137–146.
27. Alexander, A.L., J.S. Tsuruda & D.L. Parker. 1997. Elimination of eddy current artifacts in diffusion-weighted echo planar images: the use of bipolar gradients. Magn. Reson. Med. **38:** 1016–1021.
28. Gui, M., M. Lazar & K. Arfanakis. 2004. A comparison of white matter fiber-tracking results using PROPELLER and SE-EPI datasets. *In* Proceedings of the ISMRM Twelfth Scientific Meeting, Kyoto, Japan, p. 1283.
29. Pipe, J.G. 1999. Motion correction with PROPELLER MRI: application to head motion and free-breathing cardiac imaging. Magn. Reson. Med. **42:** 963–969.
30. Pipe, J.G., V.G. Farthing & K.P. Forbes. 2002. Multishot diffusion-weighted FSE using PROPELLER MRI. Magn. Reson. Med. **47:** 42–52.
31. Butts, K., A. de Crespigny, J.M. Pauly & M. Moseley. 1996. Diffusion-weighted interleaved echo-planar imaging with a pair of orthogonal navigator echoes. Magn. Reson. Med. **35:** 763–770.
32. Peters, D.C., F.R. Korosec, T.M. Grist *et al.* 2000. Undersampled projection reconstruction applied to MR angiography. Magn. Reson. Med. **43:** 91–101.
33. Forbes, K.P., J.G. Pipe, C.R. Bird & J.E. Heiserman. 2001. PROPELLER MRI: clinical testing of a novel technique for quantification and compensation of head motion. J. Magn. Reson. Imaging **14:** 215–222.

34. FORBES, K.P., J.G. PIPE, J.P. KARIS & J.E. HEISERMAN. 2002. Improved image quality and detection of acute cerebral infarction with PROPELLER diffusion-weighted MR imaging. Radiology **225:** 551–555.
35. WU, Y., A.S. FIELD & A.L. ALEXANDER. 2003. Diffusion tensor imaging of the human cervical spinal cord using PROPELLER. *In* Proceedings of the ISMRM Eleventh Scientific Meeting, Toronto, Canada, p. 2125.
36. ROBERTS, T.P. & M. HAIDER. 2004. Diffusion weighted imaging of the prostate gland in the face of magnetic susceptibility differences—parallel EPI and PROPELLER FSE approaches. *In* Proceedings of the ISMRM Twelfth Scientific Meeting, Kyoto, Japan, p. 946.
37. SUSSMAN, M.S., L.M. WHITE & T.P. ROBERTS. 2004. High-resolution diffusion-weighted imaging of cartilage using PROPELLER. *In* Proceedings of the ISMRM Twelfth Scientific Meeting, Kyoto, Japan, p. 211.
38. PIPE, J.G. 2002. Turboprop—an improved PROPELLER sequence for diffusion weighted MRI. *In* Proceedings of the ISMRM Tenth Scientific Meeting, Honolulu, Hawaii, p. 435.
39. OSHIO, K. & D.A. FEINBERG. 1991. GRASE (gradient- and spin-echo) imaging: a novel fast MRI technique. Magn. Reson. Med. **20:** 344–349.
40. CHEN, Z., S.S. LI, J. YANG *et al.* 2004. Measurement and automatic correction of high-order B_0 inhomogeneity in the rat brain at 11.7 tesla. Magn. Reson. Imaging **22:** 835–842.
41. WAN, X., G.T. GULLBERG, D.L. PARKER & G.L. ZENG. 1997. Reduction of geometric and intensity distortions in echo-planar imaging using a multireference scan. Magn. Reson. Med. **37:** 932–942.
42. JEZZARD, P. & R.S. BALABAN. 1995. Correction for geometric distortion in echo planar images from B_0 field variations. Magn. Reson. Med. **34:** 65–73.
43. ANDERSSON, J.L., S. SKARE & J. ASHBURNER. 2003. How to correct susceptibility distortions in spin-echo echo-planar images: application to diffusion tensor imaging. Neuroimage **20:** 870–888.
44. CORDES, D., K. ARFANAKIS, V. HAUGHTON & M.E. MEYERAND. 2000. Geometric distortion correction in EPI using two images with orthogonal phase-encoding directions. *In* Proceedings of the Eighth Annual Meeting of ISMRM, Denver, Colorado, p. 1712.
45. JEZZARD, P., A.S. BARNETT & C. PIERPAOLI. 1998. Characterization of and correction for eddy current artifacts in echo planar diffusion imaging. Magn. Reson. Med. **39:** 801–812.
46. PAPADAKIS, N.G., K.M. MARTIN, J.D. PICKARD *et al.* 2000. Gradient preemphasis calibration in diffusion-weighted echo-planar imaging. Magn. Reson. Med. **44:** 616–624.
47. HASELGROVE, J.C. & J.R. MOORE. 1996. Correction for distortion of echo-planar images used to calculate the apparent diffusion coefficient. Magn. Reson. Med. **36:** 960–964.
48. ARFANAKIS, K., D. CORDES, V.M. HAUGHTON *et al.* 2002. Independent component analysis applied to diffusion tensor MRI. Magn. Reson. Med. **47:** 354–363.
49. MAMATA, H., Y. MAMATA, C.F. WESTIN *et al.* 2002. High-resolution line scan diffusion tensor MR imaging of white matter fiber tract anatomy. AJNR Am. J. Neuroradiol. **23:** 67–75.
50. KOCH, M.A., V. GLAUCHE, J. FINSTERBUSCH *et al.* 2002. Distortion-free diffusion tensor imaging of cranial nerves and of inferior temporal and orbitofrontal white matter. Neuroimage **17:** 497–506.
51. PRUESSMANN, K.P., M. WEIGER, M.B. SCHEIDEGGER & P. BOESIGER. 1999. SENSE: sensitivity encoding for fast MRI. Magn. Reson. Med. **42:** 952–962.
52. PIPE, J.G. 2003. The use of parallel imaging with PROPELLER DWI. *In* Proceedings of the ISMRM Eleventh Scientific Meeting, Toronto, Canada, p. 66.
53. ARFANAKIS, K., A.A. TAMHANE, J.G. PIPE & M.A. ANASTASIO. 2005. K-space undersampling in PROPELLER imaging. Magn. Reson. Med. **53:** 675–683.

Diffusion Tensor Tractography of the Motor White Matter Tracts in Man

Current Controversies and Future Directions

ANDREI I. HOLODNY,[a] RICHARD WATTS,[b] VALERI N. KORNEINKO,[c] IGOR N. PRONIN,[c] MIKHAIL E. ZHUKOVSKIY,[d] DEVANG M. GOR,[e] AND AZIZ ULUG[b]

[a]*Functional MRI Laboratory, Department of Radiology, Memorial Sloan-Kettering Cancer Center, New York, New York 10021, USA*

[b]*Department of Radiology, Weill Medical College of Cornell University, New York, New York 10021, USA*

[c]*Department of Neuroradiology, Burdenko Institute of Neurosurgery, Moscow, Russia 125047*

[d]*Keldysh Institute for Applied Mathematics, Moscow, Russia 125047*

[e]*Department of Radiology, University of Pennsylvania, Philadelphia, Pennsylvania 19104, USA*

ABSTRACT: The anatomy of the brain is extremely complex, and certain, even large structures, such as the corticospinal tract (CST), remain poorly understood. Diffusion tractography provides an opportunity to explore the white matter tracts in a fundamentally new way. In the current paper, we show how this technique has already added to our understanding of the anatomy of the CST. We also explore the future projects involving diffusion tractography of the motor white matter tracts that will advance this method and further our understanding of brain anatomy.

KEYWORDS: brain; corticospinal tract (CST); diffusion tractography; posterior limb of the internal capsule (PLIC); white matter

INTRODUCTION

Diffusion tractography has revolutionized our ability to study the white matter tracts in the living human brain. Tractography has made it possible to study basic human anatomy on a new fundamental level.[1,2] This technique has had great impact on more practical issues such as intraoperative guidance of the resection of brain tumors adjacent to crucial white matter tracts.[3–7] A number of recent publications[8,9]

Address for correspondence: Andrei I. Holodny, M.D., Functional MRI Laboratory, Department of Radiology, Memorial Sloan-Kettering Cancer Center, 1275 York Avenue, New York, NY 10021. Voice: 212-639-3182; fax: 212-794-5863.
holodnya@mskcc.org

Ann. N.Y. Acad. Sci. 1064: 88–97 (2005). © 2005 New York Academy of Sciences.
doi: 10.1196/annals.1340.016

have demonstrated remarkable images of specific white matter tracts in normal subjects using diffusion tractography.

Nevertheless, important questions remain, including validation of diffusion tractography by other methods. This question becomes most acute when the diffusion tractography results are unexpected or contradict previously held beliefs. In order to highlight the questions and controversies surrounding diffusion tractography, we will use the example of the corticospinal tract—arguably, the most important tract in the human brain.

The corticospinal tract (CST) is the main white matter connection between the motor cortex and the spinal cord and thereby serves as the main conduit of information between the higher cortical structures and the voluntary musculature of the arms, legs, and torso. Notwithstanding the crucial function the CST plays in voluntary motion, its exact anatomical location has been the subject of some debate over the past few decades.[9,10] The anatomic detail of the internal organization of the CST, including the relationship of the hand fibers to the foot fibers as they pass through the posterior limb of the internal capsule (PLIC), is disputed.[9,10] The orientation of the homunculus of the motor cortex and a number of studies appear to support a somatotopic organization of the CST in the PLIC; however, this remains conjectural.[9,10]

Aside from the academic interest of deepening our understanding of neuroanatomy, the exact location and internal organization of the CST as it passes through the corona radiata and the PLIC have a number of important clinical applications. For example, knowledge of the exact localization of the CST would improve planning for functional neurosurgery in patients with Parkinson's disease and related disorders, for treatment of patients with strokes or lacunar infarcts in the vicinity of the PLIC, and (perhaps most significantly) for preoperative localization of the CST in brain tumor patients.[3–7]

NEUROSURGERY

Surgical resection remains the mainstay of the treatment of neurological malignancies.[11] The goal of brain tumor surgery is to maximize tumor resection while avoiding important adjacent brain structures since their inadvertent resection can lead to devastating neurological consequences. In gliomas,[12,13] the length and the quality of survival are improved with maximized tumor resection. In order to preserve vital neurological function controlled by the brain tissue directly adjacent to or involved by the tumor, it is important for the neurosurgeon to be able to identify the relationship of the tumor to the brain structure to be avoided. Functional MRI (fMRI) can identify the location of a functional area of the brain (eloquent cortex) adjacent to a brain tumor and has been shown to be useful in preoperative planning as well as in guiding the actual resection.[14,15] However, fMRI is inherently limited in that it can only show activation in the cortical gray matter.

It can be argued that preoperative identification of white matter tracts is perhaps even more important than the identification of eloquent cortices, such as the motor cortex. First, direct intraoperative white matter stimulation is much less reliable than cortical stimulation.[9] Second, the precentral gyrus (the location of the motor cortex) can usually be identified by the operating neurosurgeon by visual inspection. On the other hand, it is essentially impossible to identify separate white matter tracts, such as the CST or the thalamocortical tract, by visual inspection as they traverse the corona radiata.

Inadvertent transection of the CST leads to paralysis. Therefore, in order to improve the surgical treatment of patients with neurological malignancies, it is imperative to develop a method that could accurately depict the relationship of the CST to a tumor both preoperatively and intraoperatively. Such a method would improve neurosurgical planning as well and lessen postsurgical deficit.[3–7]

HISTORY OF THE SUBJECT OF THE LOCATION OF THE CST

The location of the CST in the PLIC has been the subject of debate and has undergone a major revision during the past few decades. One would have thought the location of arguably the most important white matter tract in the brain would have been known accurately and unambiguously since the time of Galen. However, this is not the case. Classically, the CST has been thought to lie in the anterior third of the PLIC.[16] This was first suggested by Charcot in 1883[17] and expounded upon by Dejerine in 1901[18] and Foerster in 1936.[19]

This classical localization of the CST was first challenged by neurosurgeons performing stereotactic procedures for the relief of parkinsonism.[20] Reports of (1) accidental lesions in the posterior aspect of the PLIC causing contralateral hemiplegia,[21] (2) correlation of the location of lacunar infarcts seen in the PLIC on anatomical dissection of autopsy brains to antemortem symptoms,[22] and (3) electrical stimulation by depth electrodes in patients undergoing stereotactic procedures[20] all supported the more posterior localization of the CST in the PLIC.

This has been supported by recent MRI studies not involving diffusion tractography. Yagishita *et al.*[23] compared MRI scans to brain specimens in normal subjects and patients with amyotrophic lateral sclerosis (ALS) and confirmed that the focus of T2-hyperintensity seen within the third quarter of the PLIC indeed contained the fibers of the CST.

Notwithstanding the above publications, the controversy continued to simmer.[24,25] Kumar suggested that there is a significant intersubject variability in position of the CST in the PLIC.[24] It is interesting to note that a number of major textbooks simply avoid this issue. A more recent anatomical study involving gross dissection of the brain[26] apparently showed that the more cephalad part of the CST is in the midportion of the PLIC, while the more caudad portion of the CST is in the posterior aspect of the PLIC.

Using diffusion tractography, recent publications have convincingly demonstrated that the CSTs traversed the posterior third quarter of the PLIC.[9] If the PLIC is divided into four equal quarters from anterior to posterior, the CST was shown to be in the third of these quarters.

SOMATOTOPIC ORGANIZATION OF THE CST

Given the difference of opinion regarding the *location* of the CST as it traverses the PLIC, it is not surprising that there would be disagreement regarding the *internal organization* of the CST in the PLIC.

In terms of the internal organization of the separate tracts of the CST that control movement of the hand, trunk, and foot, it is currently thought that these tracts are organized somatotopically in the PLIC, with the hand CST anterior and slightly medial to the foot along the long axis of the PLIC.[10,20] The evidence for this somatotopic organization is "rather crude".[10] The same author points out that somatotopic organization of the CST is also far from established in the crus cerebri of the cerebral peduncles.

Before the availability of tomographic imaging, there were limited ways that one could study white matter tracts in the human brain: gross dissection of postmortem specimens, correlation of autopsy brain material to antemortem clinical symptoms, and direct stimulation of the living brain by depth electrodes.

Most of what is currently believed about the internal organization of the CST in the PLIC is based on a study by Bertrand in 1965.[20] Bertrand's study suggested an organization of the fibers of the CST with the hand fiber located more anteriorly and slightly medially to the foot fiber. However, he demonstrated a large overlap between the hand and foot fibers. Indeed, Bertrand is rather modest about his claims: "Electrical stimulation within *what we have thought to be* the internal capsule itself has produced interesting information ...".[20] Unfortunately, his work was misinterpreted and translated into dogma in later publications, especially textbooks of anatomy. Although the anatomy textbook by Carpenter clearly acknowledged that the "evidence that fibers of the corticospinal tract are somatotopically arranged in this part of the internal capsule ... seems relatively crude",[10] Bertrand himself warns of "unpredictable anatomical variation".

Clearly, the limitations outlined above do not invalidate Bertrand's study. However, they do put forward an opportunity to present alternative results based on other modalities.

A recent study by our group[9] has demonstrated that 17 of the 20 fibers of the CST were organized somatotopically with the hand fibers lateral and slightly anterior to the foot fibers. In the remaining 3 (all left-sided), the hand fibers were intermixed with the foot fibers. The tumors in our study displaced the hand fibers posteriorly, but did not affect the internal organization of the CST.

The results of our study support a somatotopic organization to the separate tracts of the CST. However, from our results, it appears that the fibers are organized along the left-to-right axis (the short axis of the PLIC) as opposed to the anterior-posterior axis (the long axis), as currently believed. In 17 out of 20 cases in the present study,

---

FIGURE 1. Right (*top panel*) and left (*bottom panel*) posterior obliques of the traces of the corticospinal tracts (CSTs) superimposed on a T2-weighted image in a 40-year-old male healthy volunteer. The posterior part of the brain is towards the viewer. On the right side, the CST of the hand is in red and the CST of the foot is in green. On the left side, the CST of the hand is in green and the CST of the foot is in blue. The CST is seen to traverse the third quarter of the posterior limb of the internal capsule (PLIC). If the PLIC is divided into four equal quarters from anterior to posterior, the CST was shown to be in the third of these quarters. The CST of the hand is lateral and slightly anterior to the CST of the foot, which is the same relationship as the hand homunculus to the foot homunculus in the precentral gyrus. These findings contradict the current understanding of the relationship of the CSTs of the foot and hand in the PLIC. On the right side (*top panel*), there is some intermingling of the CSTs of the hand and the foot. Reprinted with permission from reference 9.

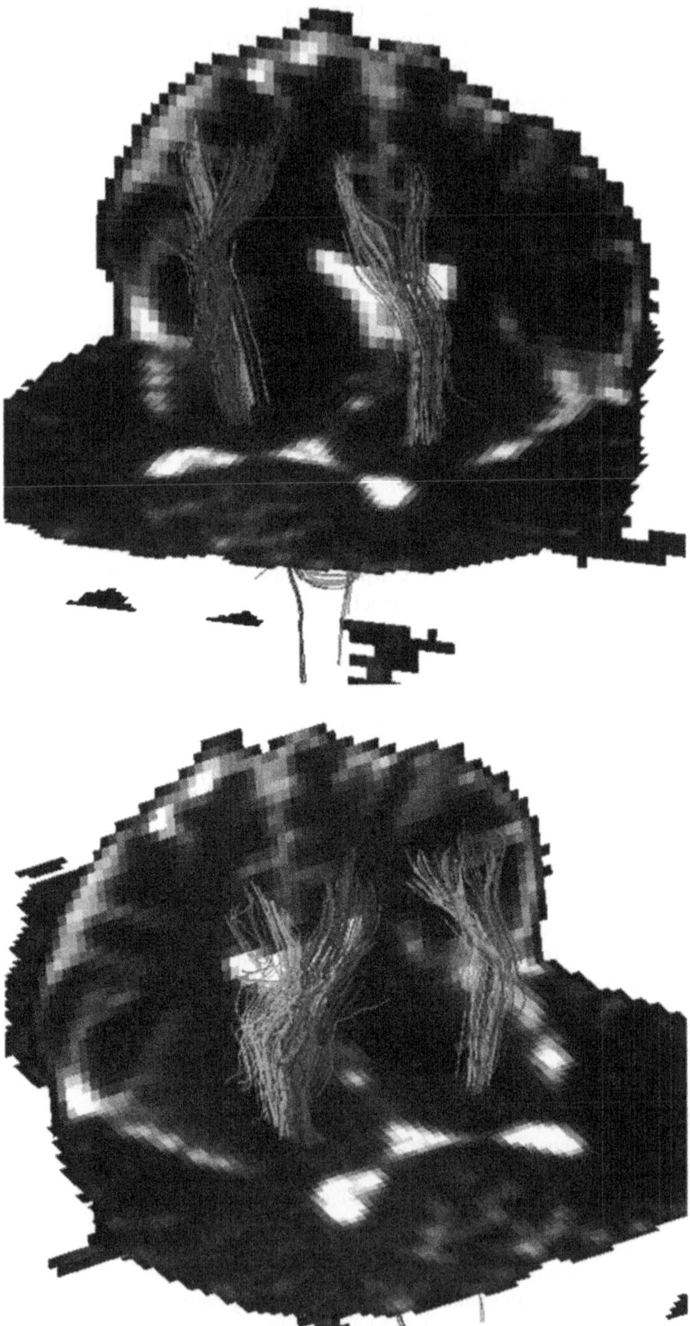

FIGURE 1. *See previous page for legend.*

the CST of the hand was lateral and slightly anterior to the CST of the foot. In the remaining 3 cases, the fibers were mixed.

According to our results, the relationship of the hand and foot CSTs is oriented exactly along the same axis as the hand and foot homunculi in the motor strip: the hand is lateral and slightly anterior to the foot. On the other hand, in the prevailing model, the orientation of the hand and foot CSTs is rotated 90° from the orientation of the hand and foot homunculi in the precentral gyrus. The location of the motor homunculus in the precentral gyrus is well characterized and confirmed routinely by intraoperative cortical mapping. Therefore, the organization of the CST in the PLIC that we have proposed appears to make much more anatomical sense than the prevailing model (FIG. 1, top and bottom panels).

FUTURE DIRECTIONS

Now that the exact location of the CST in the PLIC appears to be settled and we appear to be on our way in deciphering the somatotopic organization of the CST, a number of other important issues arise in what appears to be an inexhaustible plethora of mysteries of the anatomy of the CST.

First, the CSTs make up only part of the white matter tracts that emanate from the cortex and control movement. The CST controls the movement of the legs, torso, and arms. Control of movement of the face and tongue is executed by the corticobulbar tract (CBT). The homunculi of the tongue and face are located in the precentral gyrus, lateral to the homunculi of the leg and arm. The anatomy of the CBT in the human brain appears to be well described in the anatomy literature. These fibers originate in the lateral aspect of the precentral gyrus, and course inferiorly and laterally towards the CST fibers. The CBT fibers traverse the PLIC in the same area as the CST (although the exact location is unknown). The CBT does not proceed to the spinal cord (like the CST). Rather, the axons of the CBT terminate in the brain stem nuclei, which in turn activate movement of the face and tongue. Notwithstanding this general agreement among anatomists regarding the anatomical location of the CBT, this tract has never been successfully mapped using diffusion tractography.

The reason for this is as follows: One of the main problems currently experienced by diffusion tractography is the issue of *crossing fibers*. When a smaller white matter tract crosses a larger white matter tract, the computer programs designed to trace the white matter tracts will usually follow the crossing larger tract as opposed to the smaller tract that it was originally tracing. This problem is seen in the CBT. Caudad to the precentral gyrus and cephalad to the PLIC, the CBT intersects with the much larger white matter tract—the longitudinal fasciculus—which precludes tracing of the CBT since the computer-generated trace either terminates or follows the course of the longitudinal fasciculus. Consequently, we are usually not able to trace the CBT by conventional diffusion tractography methods.

What makes an opportune solution to the problem of crossed fibers more compelling is that, occasionally, the hand fibers of the CST themselves are also intersected by the longitudinal fasciculus. In such cases, one is not even able to trace the hand fibers of the CST. This issue becomes crucial in preoperative planning for patients with brain tumors in the vicinity of the CST.

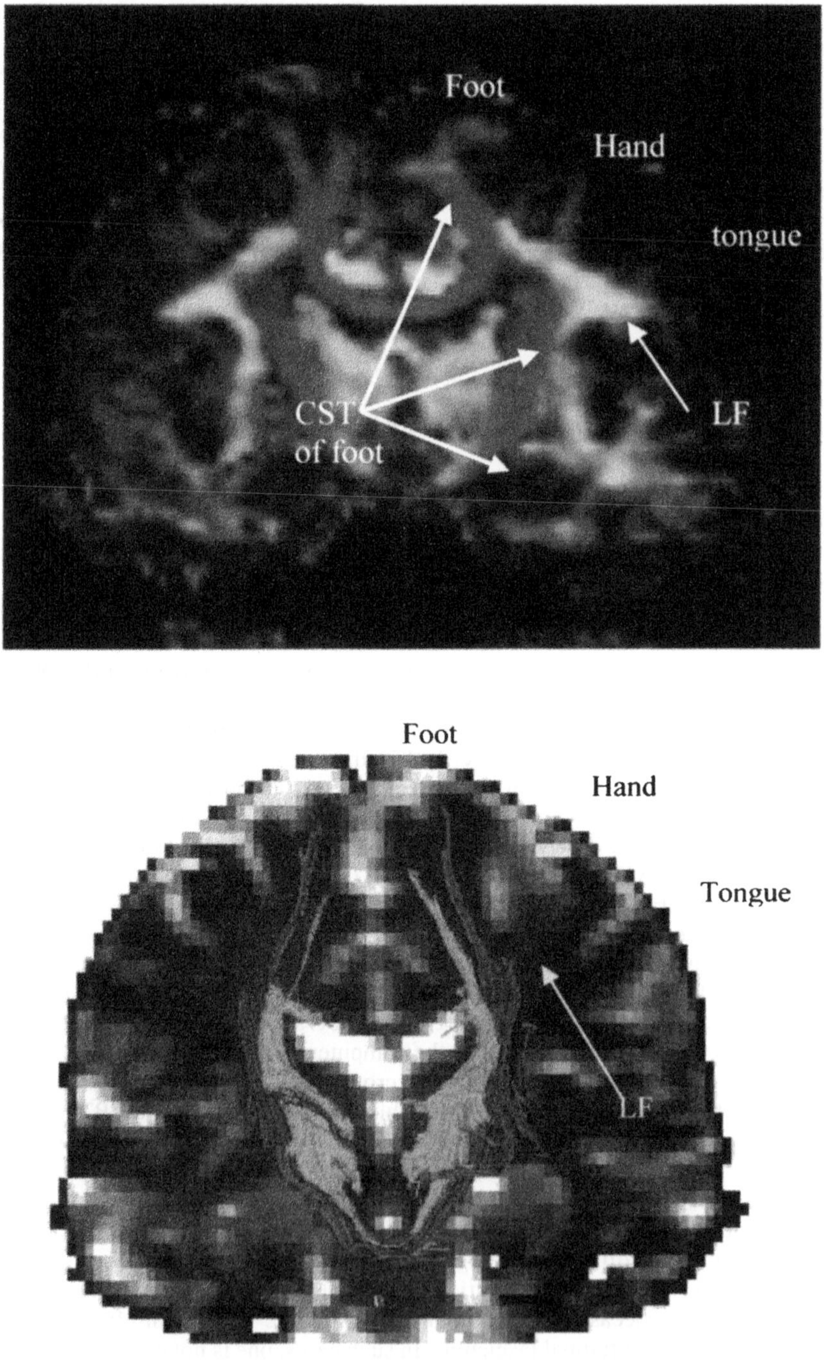

FIGURE 2. *See following page for legend.*

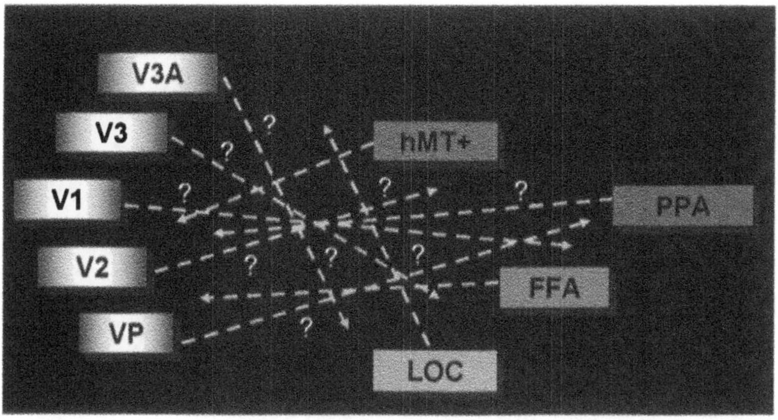

FIGURE 1. *See page 3.*

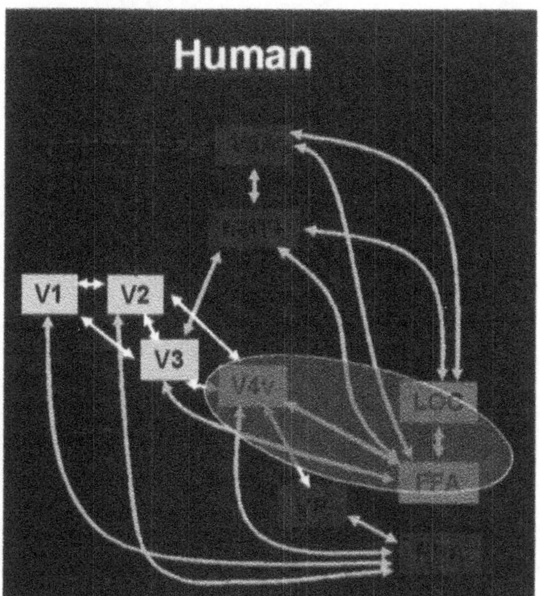

FIGURE 6. *See page 11.*

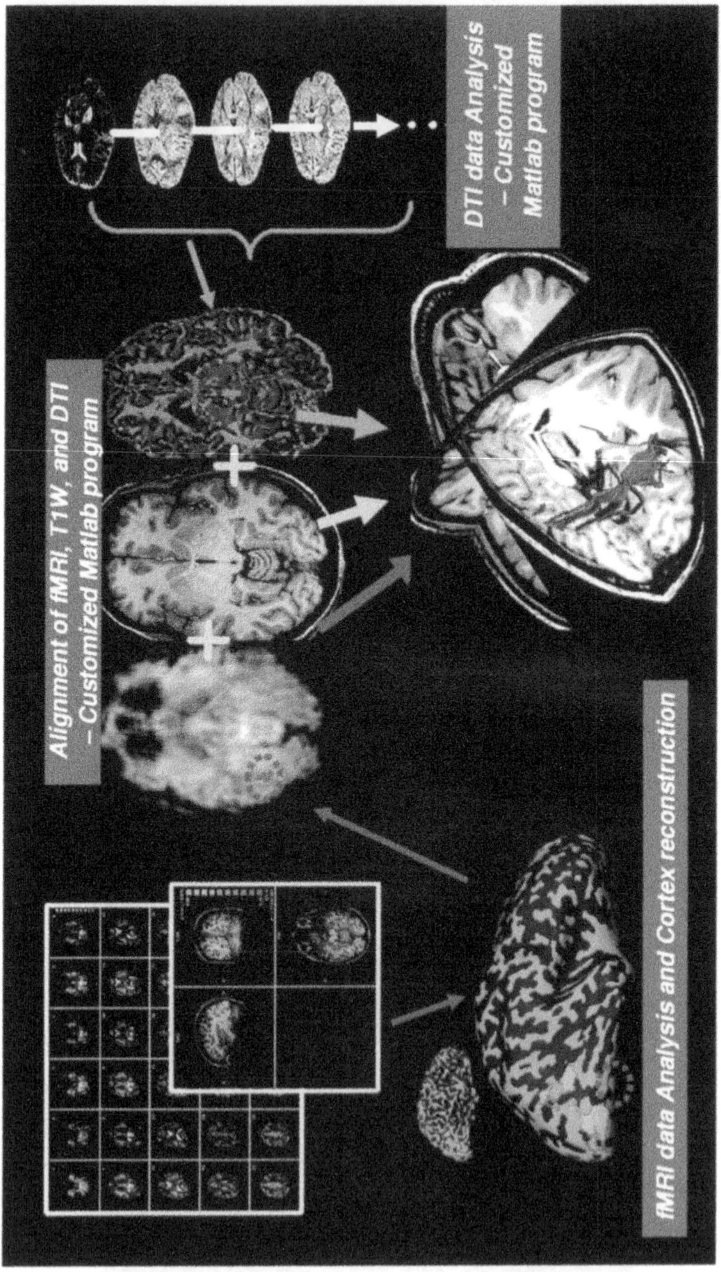

FIGURE 2. *See page 6.*

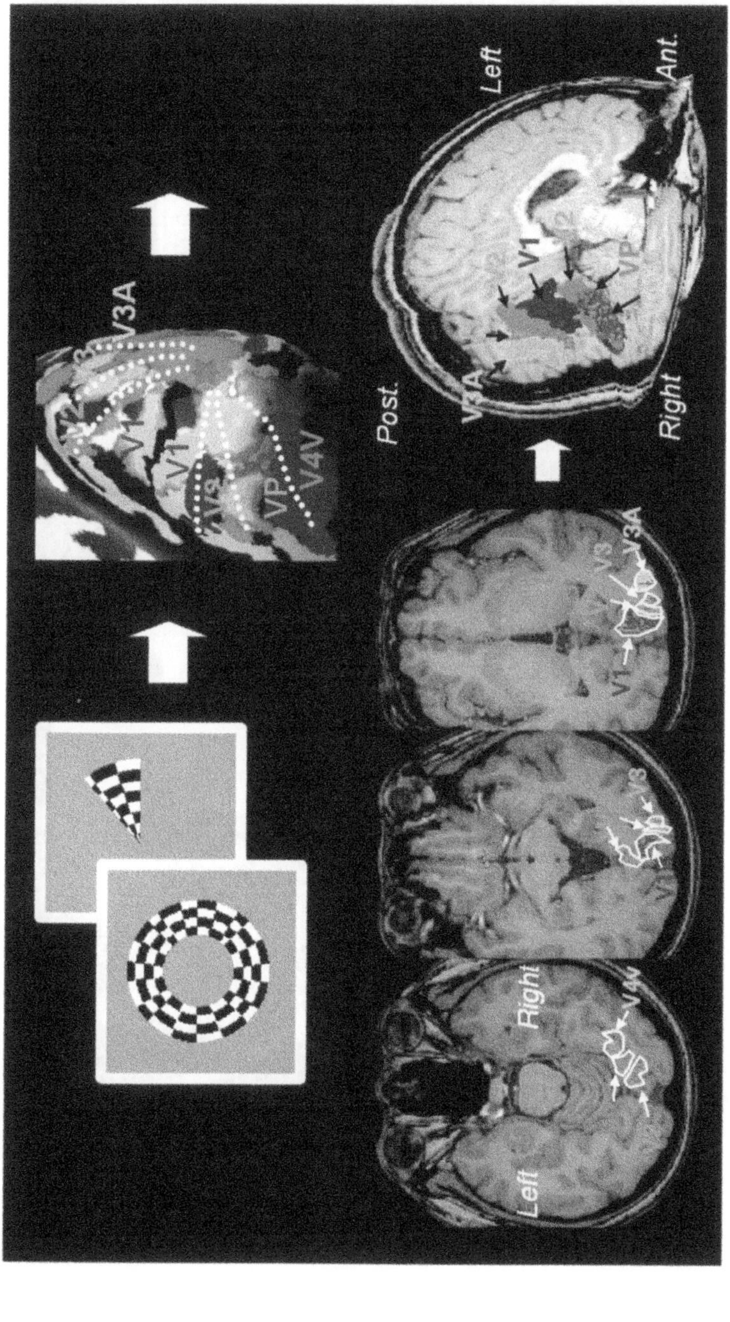

FIGURE 3. *See page 7.*

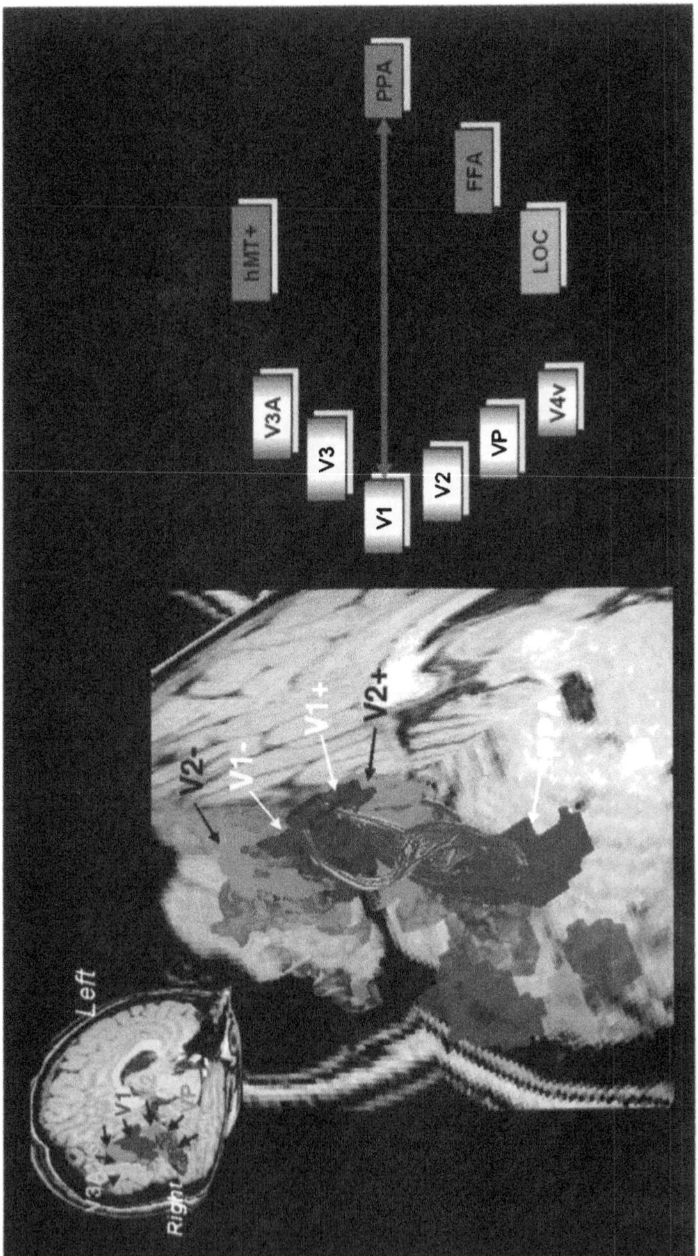

FIGURE 4. *See page 8.*

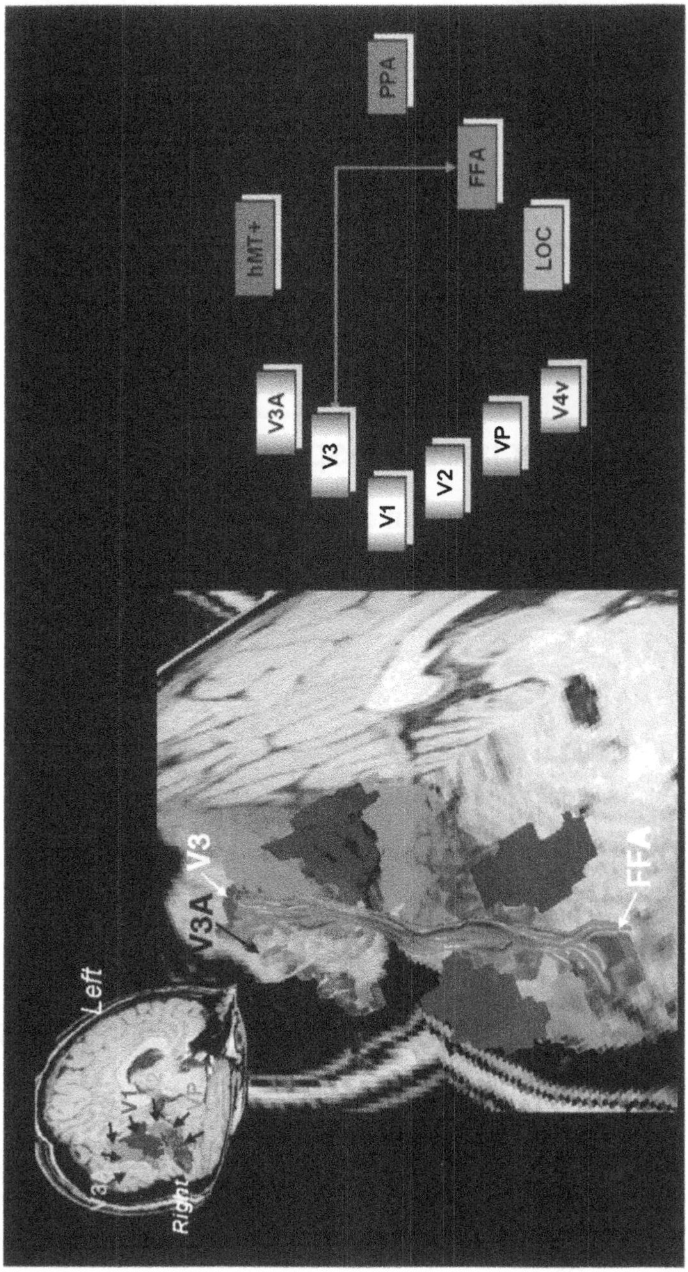

FIGURE 5. *See page 9.*

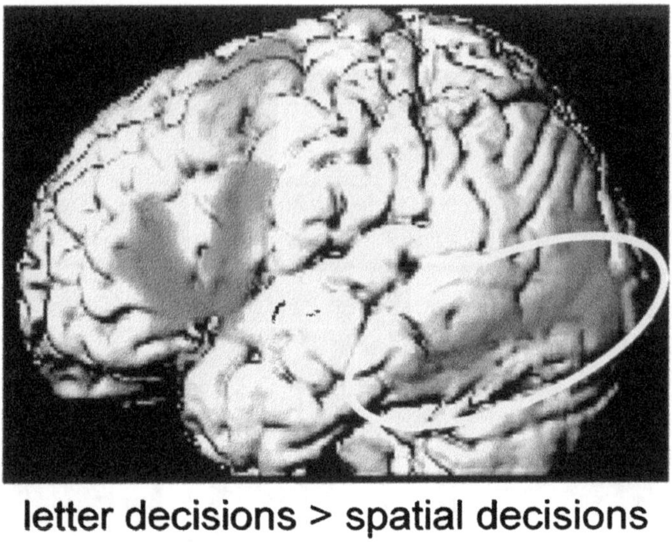

FIGURE 2. *See page 25.*

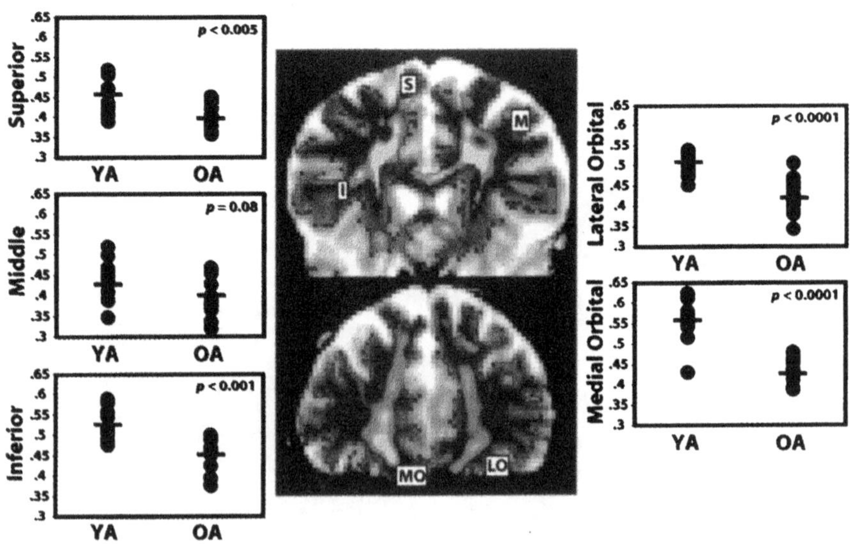

FIGURE 3. *See page 43.*

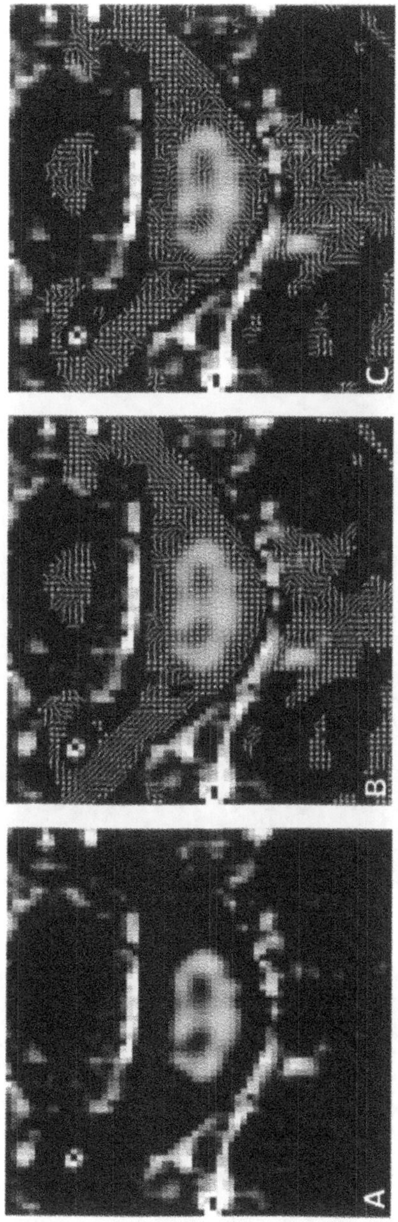

FIGURE 4. *See page 56.*

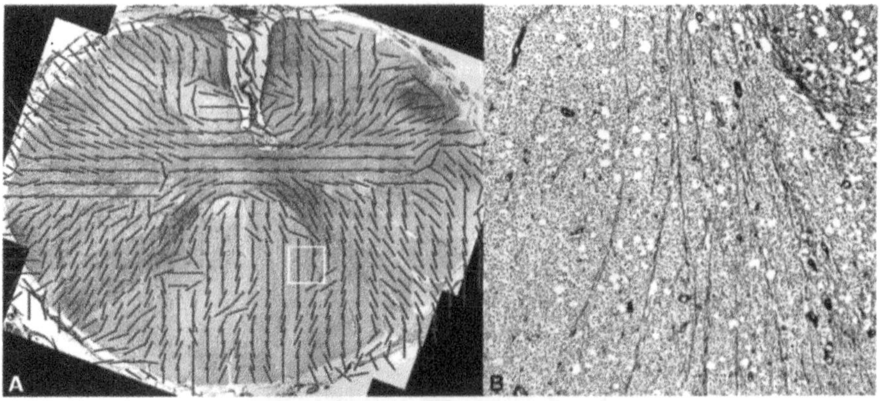

FIGURE 5. *See page 57.*

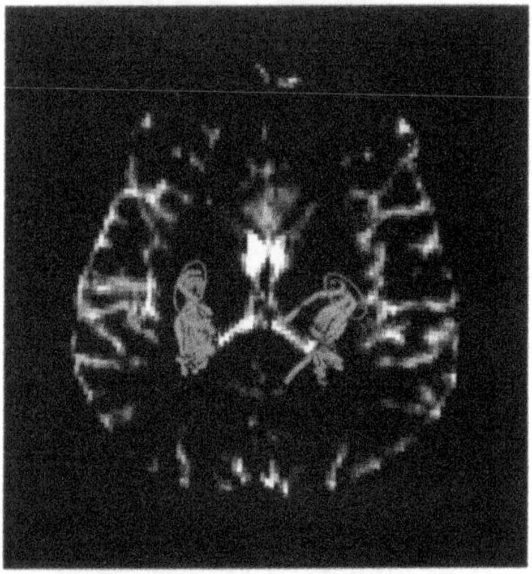

FIGURE 2. *See page 66.*

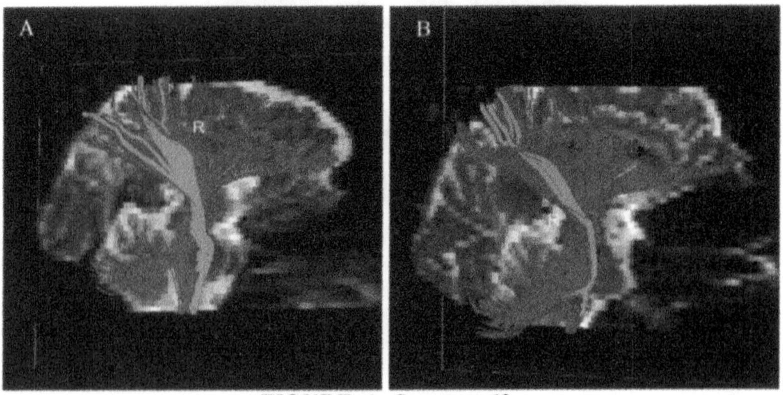

FIGURE 4. *See page 69.*

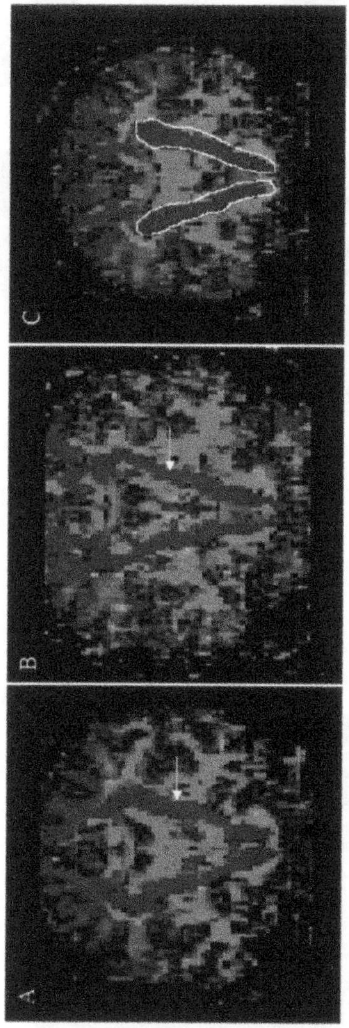

FIGURE 3. *See page 68.*

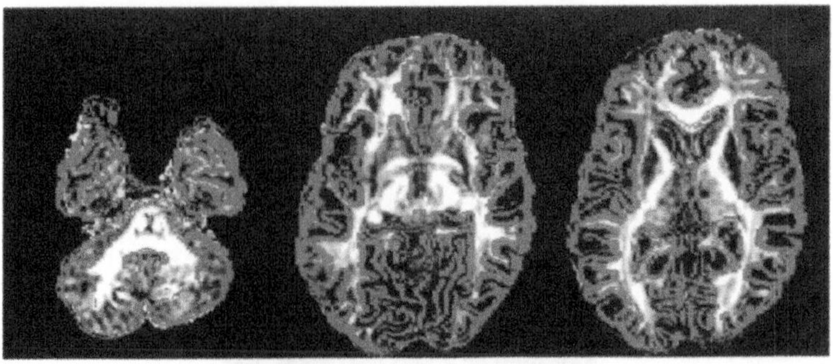

FIGURE 3. *See page 82.*

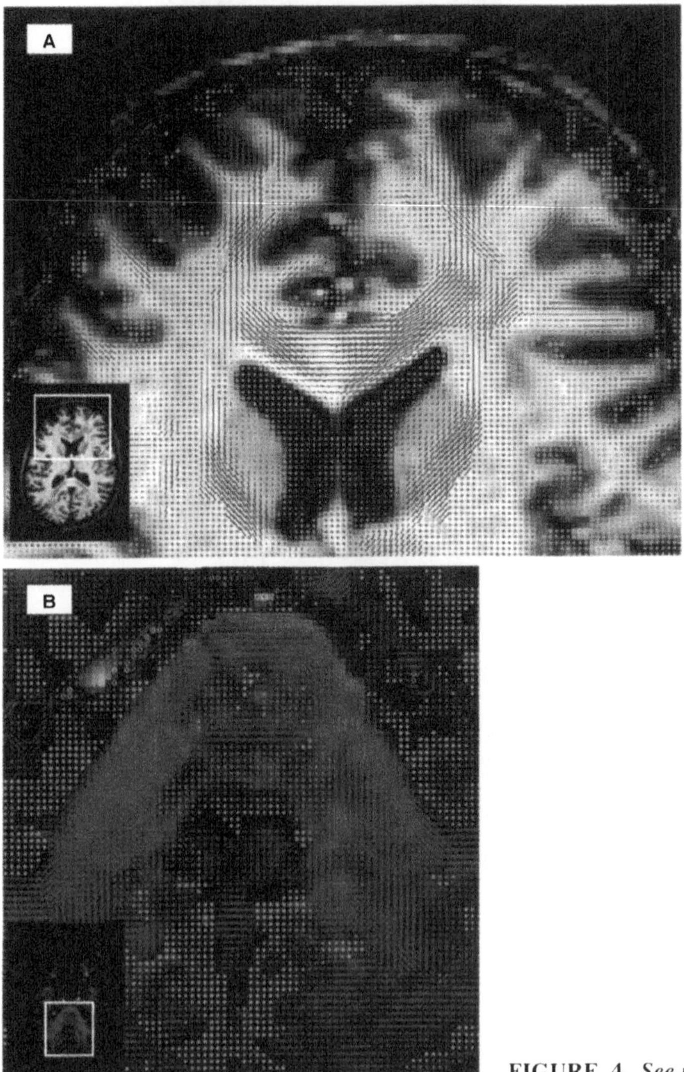

FIGURE 4. *See page 83.*

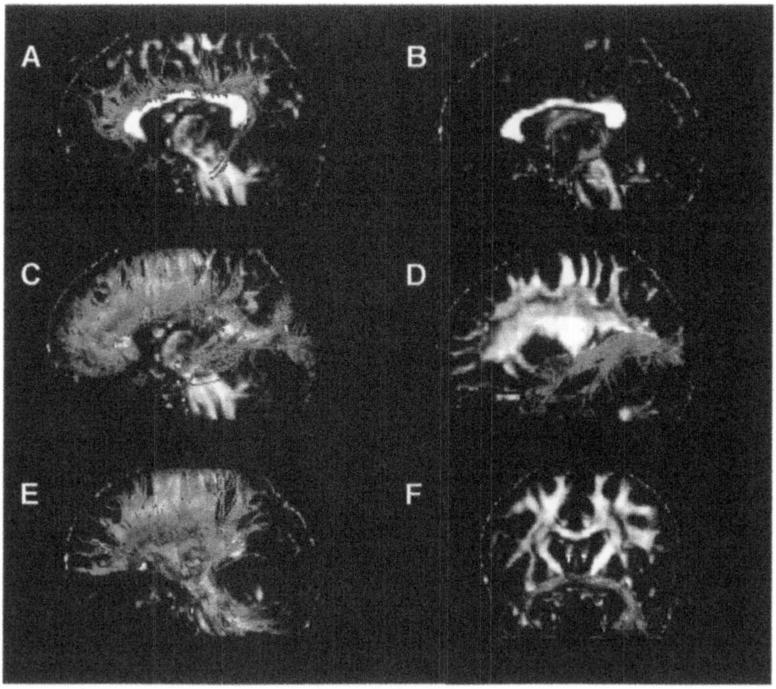

FIGURE 5. *See page 84.*

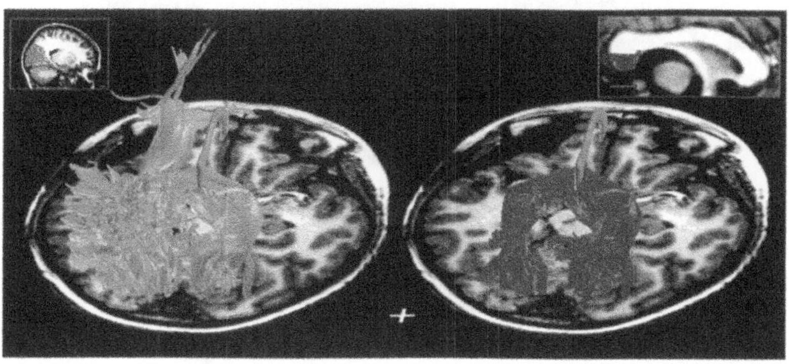

FIGURE 1. *See page 102.*

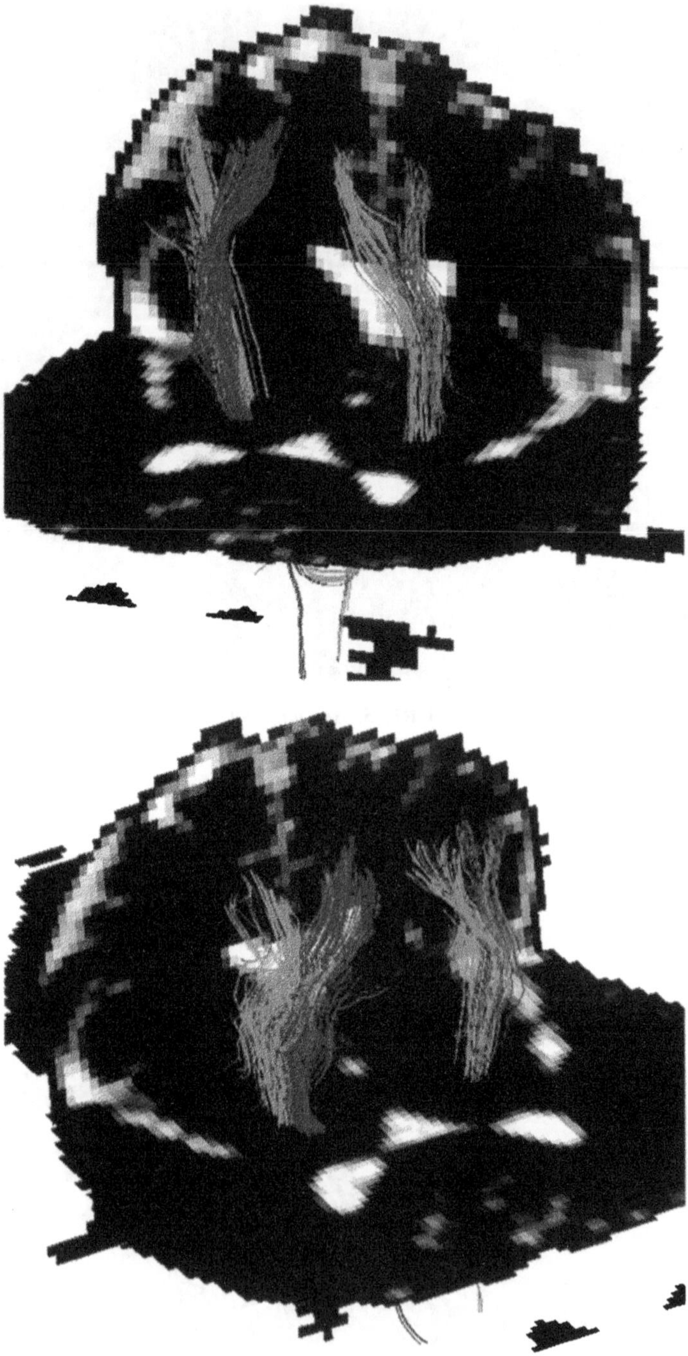

FIGURE 1. *See page 92.*

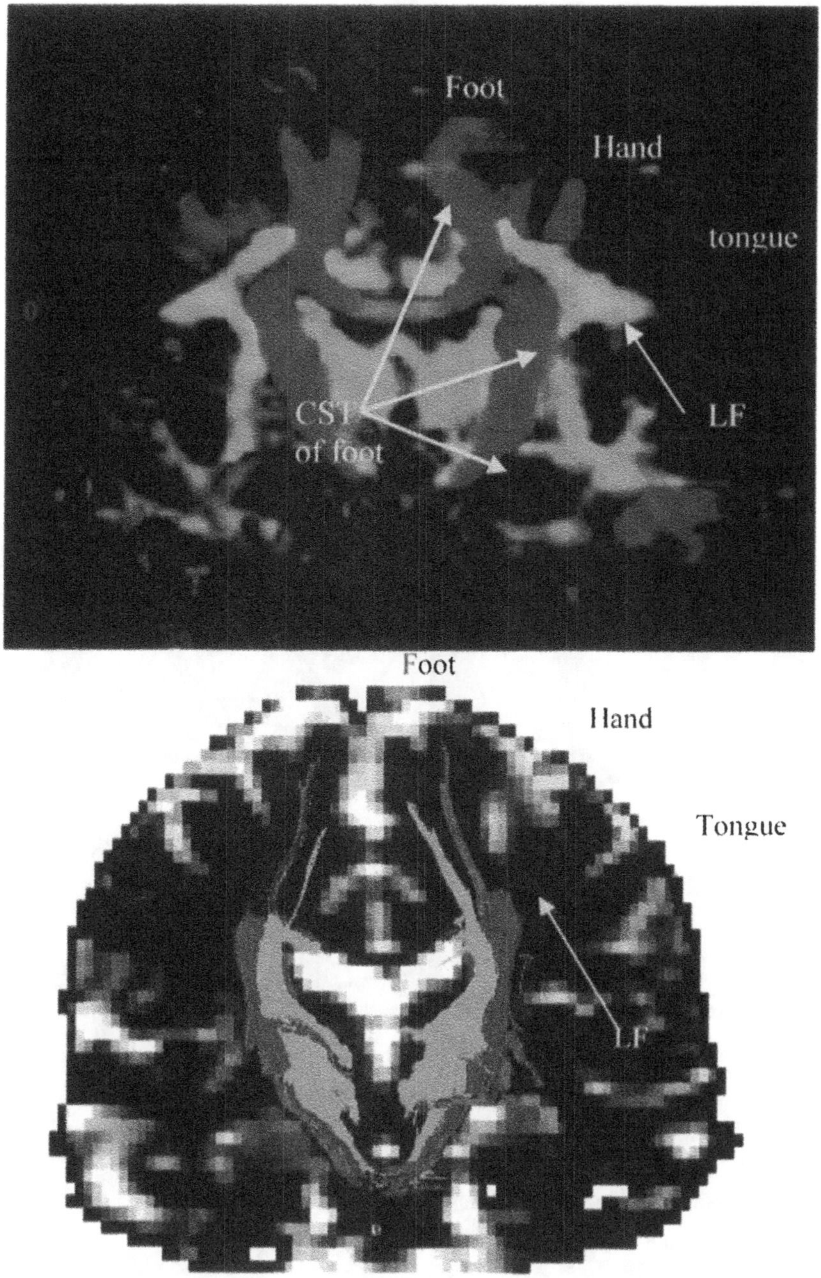

FIGURE 2. *See page 94.*

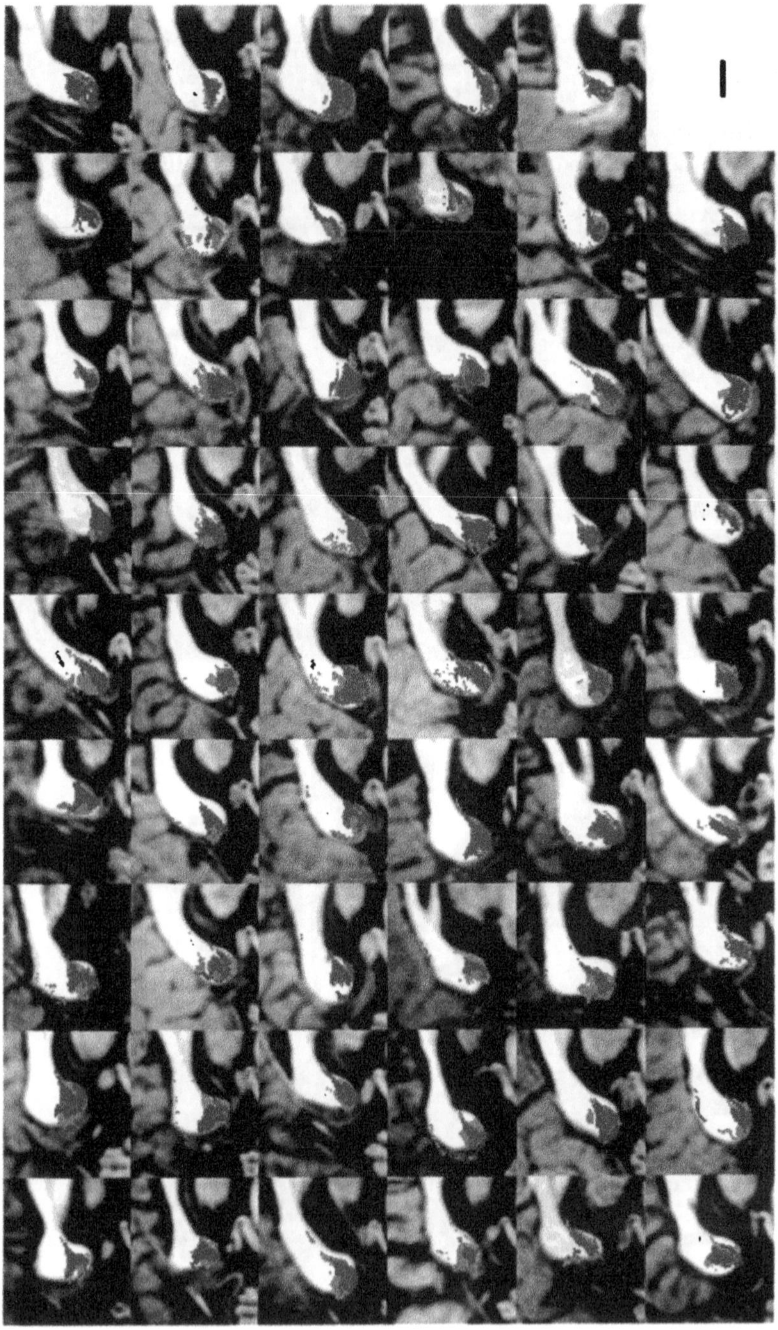

FIGURE 2. *See page 104.*

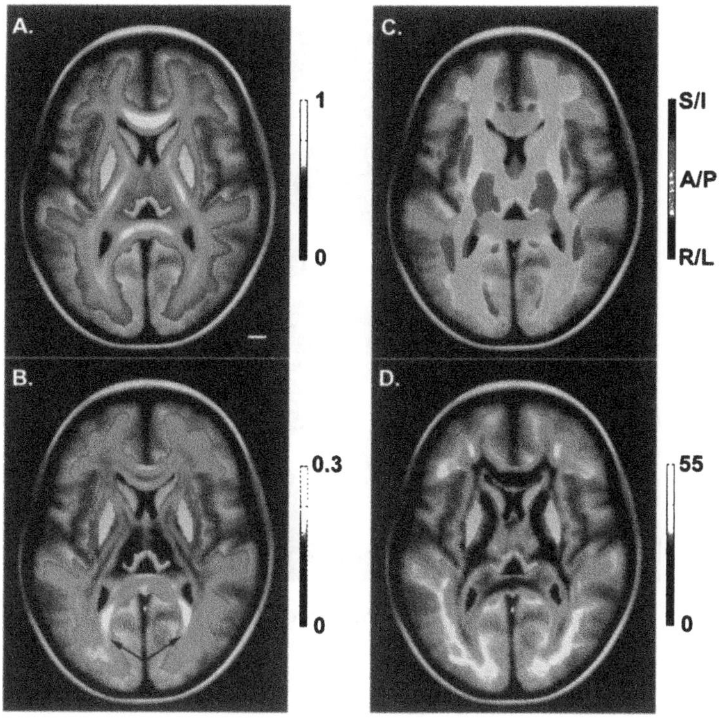

FIGURE 5. *See page 108.*

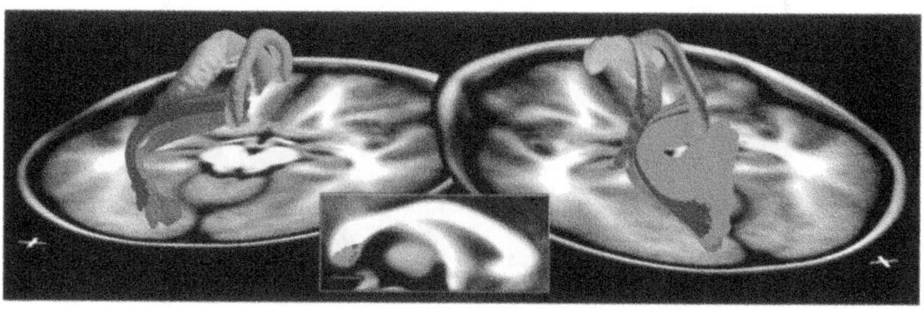

FIGURE 6. *See page 109.*

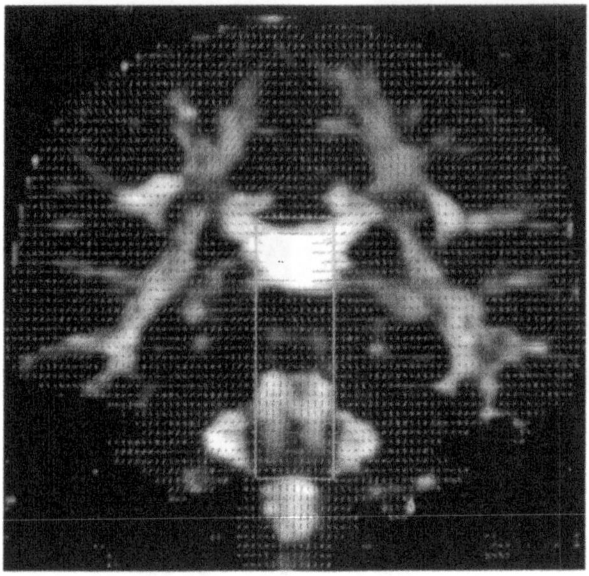

FIGURE 1. *See page 122.*

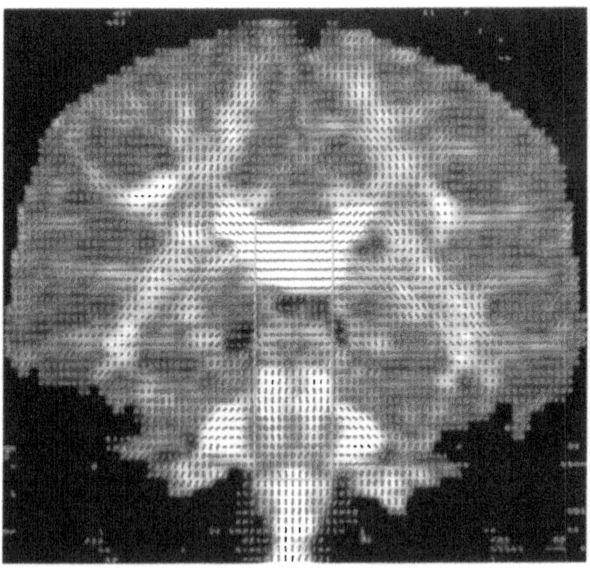

FIGURE 2. *See page 123.*

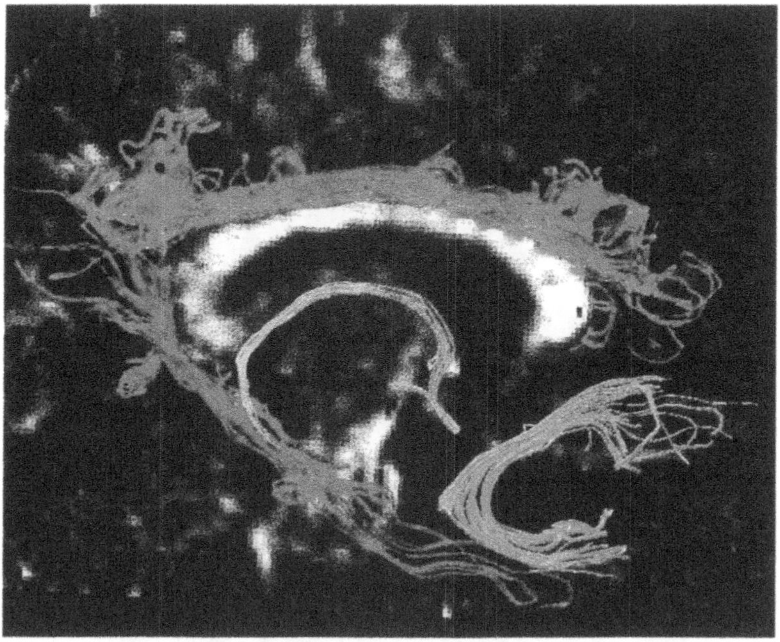

FIGURE 1. *See page 137.*

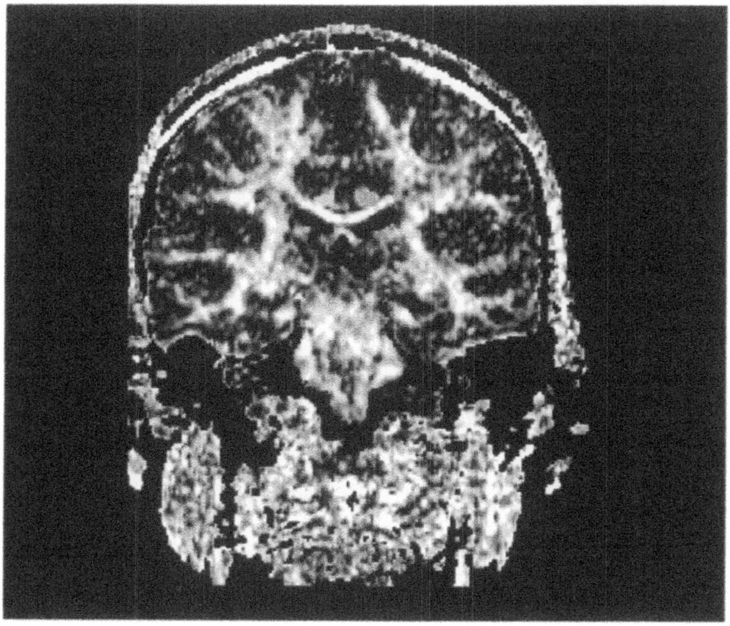

FIGURE 3. *See page 140.*

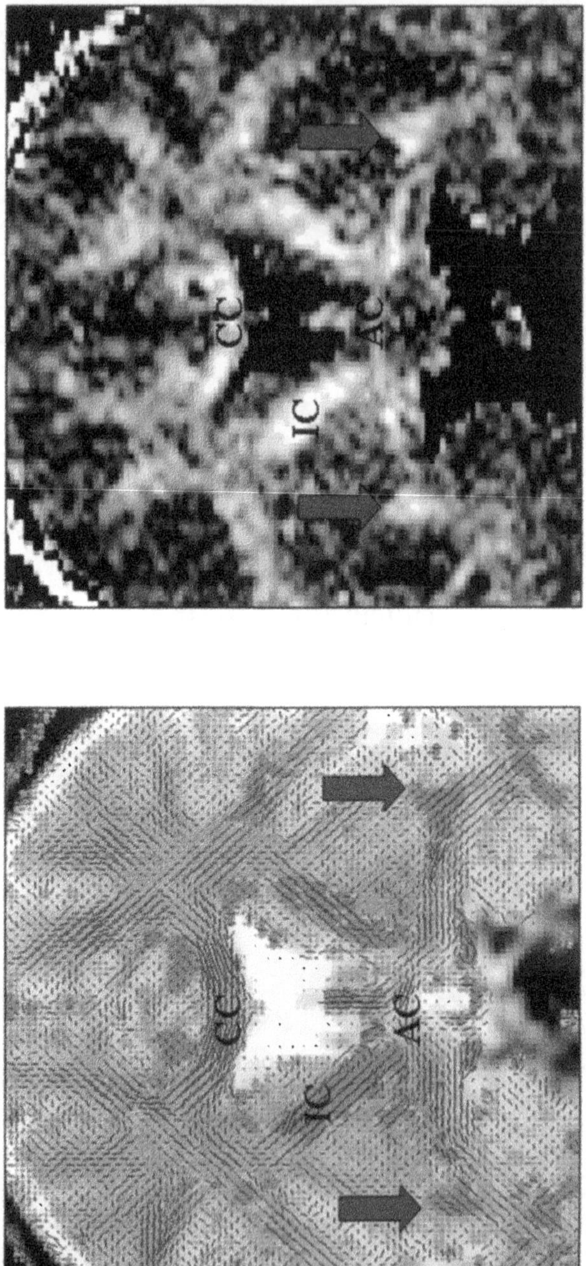

FIGURE 2. *See page 139.*

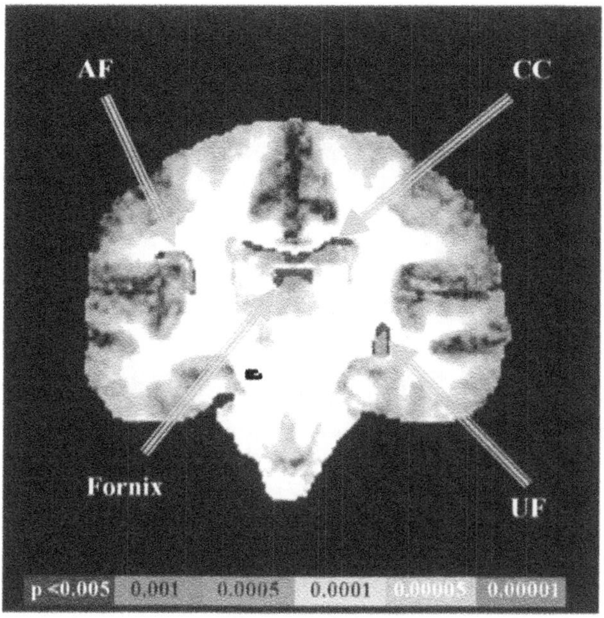

FIGURE 4. *See page 141.*

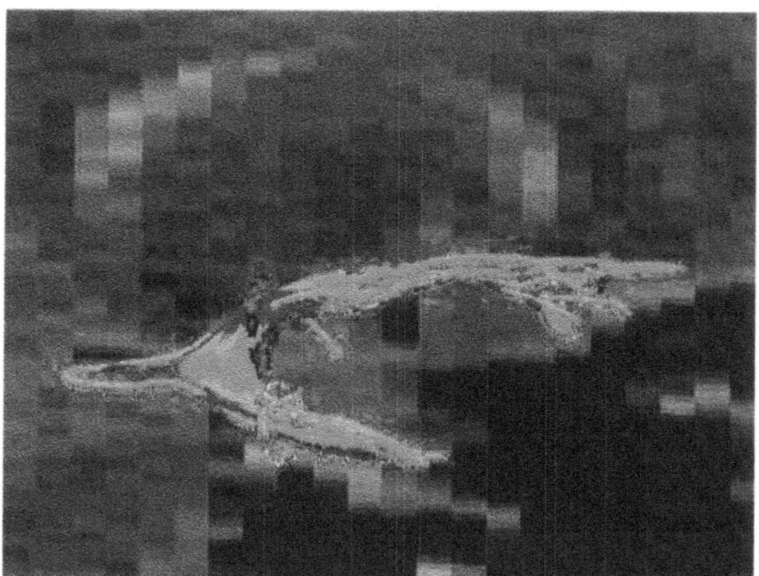

FIGURE 5. *See page 143.*

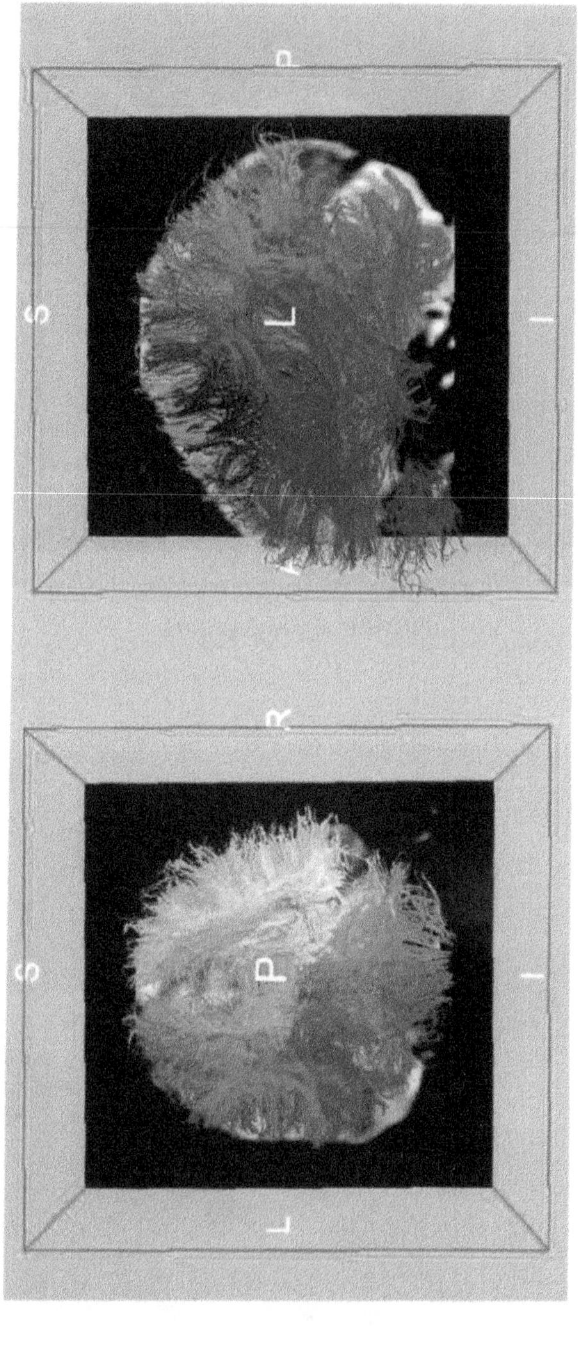

FIGURE 7. *See page 145.*

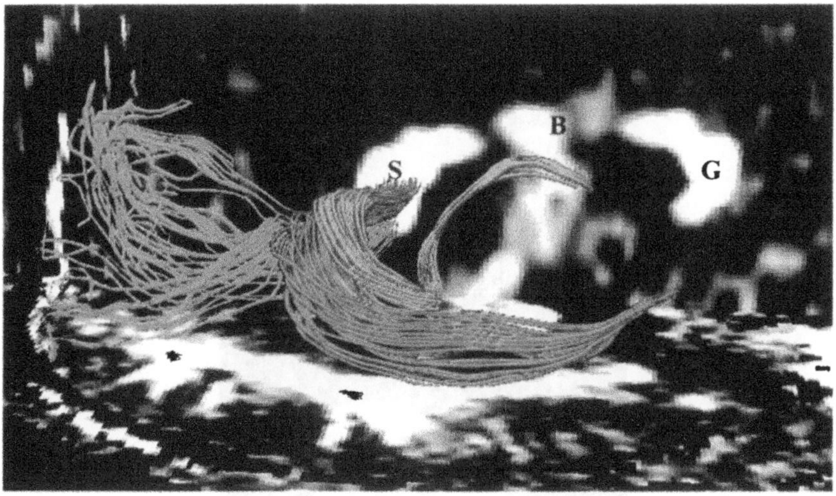

FIGURE 6. *See page 144.*

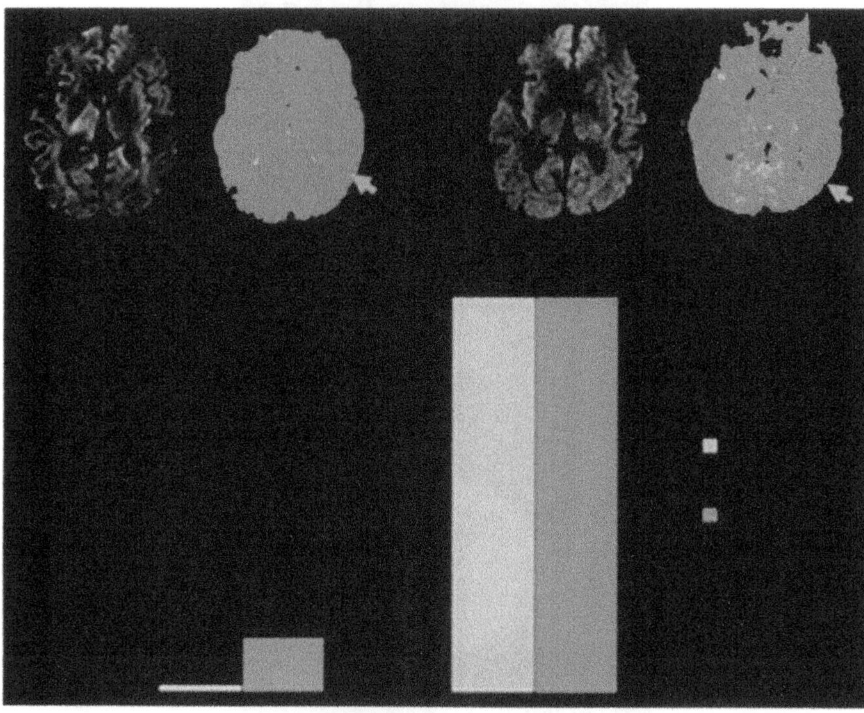

FIGURE 2. *See page 155.*

FIGURE 1. *See page 152.*

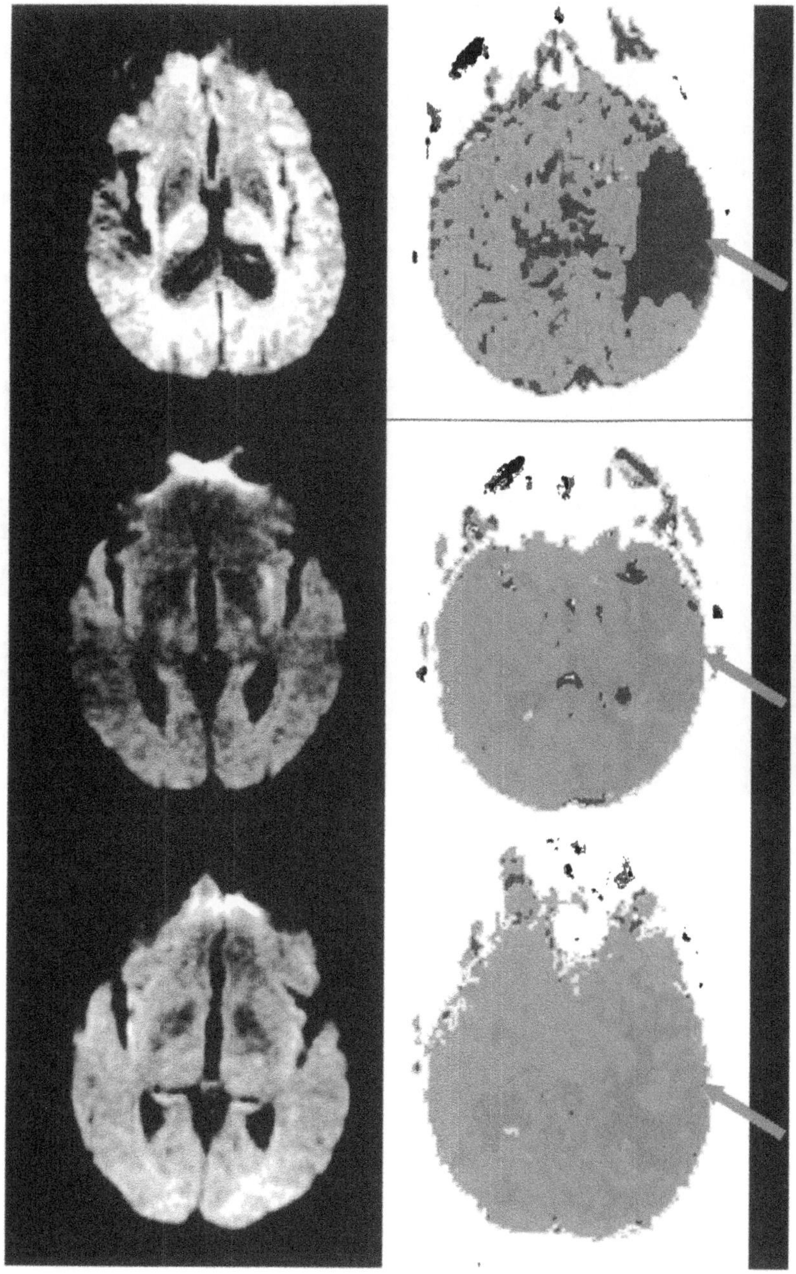

FIGURE 3. *See page 157.*

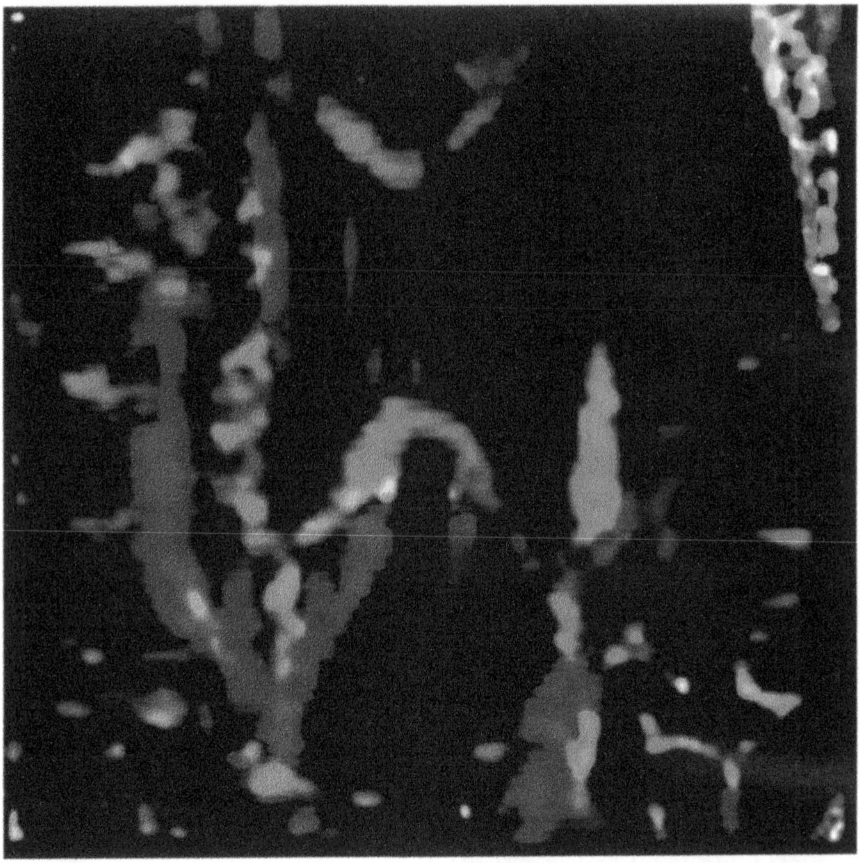

FIGURE 4. *See page 158.*

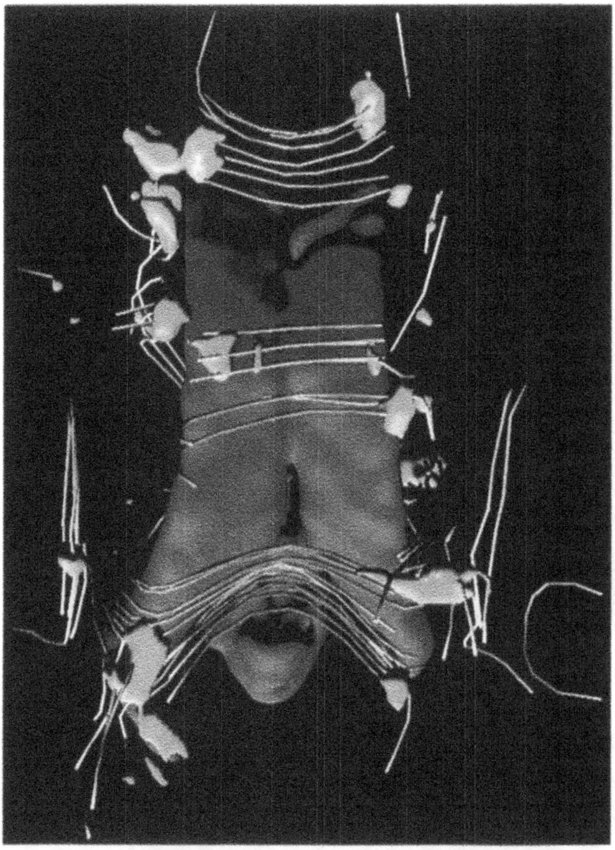

FIGURE 6. *See page 177.*

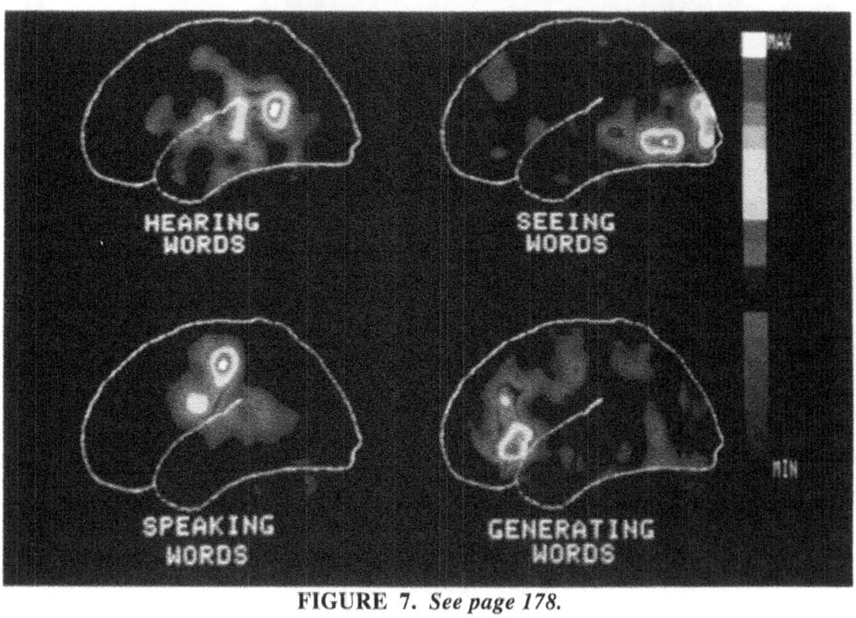

FIGURE 7. *See page 178.*

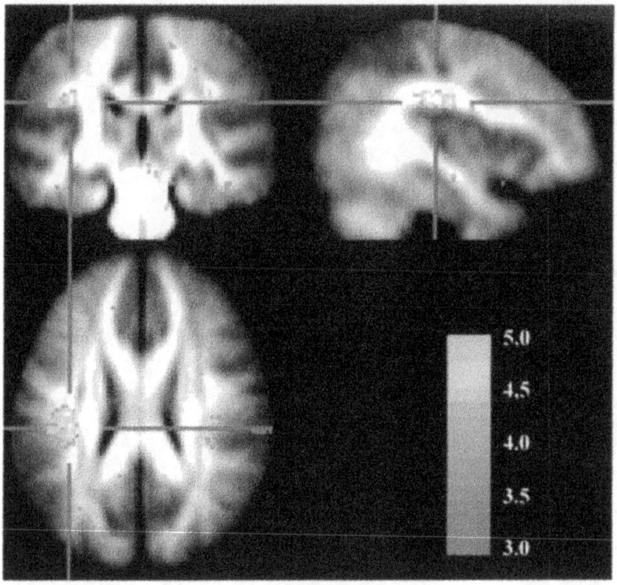

FIGURE 2. *See page 188.*

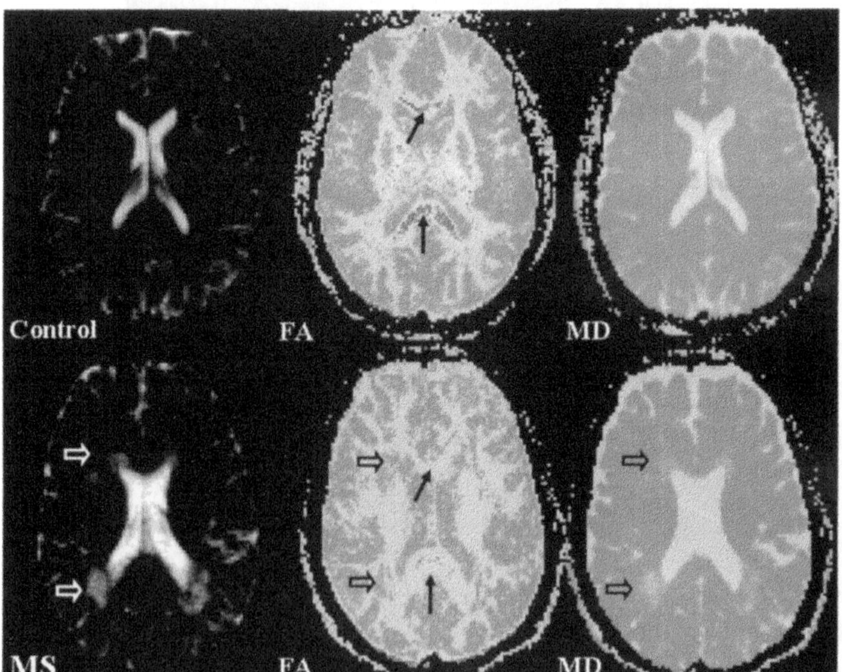

FIGURE 2. *See page 207.*

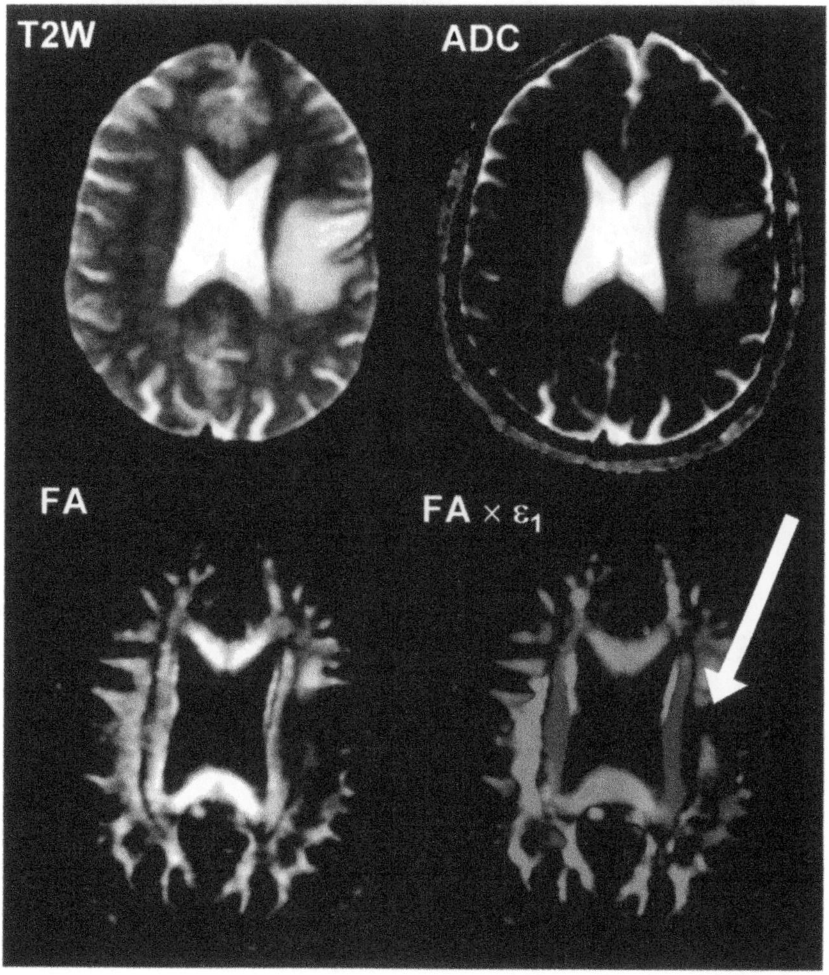

FIGURE 1. *See page 194.*

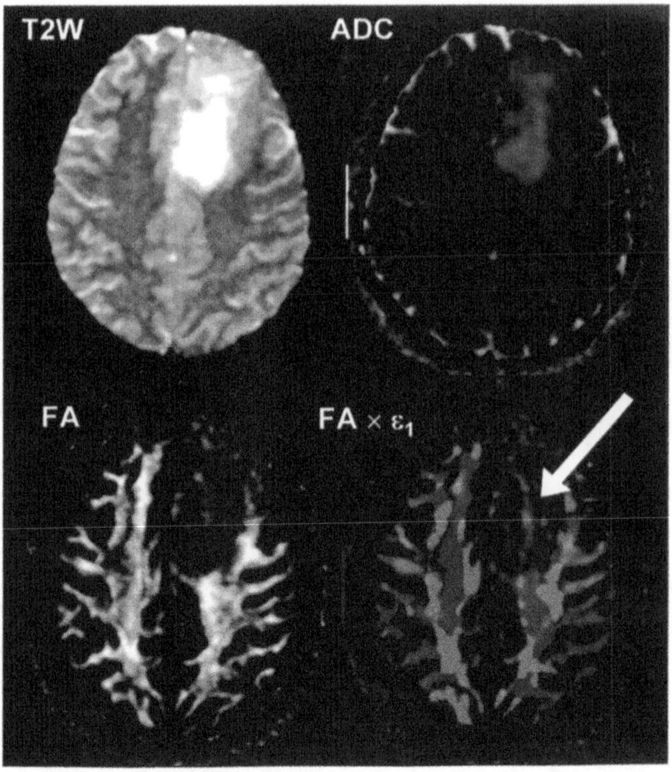

FIGURE 2. *See page 195.*

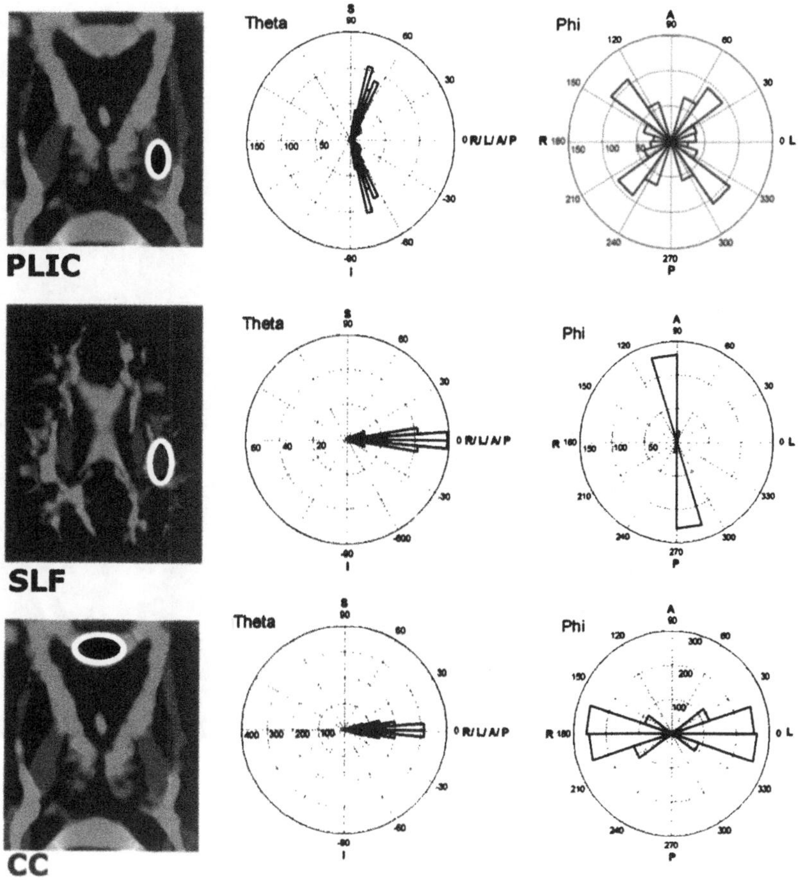

FIGURE 4. *See page 197.*

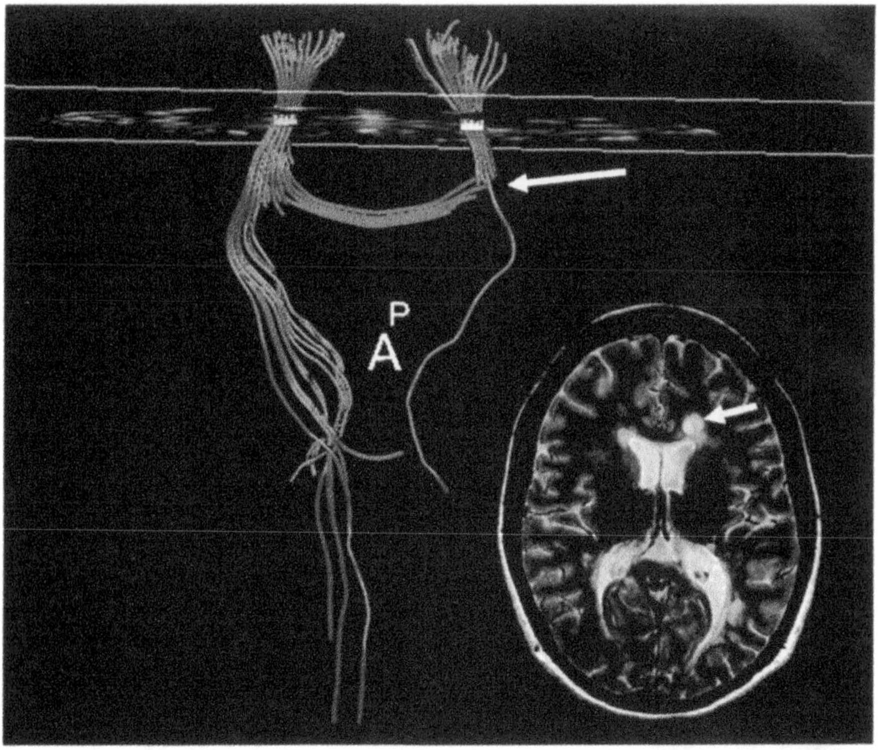

FIGURE 4. *See page 211.*

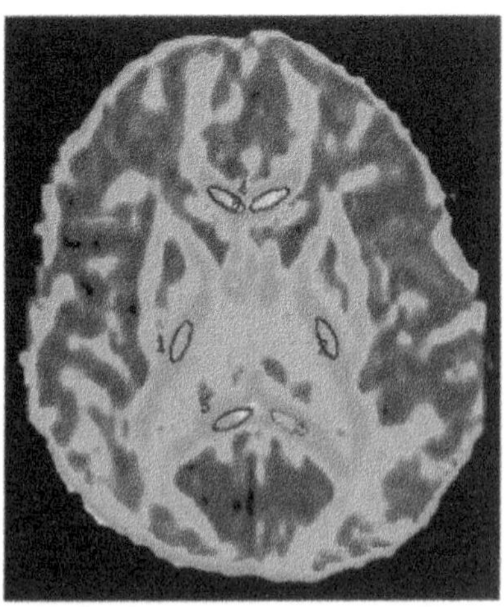

FIGURE 2. *See page 226.*

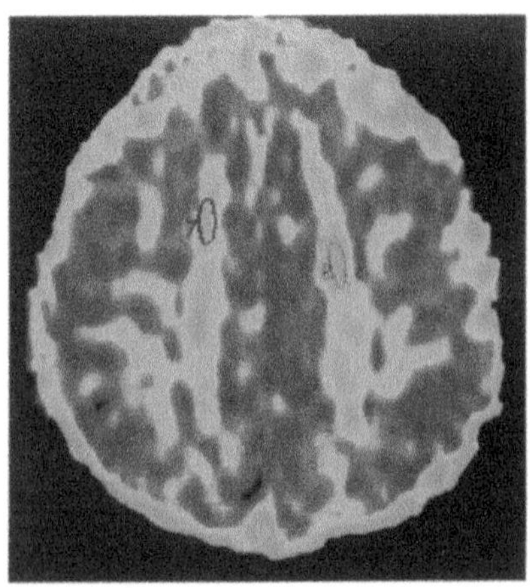

FIGURE 3. *See page 226.*

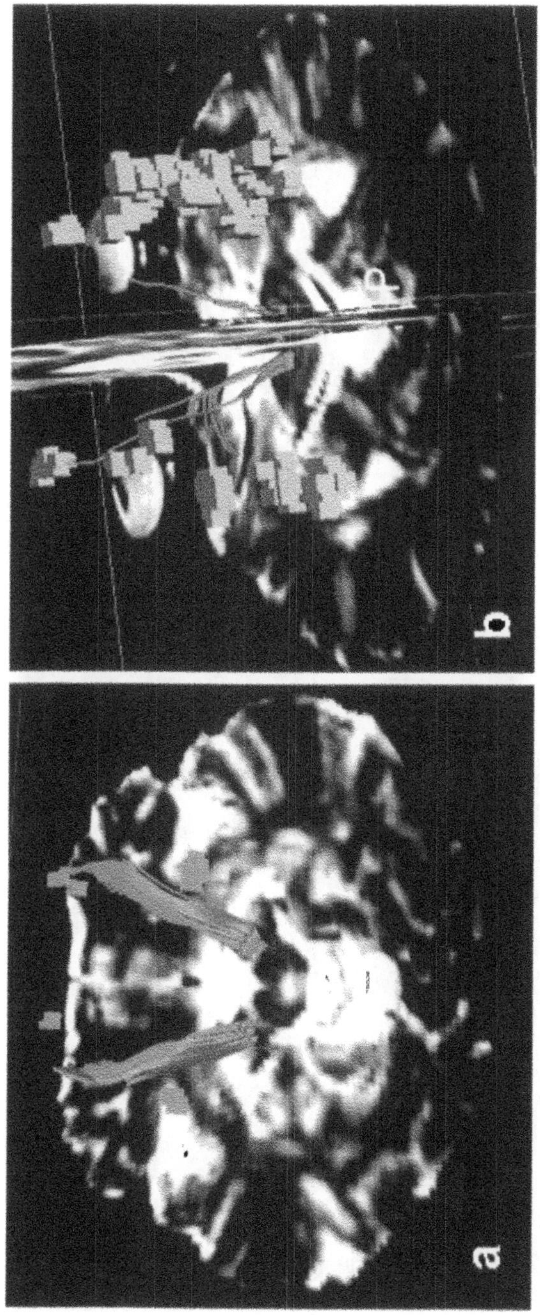

FIGURE 5. *See page 212.*

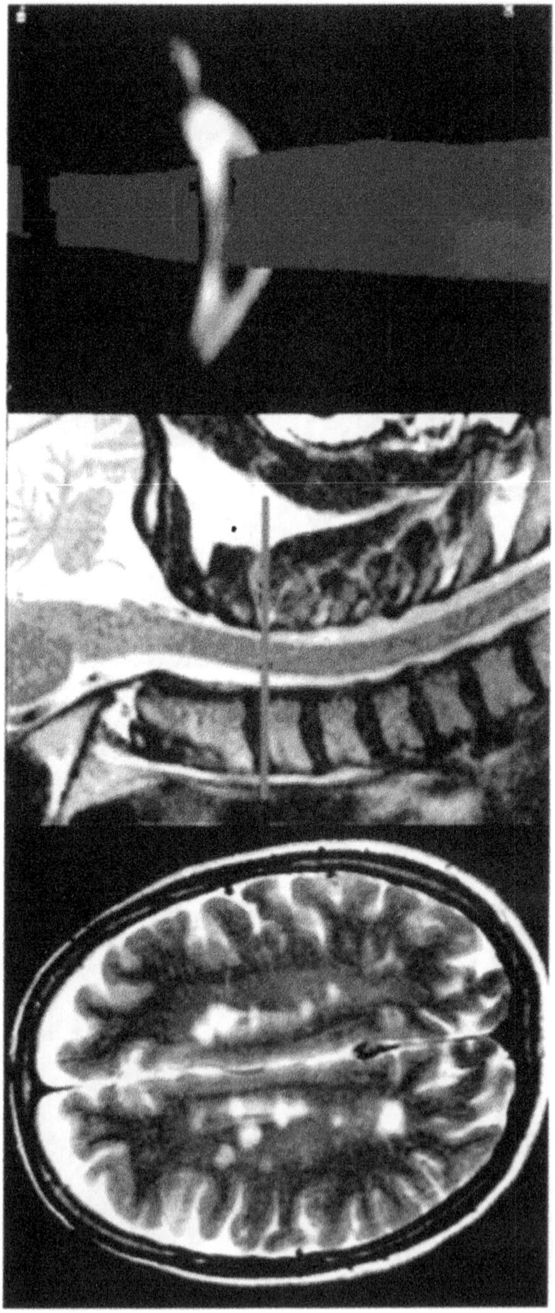

FIGURE 6. *See page 214.*

Intersection of the CST of the hand (as well as the CBT) with the longitudinal fasciculus with the consequential inability to trace these structures is illustrated in the top panel of FIGURE 2: a coronal image through the brain with a color-coded eigenvector map. The white matter tracts that run horizontally are seen in red, superior-inferior in blue, and anterior-posterior in green. The CST of the foot can be traced uninterrupted from the area of the foot homunculus in the precentral gyrus to the midbrain. This is seen as a continuous blue line in the top panel of FIGURE 2. The trace of the CST of the foot is depicted in green and red in the bottom panel of FIGURE 2. However, the trace of the CST of the hand and the CBT is interrupted by the longitudinal fasciculus (LF) that runs perpendicular to the CST of the hand and the CBT (top and bottom panels of FIGURE 2). Due to this interruption, one cannot trace the CST of the hand and the CBT using currently available methods. The location of the hand homunculus is identified in the bottom panel of FIGURE 2 by BOLD fMRI. Therefore, in this example, one would not be able to provide information to the operating neurosurgeon about the location of the CST of the hand—arguably, the most important part of the CST.

Since diffusion tractography is heavily dependent upon mathematics, we do not consider it inappropriate to resort to a mathematical analogy. At the Paris International Congress of 1900, a German mathematician, David Hilbert, proposed 23 outstanding problems in mathematics. These problems have come to be known as Hilbert's problems and were designed to serve as examples for the kinds of problems whose solutions would lead to the furthering of disciplines in mathematics. To a large part, search for the solutions to these problems, some of which remain unsolved today, has defined the field of mathematics during the 20th century and beyond.

We would like to propose a problem in the Hilbertian motif. The solution to this problem will not only answer a specific question. Rather, the search itself, as well as the methodological advances that will proceed for the solution, will serve to advance the understanding of the fields of neuroanatomy, mathematical modeling of complex biological systems, and preoperative and intraoperative neurosurgical planning.

Therefore, the Hilbertian problem that we propose is to trace the CBT from its origin in the precentral gyrus to its terminus in the brain stem using diffusion tractography. In our opinion, successful solution to this problem will necessarily entail fundamental advances in a number of fields of endeavor.

Why is the anatomy of the CBT an optimal paradigm to study? First, the anatomy of the CBT is generally understood. We know that it originates in the lateral aspect

FIGURE 2. (*Top panel*) A coronal image through the brain with a color-coded eigenvector map. The white matter tracts that run horizontally are seen in red, superior-inferior in blue, and anterior-posterior in green. The CST of the foot can be traced uninterrupted from the area of the foot homunculus in the precentral gyrus to the midbrain. This is seen as a continuous blue line. (*Bottom panel*) The trace of the CST of the foot is depicted in green and red. However, the trace of the CST of the hand and the CBT is interrupted by the longitudinal fasciculus (LF) that runs perpendicular to the CST of the hand and the CBT (*top and bottom panels*). Due to this interruption, one cannot trace the CST of the hand and the CBT using currently available methods. The location of the hand homunculus is identified in the bottom panel by BOLD fMRI. Therefore, in this example, one would not be able to provide information to the operating neurosurgeon about the location of the CST of the hand—arguably, the most important part of the CST.

of the precentral gyrus, traverses medially and inferiorly, crosses the LF and PLIC, and terminates in the nuclei of the pons and medulla oblongata. However, the *exact* location of the CBT is poorly understood. The situation is analogous to the CST as described in the first part of this opus. Specifically, the somatotopic organization of the motor white matter tracts as they traverse the PLIC has recently been defined for the foot and hand CSTs.[9] However, the location of the CBT as it traverses the PLIC and the relationship of the CBT to the CST, including the questions of somatotopic organization of these structures, remain unknown.

Second, the exact location of the CBT is more than just an anatomic curiosity. As with the CST, the exact location of the CBT is crucial in neurosurgical planning as well as intraoperative guidance. Accidental transection of the CBT can lead to paralysis. Neurosurgery also presents an advantage since one can confirm the location of the CBT, as predicted by diffusion tractography methods, using stereotactic neurosurgical guidance systems and intraoperative stimulation.

Third, the anatomy of the CBT is an ideal paradigm to test new mathematical methods to overcome the "crossing fibers" problem in diffusion tractography, which constitutes the crucial next step in the advance of this technology. The CBT is a small white matter tract that intersects a large white matter tract. Therefore, when someone proposes a method that purportedly can trace a small white matter tract through a large intersecting tract, one should be able to use the paradigm of the CBT (and occasionally the CST of the hand) intersecting the longitudinal fasciculus to determine if indeed the method lives up to its billing.

REFERENCES

1. Basser, P.J., J. Mattiello & D. Le Bihan. 1994. MR diffusion tensor spectroscopy and imaging. Biophys. J. **66:** 259–267.
2. Basser, P.J. & C. Pierpaoli. 1996. Microstructural and physiological features of tissue elucidated by quantitative-diffusion-tensor MRI. J. Magn. Reson. **B111:** 209–219.
3. Holodny, A.I., T.H. Schwartz, M. Ollenschleger *et al.* 2001. Tumor involvement of the corticospinal tract: diffusion magnetic resonance tractography with intraoperative correlation. J. Neurosurg. **95**(6)**:** 1082.
4. Holodny, A.I. & M. Ollenschlager. 2002. Diffusion imaging in brain tumors. Neuroimaging Clin. N. Am. **12**(1)**:** 107–124.
5. Field, A.S. & A.L. Alexander. 2004. Diffusion tensor imaging in cerebral tumor diagnosis and therapy. Top. Magn. Reson. Imaging **15**(5)**:** 315–324.
6. Kamada, K., T. Todo, Y. Masutani *et al.* 2005. Combined use of tractography-integrated functional neuronavigation and direct fiber stimulation. J. Neurosurg. **102**(4)**:** 664–672.
7. Laundre, B.J., B.J. Jellison, B. Badie *et al.* 2005. Diffusion tensor imaging of the corticospinal tract before and after mass resection as correlated with clinical motor findings: preliminary data. AJNR Am. J. Neuroradiol. **26**(4)**:** 791–796.
8. Jellison, B.J., A.S. Field, J. Medow *et al.* 2004. Diffusion tensor imaging of cerebral white matter: a pictorial review of physics, fiber tract anatomy, and tumor imaging patterns. AJNR Am. J. Neuroradiol. **25**(3)**:** 356–369.
9. Holodny, A.I., D.M. Gor, R. Watts *et al.* 2005. Diffusion tensor tractography of the somatotopic organization of the corticospinal tract in the internal capsule: controversial new anatomical findings. Radiology **234**(3)**:** 649–653.
10. Carpenter, M.B. 1983. Human Neuroanatomy, pp. 537–538. Williams & Wilkins. Baltimore.
11. DeAngelis, L.M. 2001. Brain tumors. N. Engl. J. Med. **344**(2)**:** 114–123.
12. Ammirati, M., N. Vick, Y. Liao *et al.* 1987. Effect of the extent of surgical resection on survival and quality of life in patients with supratentorial glioblastomas and anaplastic astrocytomas. Neurosurgery **21:** 201–206.

13. DEVAUX, B.C., J.R. O'FALLON & P.J. KELLY. 1993. Resection, biopsy, and survival in malignant glial neoplasms: a retrospective study of clinical parameters, therapy, and outcome. J. Neurosurg. **78:** 767–775.
14. SCHULDER, M., J.A. MALDJIAN, W.C. LIU *et al.* 1998. Functional image guided surgery of intracranial tumors located in or near the sensorimotor cortex. J. Neurosurg. **89:** 412–441.
15. HIRSCH, J., M.I. RUGE, K.H. KIM *et al.* 2000. An integrated functional magnetic resonance imaging procedure for preoperative mapping of cortical areas associated with tactile, motor, language, and visual functions. Neurosurgery **47**(3)**:** 711–721.
16. CARPENTER, M.B. 1976. Human Neuroanatomy, pp. 463–464. Williams & Wilkins. Baltimore.
17. CHARCOT, J.M. 1883. Lectures on the Localisation of Cerebral and Spinal Diseases, pp. 186–189. New Sydenham Society. London.
18. DEJERINE, J. 1901. Anatomie des Centres Nerveaux, p. 720. Rueff. Paris.
19. FOERSTER, O. 1936. Motorische felder und bahnen. *In* Handbuch der Neurologie. Vol. 6, p. 357. Springer-Verlag. Berlin.
20. BERTRAND, G., J. BLUNDELL & R. MUSELLA. 1965. Electrical exploration of the internal capsule and neighboring structures during stereotactic procedures. J. Neurosurg. **22:** 333–343.
21. KRETSCHMANN, H.J. 1988. Localization of the corticospinal fibers in the internal capsule in man. J. Anat. **160:** 219–225.
22. HANAWAY, J. & R.R. YOUNG. 1977. Localization of the pyramidal tract in the internal capsule of man. J. Neurol. Sci. **34:** 63–70.
23. YAGISHITA, A., I. NAKANO, M. ODA & A. HIRANO. 1994. Location of the corticospinal tract in the internal capsule at MR imaging. Radiology **191:** 455–460.
24. KUMAR, A.J., W. KOHLER, B. KRUSE *et al.* 1995. MR findings in adult-onset adrenoleukodystrophy. AJNR Am. J. Neuroradiol. **16:** 1227–1237.
25. YAGISHITA, A. & I. NAKANO. 1997. Location of the pyramidal and spinothalamic tracts [letter]. AJNR Am. J. Neuroradiol. **18:** 195.
26. ROSS, E.D. 1980. Localization of the pyramidal tract in the internal capsule by whole brain dissection. Neurology **30:** 59–64.

Occipital-Callosal Pathways in Children

Validation and Atlas Development

ROBERT F. DOUGHERTY,[a,b] MICHAL BEN-SHACHAR,[a,b] GAYLE DEUTSCH,[a] POLINA POTANINA,[a] ROLAND BAMMER,[c] AND BRIAN A. WANDELL[a,b,d]

[a]*Stanford Institute for Reading and Learning,* [b]*Psychology Department,* [c]*Radiology Department,* [d]*Neurosciences Program, Stanford University, Stanford, California, USA*

ABSTRACT: Diffusion tensor imaging and fiber tracking were used to measure fiber bundles connecting the two occipital lobes in 53 children of 7–12 years of age. Independent fiber bundle estimates originating from the two hemispheres converge onto the lower half of the splenium. This observation validates the basic methodology and suggests that most occipital-callosal fibers connect the two occipital lobes. Within the splenium, fiber bundles are organized in a regular pattern with respect to their cortical projection zones. Visual cortex dorsal to calcarine projects through a large band that fills much of the inferior half of the splenium, while cortex ventral to calcarine sends projections through a band at the anterior inferior edge of the splenium. Pathways projecting to the occipital pole and lateral-occipital regions overlap the dorsal and ventral groups slightly anterior to the center of the splenium. To visualize these pathways in a typical brain, we combined the data into an atlas. The estimated occipital-callosal fiber paths from the atlas form the walls of the occipital horn of the lateral ventricle, with dorsal paths forming the medial wall and the ventral paths bifurcating into a medial tract to form the inferior-medial wall and a superior tract that joins the lateral-occipital paths to form the superior wall of the ventricle. The properties of these fiber bundles match those of the hypothetical pathways described in the neurological literature on alexia.

KEYWORDS: diffusion tensor imaging; fiber tracking; splenium; alexia; white matter atlas

INTRODUCTION

The white matter fiber bundles that connect distant cortical regions are essential for neural computations, and damage to these bundles produces striking neuropsychological deficits. Proper fiber bundle development requires remarkable coordination between source and target neurons in the developing brain. Variations in the precision or stability of these long-range connections may account for important aspects of behavioral and cognitive performance.[1,2]

The magnetic resonance imaging method of diffusion tensor imaging (DTI), coupled with fiber-tracking (FT) algorithms, promises to provide substantial infor-

Address for correspondence: Robert F. Dougherty, Psychology Department, Jordan Hall, Building 420, Stanford University, Stanford, CA 94305-2130. Voice: 650-725-2466. bobd@stanford.edu

Ann. N.Y. Acad. Sci. 1064: 98–112 (2005). © 2005 New York Academy of Sciences. doi: 10.1196/annals.1340.017

mation about human fiber bundles and their development.[3–12] An important objective of DTI-FT methods is to clarify the specific behavioral consequences associated with fiber bundles in individual subjects. For example, DTI methods make it possible to compare an individual's behavior or skill level with white matter structure.[13,14] It is also possible to study white matter fiber bundles in individual subjects with specific behavioral disorders.[15]

A quantitative description of typical white matter is essential for understanding how deviations from the norm impact behavior. One useful description of white matter properties takes the form of a template that describes the typical DTI values and their variance in a normal population. In those portions of the white matter where fiber tracts can be reliably estimated, it is further useful to label template structures, including fiber bundles and the paths that they traverse. A number of groups are developing such atlases.[16]

Here, we describe our measurements of typical white matter properties in children. We concentrate on fiber bundles that pass through the splenium of the corpus callosum and connect the occipital lobes (occipital-callosal fiber tracts). In adults, we previously found that these fiber pathways can be estimated reliably. Here, we extend those findings to a larger population of children ($N = 53$; 7–12 years old).

Finally, we use these measurements to identify the fiber bundles likely to be involved in a specific reading deficit, pure alexia.[15,17–20] Specifically, we analyze the positions of the white matter lesions described in the literature with respect to the fiber bundles in the atlas. This analysis identifies specific long-range splenial fiber bundles that are important for fluent reading and probably for proper reading development.

METHODS

Subjects

Children were recruited from the San Francisco Bay region of California ($N = 53$; 7–12 years old). Written informed consent/assent was obtained from all children and parents. All subjects were physically healthy and had no history of neurological disease, head injury, attention deficit/hyperactivity disorder, or psychiatric disorder. The Stanford Panel on Human Subjects in Medical and Non-Medical Research approved all procedures.

The data were collected as part of a longitudinal study of reading development and represent the first measurement time point. All subjects also participated in neuropsychological testing and functional MRI scan sessions, but these data will not be discussed here.

Magnetic Resonance Imaging Protocols

MR data were acquired on 1.5T Signa LX (Signa CVi, GE Medical Systems, Milwaukee, WI) using a self-shielded, high-performance gradient system capable of providing a maximum gradient strength of 50 mT/m at a gradient rise time of 268 μs for each of the gradient axes. A standard quadrature head coil, provided by the vendor, was used for excitation and signal reception. Head motion was minimized by placing cushions around the head and securing a Velcro strap across the forehead.

Children watched and listened to cartoons via a video projection system and pneumatic headphones (Resonance Technologies, Northridge, CA) during the scan session to occupy their attention and reduce anxiety.

The DTI protocol involved 8–10 repeats of a 90-s whole-brain scan. These repeats were averaged to improve signal quality. The pulse sequence was a diffusion-weighted single-shot spin-echo, echo-planar imaging sequence (TE = 63 ms; TR = 6 s; FOV = 260 mm; matrix size = 128 × 128; bandwidth = ±110 kHz; partial *k*-space acquisition). We acquired 48–54 axial, 2-mm-thick slices (no skip) for two *b*-values: $b = 0$ and $b = {\sim}800$ s/mm^2. The high *b*-value was obtained by applying gradients along 12 different diffusion directions. Two gradient axes were energized simultaneously to minimize TE: $G0 = (0\ 0\ 0)^T$, $G1 = 1/\sqrt{2}\ (1\ 1\ 0)^T$, $G2 = 1/\sqrt{2}\ (1\ 0\ 1)^T$, $G3 = 1/\sqrt{2}\ (0\ 1\ 1)$, $G4 = 1/\sqrt{2}\ (-1\ 1\ 0)^T$, $G5 = 1/\sqrt{2}\ (-1\ 0\ 1)^T$, and $G6 = 1/\sqrt{2}\ (0\ -1\ 1)$. This pattern was repeated twice for each slice. The polarity of the effective diffusion-weighting gradients was reversed for odd repetitions to remove cross-terms between diffusion gradients and both imaging and background gradients.

DTI data were preprocessed using a custom program based on normalized mutual information that removed eddy current distortion effects and determined a constrained nonrigid image registration.[21] The six elements of the diffusion tensor were determined by multivariate regression using matrix calculations on a per voxel basis.[22–24]

We also collected high-resolution T1-weighted anatomical images using a 3D-SPGR sequence (~1 × 1 × 1 mm voxel size). We routinely collected 3–4 3D-SPGR scans (both axial and sagittal slice orientations) and coregistered and averaged all scans that were relatively artifact-free.

Fiber-Tracking Algorithms and Software

For each subject, the T2-weighted ($b = 0$) images were aligned to the T1-weighted 3D-SPGR anatomical images. The alignment was initiated using the scanner coordinates stored in the image headers. This alignment was refined using a mutual-information 3D rigid-body algorithm from SPM2.[25] Several anatomical landmarks, including the anterior commissure (AC), the posterior commissure (PC), and another point in the midsagittal plane, were identified by hand in the T1 images. With these landmarks, we computed a rigid-body transform from the native image space to the conventional AC–PC aligned space, with the AC at the origin, the AC–PC line forming the y-axis, the AC–midsagittal line forming the z-axis, and the remaining orthogonal (left–right) forming the x-axis. The DTI data were then resampled to this AC–PC aligned space with 2-mm isotropic voxels using a spline-based tensor interpolation algorithm,[26] taking care to rotate the tensors to preserve their orientation with respect to the anatomy.[27] The T1 images were resampled to AC–PC aligned space with 1-mm isotropic voxels. We confirmed that this coregistration technique aligns the DTI and T1 images to within 1–2 mm in the occipital lobes. However, the well-known geometric distortions inherent in EPI acquisition limit the accuracy in regions prone to susceptibility artifacts, such as orbitofrontal and anterior temporal regions.

Fiber bundles were estimated using a streamlines tracking algorithm[3,6,10] with a fourth-order Runge-Kutta path integration method[28] and 1-mm fixed step size. A continuous tensor field was estimated using trilinear interpolation of the tensor elements. Path tracing proceeded until the FA fell below 0.15, or until the minimum angle between the current and previous path segments was larger than 30°. Both directions from the initial seed point principal diffusion axis were traced.

The seed points were laid down in a grid that filled the white matter of the left or right occipital lobe (defined in each subject by hand) at 1-mm spacing. About 16,000 estimated fiber paths were identified in each hemisphere. As we focused on callosal projections for these analyses, only fiber bundles that passed through the corpus callosum were retained. This was typically about 20% of the total left or right fibers.

Creating Average Tensor Maps

To facilitate individual subject analyses and allow averaging of the diffusion tensor field across subjects, we computed a smooth deformation of each subject's brain to a custom brain template. This deformation was used in the dorsal/ventral/lateral segmentation analysis to warp regions of interest (defined on the cortical surface of one subject) to all subjects. These deformations were also used to create an average tensor map for visualizing the expected path of the occipital-callosal fibers.

The deformation was computed based on the high-resolution T1-weighted images. A brain mask was computed for each of these using the brain extraction tool (BET, version 2.1) from the Oxford FMRIB Analysis Group.[29] By stripping the skull with this brain mask, we avoid the false-matches between brain-edge and scalp-edge that occasionally occur when the skull and scalp are included in the images.

The final template and average tensor map was created by an iterative process. An initial T1-weighted template was created by resampling each brain to a common space (similar to that of Talairach and Tournoux) based on landmarks that were defined by hand in each subject, and then averaging all the resampled, skull-stripped, brains. This avoids the biases inherent in basing the final template on a single subject or on a template comprising a different sample. This method also ensures that the final template is representative of the group in shape and size. As described above, the T1-weighted images were initially transformed with a rotation and translation into AC–PC space, with 1-mm isotropic voxels. To further align the brains for creating the initial atlas, we also marked six points in each brain that defined the maximal extent of the cortical surface along each of the three axes. These six landmarks were used to define seven different scale factors: two along the x-axis (left of the AC and right of the AC), two along the z-axis (inferior of the AC and superior of the AC), and three along the y-axis (anterior of the AC, posterior of the PC, and the AC-to-PC section were all scaled separately). The scale factors were chosen so that the maximal extent of each brain would end up at the mean position of all 53 brains. The mean extents for the 53 brains were 71.8 mm for superior of AC, 40.5 mm for inferior of AC, 66.1 mm for right of AC, 66.7 mm for left of AC, 68.0 mm for anterior of AC, 25.1 mm for AC-to-PC, and 75.4 mm for posterior of PC.

Next, an intermediate template was created by spatially normalizing the original (skull-stripped) brains to this initial template using the nonlinear deformation algorithm from SPM2.[30] These intermediate images were averaged to create a refined template, which had a more sharply defined brain-edge and ventricles. Finally, we spatially normalized the original (skull-stripped) brains (again, using the SPM2 tools) to this refined template. For each brain, this deformation—mapping the T1-weighted images to the intermediate template—was also applied to the diffusion tensor map, taking care to rotate the tensors to preserve their orientation.[31] When applying the deformations, the T1-weighted volume was resampled to a 1-mm grid and the tensor map was resampled to a 2-mm grid, both using trilinear interpolation. Finally, all of these resampled T1-weighted images and diffusion tensor maps were then averaged to produce the mean tensor map and T1-weighted template.

These templates, along with the custom software used for all the analysis and visualization, can be obtained from the authors upon request.

RESULTS

Comparison of Right and Left Occipital-Callosal Bundle Positions

We evaluate the reliability of fiber bundle estimates by comparing estimates derived from two independent data sets. We first define two independent volumes of interest (VOI) comprising all the white matter in the left and right occipital lobes. We then compare where fibers from these two sets cross the callosum. Any occipital-occipital fiber tract has a single crossing point in the callosum. Hence, if occipital-occipital fibers are properly identified by DTI-FT applied to the left and the right, then the estimated fibers from these two data sets should meet at a common position in the callosum.

There are two reasons why a callosal location would be identified in one estimate, but not the other. First, a fiber tract may pass from the occipital lobe through the callosum and project somewhere other than the occipital lobe. Such a fiber would not have a matching counterpart from the other occipital lobe. Second, the DTI-FT algorithm may make a type I or II error. In the case of poor sensitivity, the algorithm correctly estimates the portion of the fiber tract in, say, the left occipital lobe, but fails to identify the matching portion in the right lobe. Alternatively, the estimated occipital-callosal tract may not exist.

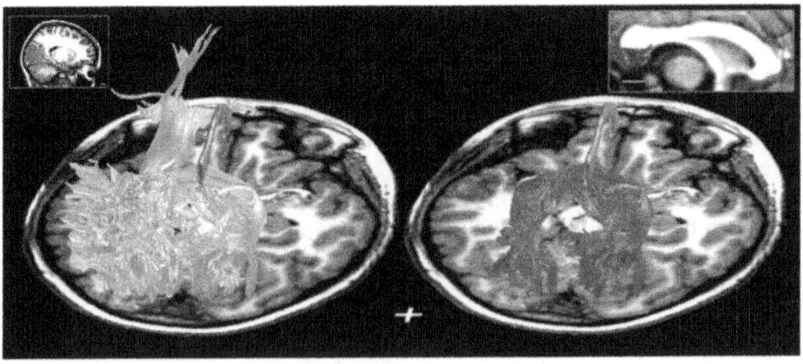

FIGURE 1. Method for estimating occipital-callosal fiber bundles. A region of interest is selected in the occipital lobe of one hemisphere (sagittal image showing red overlay; upper left inset). The points in this region of interest serve as seed locations to estimate all fiber bundles that pass through the occipital lobe (yellow fibers in the left image). The subset of occipital-lobe fibers that pass through the corpus callosum (shown in cyan) are identified (blue fiber bundles in the right image). The location of these fibers in the plane of the corpus callosum is shown in the upper right inset. The process is repeated for both the left and right hemispheres. Scale bar: 1 cm.

The estimated fiber bundles passing through seed points in the VOI of the left occipital lobe are shown in the left panel of FIGURE 1 for one representative child. The inset image shows a coronal view of the VOI. The subset of these fiber bundles that also pass through the corpus callosum are shown in the right panel of FIGURE 1. These comprise about 15% of the total left occipital lobe fiber bundles. The inset image shows the location within the splenium (the posterior callosum) of the fiber bundle estimates from the left hemisphere. Using these methods, DTI-FT identifies mainly callosal bundles projecting to the dorsal and medial cortical surface. We suspect that callosal tracts projecting to the lateral and ventral surfaces exist,[32,33] but they are missed by current methods.[7]

FIGURE 2 is a qualitative overview of the overlap between the right and left occipital-callosal fiber bundles in the individual brains. Each image section shows estimates from a different child. The images are cropped to show the posterior portion of the corpus callosum, including the splenium and the isthmus.

In these children, just as in adults, the left and right occipital-callosal fiber positions overlap substantially in the splenium. Fiber bundles from the occipital lobe in the right that travel to, say, the temporal lobe in the left would not be part of this overlap. Hence, the high degree of overlap confirms the hypothesis that the majority of occipital lobe callosal fibers project to the contralateral occipital lobe.

FIGURE 2 also illustrates the significant size and shape variability of the splenium and isthmus. For example, consider the first image section. That child has a nearly circular splenium and a very narrow isthmus that is easy to identify. The last two subjects have a long, cylindrical splenium, and it is hard to identify the isthmus.

The traditional method of segmenting the callosum assigns a much larger splenial area to the last subject than the first. Yet, the estimated occipital-callosal fiber bundles are very similar in cross-sectional area. Hence, DTI-FT should provide a more secure method for segmenting the callosum into bands that correspond to functional units compared to conventional segmentation definitions based on callosal shape.[34]

The agreement between subjects in the shape and position of the occipital-callosal fiber bundles at the midsagittal plane encouraged us to pool the data into a single image. We did this in two ways. First, we transformed the individual subject data into a common spatial reference frame based on the position of the anterior and posterior commissures (see METHODS). To quantify the left–right convergence, we defined a grid comprising 0.5×0.5 mm bins. The grid was placed at a position that was 36 mm posterior to the AC and 11 mm above the AC–PC line in the midsagittal plane. For each bin, we calculated the fiber bundle density estimates (bundles per square mm per subject) from the left and right occipital lobes. The correlation between these two values is shown in a scatter plot (FIG. 3A). These estimates obtained from the left and right data sets are highly correlated, with a linear regression accounting for 86% of the variance. Within each bin, we estimated on average 10% more fiber bundles arriving from the left occipital lobe. The spatial distributions within the splenium of the left and right fiber bundle density estimates are shown in the images of FIGURE 3B.

As noted above, across individuals there is considerable variation in the shape and position of the splenium with respect to the AC. For example, in this co-registered coordinate frame, it is still possible for the center of the splenium between two subjects to differ by a centimeter. To improve the coarse alignment based on the AC and PC positions, we placed the center of mass of each subject's occipital-

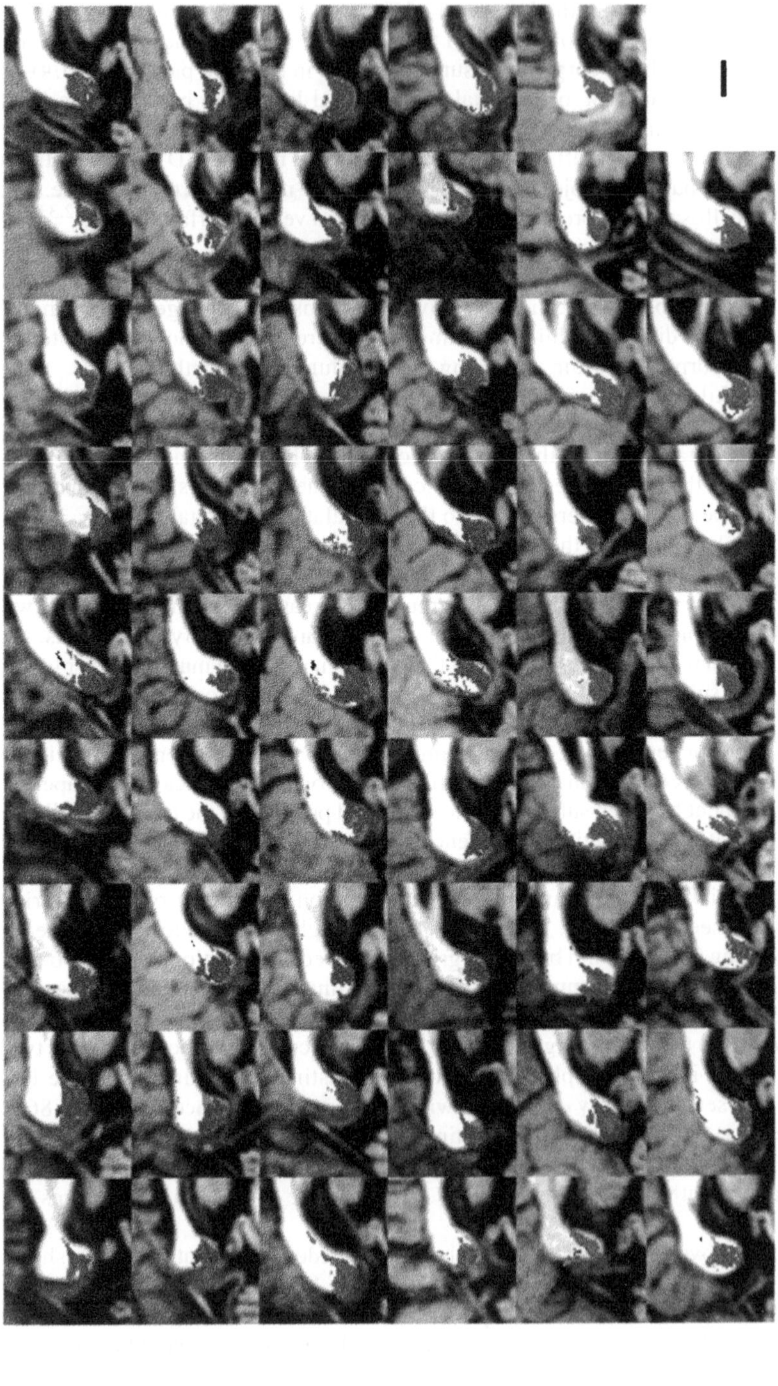

FIGURE 2. Left and right occipital-callosal fiber bundle positions in 53 children. The positions of fiber bundles estimated from left (red) and right (blue) occipital lobes are indicated by the color overlay. A pixel is colored if there is at least one estimated fiber bundle at that location. Scale bar: 1 cm.

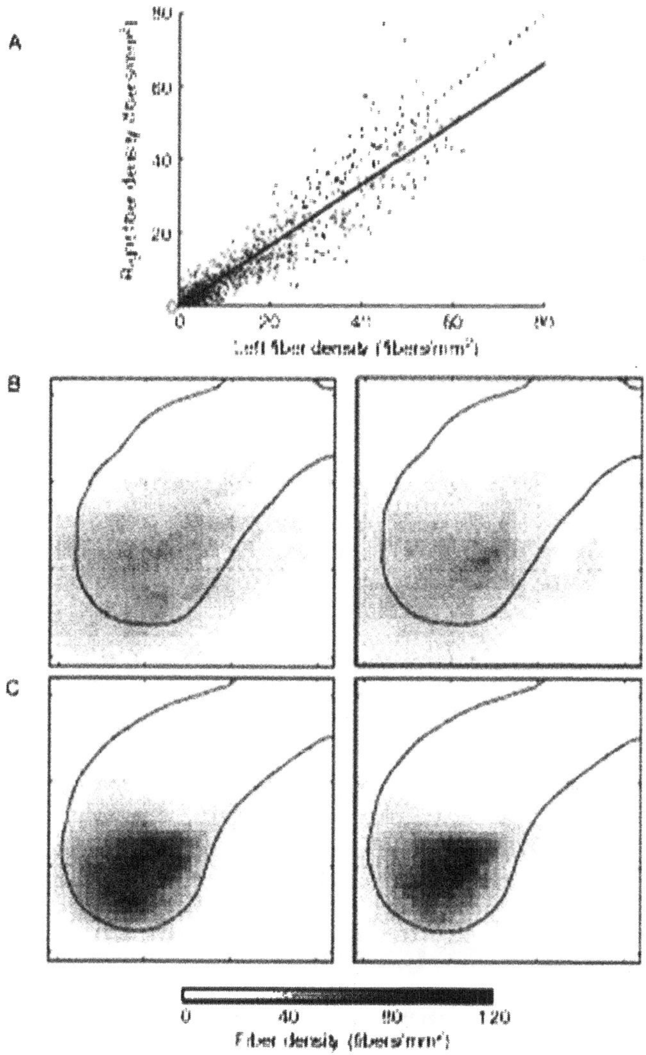

FIGURE 3. Convergence of occipital-callosal fiber bundle densities estimated independently in the left and right hemispheres. The data are cumulated from all 53 children. **(A)** The data from each child were translated and scaled into a common coordinate frame, aligned with the AC–PC axis. The graph compares the number of fiber bundles per square mm (per subject) from the left (horizontal axis) and right (vertical axis) at locations within a grid drawn in the corpus callosum. **(B)** The images show the fiber bundle density estimates overlaid on a contour that represents the average splenium in AC–PC aligned coordinates. These data form the basis of the graph in panel A. The two images show the spatial distribution of the fibers derived from the left (*left panel*) and right (*right panel*) occipital lobes. **(C)** The data from each child were translated in two dimensions so that the combined center of mass of left and right occipital-callosal fiber bundles is at a common location. This translation reduces the variance caused by errors in registering the brains. Grid lines are spaced at 5 mm.

callosal fiber bundles at a common position. The images in FIGURE 3C show the fiber bundle density estimates overlaid after aligning each subject in this way. Again, the right and left fiber density bundle estimates agree quite closely (the percent variance explained increases to 94%). The shape of the occipital-callosal fiber bundles is more apparent, and with this alignment it becomes clear that the fiber bundles pass mainly through the most ventral portion of the splenium.

The occipital-callosal fibers can be further subdivided according to their cortical projection zones. We defined three regions of interest (ROIs) corresponding to the anatomical positions of several visual field maps on the cortical surface of one

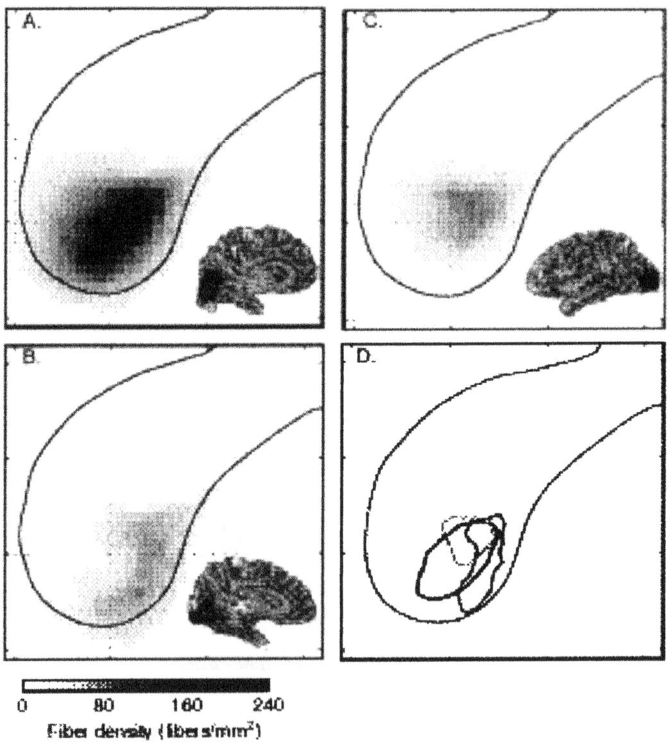

FIGURE 4. Segmentation of the occipital-callosal fiber bundles according to cortical projection zone. The four panels show the distribution of the fibers that terminate near three distinct occipital projection zones. The inset figures show a smoothed version of the cortex at the boundary between white and gray matter. The shading describes the gyri (light) and sulci (dark). The cortical region of interest is shown by the dark overlay. **(A)** The largest group of estimated fibers terminate in the dorsal region of the occipital cortex. **(B)** Fibers located in the anterior-ventral splenium terminate near the ventral occipital surface. **(C)** Fibers terminating in the posterior and lateral aspect of the occipital lobe, which include mainly foveal representations, form a small localized group near the middle of the spenium. **(D)** The contour lines that capture 60% of the fibers within each group are shown with respect to the outline of the splenium (dorsal is black, ventral is dark gray, and lateral is light gray). Grid lines are spaced at 5 mm.

representative subject. The ROIs were warped to the native space of each individual subject by first warping them to the standard space of our template brain and then warping them to each individual brain (using the reverse of the template deformation). The first region is dorsal to the fundus of the calcarine sulcus and roughly corresponds to the dorsal portions of V1, V2, V3, V3A, and V3B maps. A second ROI was defined ventral to calcarine and includes the ventral portions of V1, V2, V3, and all of hV4. Finally, a third ROI was defined on the cortical surface covering the occipital pole and extending onto the lateral surface into the lateral-occipital complex.[35,36] We confirmed that these warped ROIs were anatomically reasonable in each of the individual brains (e.g., the dorsal/ventral split fell within the calcarine sulcus).

We subdivided the center of mass–aligned occipital-callosal fiber bundles based on the cortical ROIs closest to each fiber bundle end point (see ref. 7 for details). The fiber bundle density plots for the three ROIs are shown in FIGURES 4A, 4B, and 4C. The largest percentage of the occipital-callosal fiber bundles project to dorsal cortex; the fiber bundles projecting to ventral cortex are slightly anterior and ventral. The lateral-occipital projecting fibers overlap the other two groups in the superior-anterior region. Damage near this position can produce alexia.[17] The contours in FIGURE 4D outline the regions containing 60% of the estimated fiber densities from each of the subgroups.

We computed an atlas to summarize the regularities across the entire group of children. This atlas includes the diffusion tensor data from all 53 children; it was created by finding smooth deformations of each individual brain via a common reference frame (see METHODS). The atlas data from a single axial slice through the splenium are shown in FIGURE 5. The atlas includes both fractional anisotropy (FA) and principal diffusion direction (PDD) measures.

Mean and standard deviation of the measured FA are overlaid on the average anatomical image (FIGS. 5A and 5B). Only locations with a mean FA value above 0.2 are shown in the overlay. The FA values are very high in the callosal pathways (~0.6–0.8) and in general in the dense medial white matter regions of the brain. The typical FA value decreases steadily as the measurements approach the cortical surface. Also, note that the FA standard deviation tends to be lowest at the center of a major pathway, increasing towards the edge. This tendency can be explained by imperfections in the alignment of the multiple brains. Small misalignments in the middle of a major white matter tract do not introduce significant FA variation, while misalignments at the edge where FA contrast is high will produce significant variance in the atlas.

The FA standard deviation is also very high in the splenium and forceps major; the standard deviation is particularly high where the forceps major passes adjacent to the lateral horn of the occipital ventricle. The standard deviation in this region is approximately that of a uniform distribution on the unit interval (0.2887). Hence, atlas-based methods are insensitive for identifying FA group differences in this region.

The PDD mean and dispersion in the same slice are shown in FIGURES 5C and 5D. The dispersion is measured using the bipolar Watson distribution, which describes an antipodal symmetric distribution of unsigned random vectors on a unit sphere.[37] The mean direction is quite stable within the internal capsule and corpus callosum. Approaching the edges of the white matter (e.g., in the posterior occipital lobe), the PDD becomes very variable. In this case, the maximum dispersion for a uniformly distributed set of directions is 55°. Hence, the PDD at this slice in the posterior occipital lobe is almost randomly distributed.

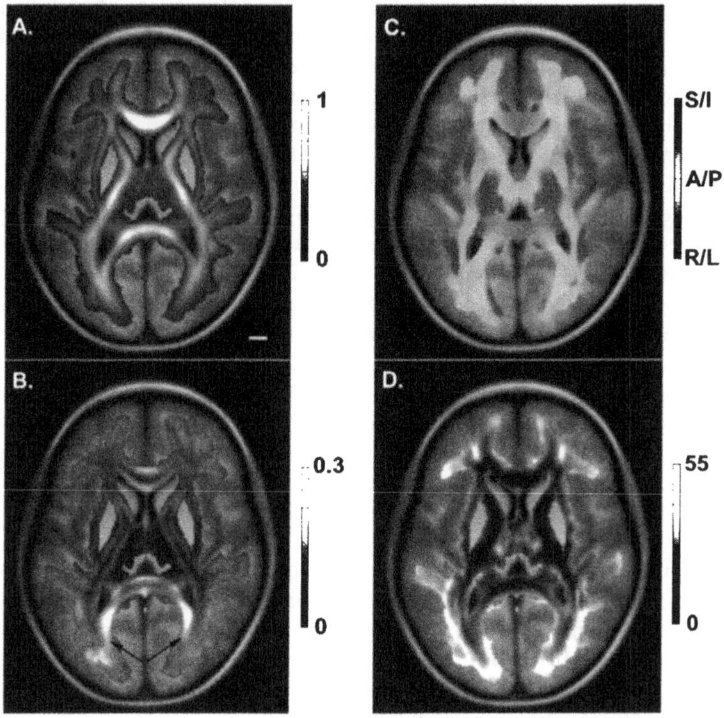

FIGURE 5. A single slice from the atlas summarizing the children's DTI measurements. The DTI information is superimposed on the mean anatomical image from coregistering the 53 brains. **(A)** The color overlay represents the mean fractional anisotropy (FA). **(B)** The standard deviation of the FA. **(C)** The principal diffusion direction (PDD): S/I, superior/inferior; A/P, anterior/posterior; R/L, right/left. **(D)** The PDD reliability is represented by the angular dispersion (in degrees). This statistic represents the angular scatter of the PDD on the unit sphere. Scale bar: 1 cm.

These images clarify that atlas-based methods have much higher sensitivity (lower variance) in some brain regions than others. The group data have relatively low variance near the sagittal midline, but the reliability of both PDD and FA maps declines severely near the cortical surface. This uneven sensitivity pattern should be part of the interpretation of any DTI result. For example, two recent studies demonstrated a covariation of reading performance with FA in a portion of the atlas with very low FA variance.[13,14] The atlas images remind us that such covariation between DTI data and behavior may be present elsewhere, but missed because of high group variance in those regions.

There is good agreement in the estimates of occipital-callosal fibers tracked independently from the two hemispheres in individual subjects (FIGS. 2 and 3). Considering the atlas data, we find that it should be possible to track reliably fiber bundle paths that course through the corpus callosum and pass near the ventricles. This is because the reliability of the PDD in this portion of the brain is relatively high; while the FA reliability is low in some of these regions, the mean FA value is high enough

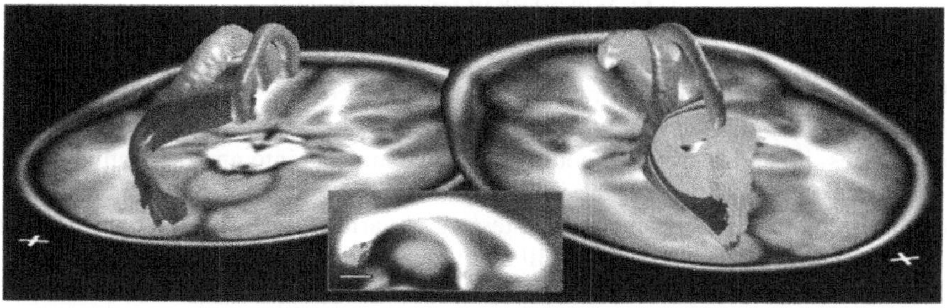

FIGURE 6. Fiber bundle pathways between the callosum and three regions of interest on the cortical surface. The fiber bundles are derived from the average tensor maps. The image on the left shows the pathways followed by the fibers that terminate near the ventral (red) and lateral occipital (blue) cortical regions of interest (see FIG. 4). The ventral fibers pass on both sides of the lateral ventricle (shown in dark yellow), forming part of the tapetum. The image on the right shows these same fibers from a different viewpoint and adds the largest group of fibers, those that terminate near the dorsal region of interest (green). The inset image shows the location of these fiber groups within the splenium. The corpus callosum is shown in cyan. Scale bar: 1 cm.

to support estimation of the mean PDD. We estimated the path followed by the occipital-callosal bundles that connect the callosum and the three cortically defined ROIs (FIG. 6). The lateral occipital fiber bundles follow a path that is mainly superior to the lateral occipital ventricle. The fibers headed towards the ventral occipital lobe divide, with some fibers passing superior and others inferior to the ventricle. The division of these fibers around the ventricle is present in the atlas, and it also can be seen in many individual subjects. The spatial relationship between these fiber bundles, the lateral occipital ventricle, and the callosum are illustrated by the image on the left. The dorsal fiber bundles are the largest group. These are added in to the image on the right. The positions of these different fiber bundles within the splenium are shown by the inset image.

DISCUSSION

The occipital-callosal fiber bundles estimated in children are quite similar to previous estimates from a much smaller adult data set. In both cases, the independent estimates from right and left hemisphere data converge into a small region within the splenium of the corpus callosum. Further, the data from the larger sample of children can be used to quantify the precision of the estimates. Positions of the occipital-callosal fiber bundles within the splenium appear to be accurate to within 1 mm in individual subjects (FIGS. 2–4).

DTI, like any measurement modality, has strengths and limitations. The measurements of these occipital-callosal fibers are reliable and can be made at very high spatial resolution. On the other hand, we find very limited ability to identify fiber bundles that pass through the corpus callosum to the lateral surface. These pathways interdigitate through much larger pathways that form the inferior longitudinal fasciculus (occipitofrontal and occipitotemporal pathways[4]). The current data and analysis methods cannot resolve the smaller pathways in this region of fiber crossing.

DTI Sensitivity in Group Analyses

In some cases, it is essential to analyze data from individual subjects, say for diagnosis and treatment. The measurements described here confirm the validity of the DTI-FT methods for identifying such pathways in certain parts of the brain. In other cases, it can be useful to pool data across subjects into an atlas, which represents the expected pattern in the "average" brain for a given population. The atlas identifies common points and permits testing of various statistical hypotheses, such as a covariation between reading skill and white matter anisotropy.[13,14,38]

While atlases can increase statistical power, they also have significant limitations. Some of these limitations are made explicit by including both mean and variance values in the atlas (FIG. 5). The group atlas for DTI shows that the reliability of FA and PDD values is inhomogeneous across the brain. In general, measurements from the internal capsule and inferior longitudinal fasciculus are quite reliable; group averages in other regions, such as near the lateral occipital or frontal regions, are much less so. Consequently, smaller differences can be detected in some regions compared to others. This may be a factor in explaining the tendency to find significant covariation with other factors in a small number of low variance regions (e.g., ref. 39).

The reliability of the atlas data also has implications for the ability to track fiber bundles from group data (e.g., ref. 9). A fiber bundle may be reliably detected in each individual, but coregistration of many individuals may introduce FA and PDD variance that masks the bundle in the average. Moreover, if there is a systematic bias in the individual data, such as fibers taking a wrong turn into a major bundle of crossing fibers, this bias will not be eliminated (and may be accentuated) in the group average. It is essential to be alert to these uncertainties and limitations when interpreting DTI-FT in group data. In general, if care is taken to ensure that the major pathways are present in the group average data, then fiber bundles identified in the atlas are a convenient way to summarize expected pathway positions in a "typical" brain (FIG. 6).

Splenial Pathways Necessary for Reading

More than a hundred years of neurology has focused attention on the loss of reading (alexia) following white matter damage in the splenium and nearby occipital and temporal pathways.[1,40] Careful neurological analyses show that alexia is closely coupled to damage in the splenium, forceps major, and white matter above the occipital horn.[17,19] On the other hand, the fibers passing ventromedial to the occipital ventricle are "of little importance for reading" (p. 1822 of ref. 19). Concerning the location of these fibers in the splenium, it appears that "only one group … seems to be pertinent for reading. That corresponds roughly to the middle component of fibers of the splenium, considered on the dorso-ventral axis …" (p. 1580 of ref. 17).

Thus, the region of the splenium containing fiber bundles essential for reading likely includes the fibers we have identified that form the superior and lateral wall of the lateral ventricle (FIG. 6). These bundles also pass through the central part of the dorsal-ventral axis in the splenium. The fibers in the splenium that are slightly superior to the ones we have identified connect with the lateral temporal and medial parietal pathways.[34] The parietal fibers can be identified, and with additional work it may be possible to follow the fibers that pass to the lateral aspect of the temporal lobe. The identification of these splenial pathways builds upon the classic neurological findings and, when coupled with functional measurements, may contribute to a better understanding of the reading pathways in normal developing children.

ACKNOWLEDGMENTS

We thank Jan Ruby, Adele Behn, and Sweta Patanaik for their work recruiting subjects and organizing the data. We also gratefully acknowledge the cheerful and helpful cooperation of the parents and children who participated in this study. Funding was provided by NIH Grant No. EY015000 and the Schwab Foundation for Learning.

REFERENCES

1. DEJERINE, J. 1891. Sur un cas de cecite verbale avec agraphie, suivi d'autopsie. Mém. Soc. Biol. **3:** 197–201.
2. JUST, M.A. *et al.* 2004. Cortical activation and synchronization during sentence comprehension in high-functioning autism: evidence of underconnectivity. Brain **127:** 1811–1821.
3. BASSER, P.J. *et al.* 2000. *In vivo* fiber tractography using DT-MRI data. Magn. Reson. Med. **44:** 625–632.
4. CATANI, M. *et al.* 2002. Virtual *in vivo* interactive dissection of white matter fasciculi in the human brain. Neuroimage **17:** 77–94.
5. CATANI, M. *et al.* 2003. Occipito-temporal connections in the human brain. Brain **126:** 2093–2107.
6. CONTURO, T.E. *et al.* 1999. Tracking neuronal fiber pathways in the living human brain. Proc. Natl. Acad. Sci. USA **96:** 10422–10427.
7. DOUGHERTY, R.F. *et al.* 2005. Functional organization of human occipital-callosal fiber tracts. Proc. Natl. Acad. Sci. USA **102:** 7350–7355.
8. HAGMANN, P. *et al.* 2003. DTI mapping of human brain connectivity: statistical fibre tracking and virtual dissection. Neuroimage **19:** 545–554.
9. JONES, D.K. *et al.* 2002. Spatial normalization and averaging of diffusion tensor MRI data sets. Neuroimage **17:** 592–617.
10. MORI, S. *et al.* 1999. Three-dimensional tracking of axonal projections in the brain by magnetic resonance imaging. Ann. Neurol. **45:** 265–269.
11. MORI, S. *et al.* 2002. Imaging cortical association tracts in the human brain using diffusion-tensor-based axonal tracking. Magn. Reson. Med. **47:** 215–223.
12. MOSELEY, M., R. BAMMER & J. ILLES. 2002. Diffusion-tensor imaging of cognitive performance. Brain Cogn. **50:** 396–413.
13. KLINGBERG, T. *et al.* 2000. Microstructure of temporo-parietal white matter as a basis for reading ability: evidence from diffusion tensor magnetic resonance imaging. Neuron **25:** 493–500.
14. DEUTSCH, G.K. *et al.* 2005. Children's reading performance is correlated with white matter structure measured by diffusion tensor imaging. Cortex **41:** 354–363.
15. MOLKO, N. *et al.* 2002. Visualizing the neural bases of a disconnection syndrome with diffusion tensor imaging. J. Cogn. Neurosci. **14:** 629–636.
16. WAKANA, S. *et al.* 2004. Fiber tract-based atlas of human white matter anatomy. Radiology **230:** 77–87.
17. DAMASIO, A.R. & H. DAMASIO. 1983. The anatomic basis of pure alexia. Neurology **33:** 1573–1583.
18. LE, T.H. *et al.* 2005. Diffusion tensor imaging with three-dimensional fiber tractography of traumatic axonal shearing injury: an imaging correlate for the posterior callosal "disconnection" syndrome—case report. Neurosurgery **56:** 189.
19. BINDER, J.R. & J.P. MOHR. 1992. The topography of callosal reading pathways: a case-control analysis. Brain **115**(part 6): 1807–1826.

20. MAO-DRAAYER, Y. & H. PANITCH. 2004. Alexia without agraphia in multiple sclerosis: case report with magnetic resonance imaging localization. Mult. Scler. **10:** 705–707.
21. BAMMER, R. & M. AUER. 2001. Correction of eddy-current induced image warping in diffusion-weighted single-shot EPI using constrained non-rigid mutual information image registration. *In* Proceedings of the Ninth Annual ISMRM Meeting. Vol. 9, p. 508.
22. BASSER, P.J. 1995. Inferring microstructural features and the physiological state of tissues from diffusion-weighted images. NMR Biomed. **8:** 333–344.
23. BASSER, P.J. & C. PIERPAOLI. 1996. Microstructural and physiological features of tissues elucidated by quantitative-diffusion-tensor MRI. J. Magn. Reson. **B111:** 209–219.
24. HEDEHUS, M. & S. SKARE. 2004. DTI post-processing using "tensorcalc" software.
25. FRISTON, K., J. ASHBURNER & WELLCOME DEPARTMENT OF IMAGING NEUROSCIENCE. 2004. Statistical parametric mapping.
26. PAJEVIC, S., A. ALDROUBI & P.J. BASSER. 2002. A continuous tensor field approximation of discrete DT-MRI data for extracting microstructural and architectural features of tissue. J. Magn. Reson. **154:** 85–100.
27. ALEXANDER, D.C. *et al.* 2001. Spatial transformations of diffusion tensor magnetic resonance images. IEEE Trans. Med. Imaging. **20:** 1131–1139.
28. PRESS, W.H. *et al.* 2002. Numerical Recipes in C++: The Art of Scientific Computing. Cambridge University Press. London/New York.
29. SMITH, S.M. 2002. Fast robust automated brain extraction. Hum. Brain Map. **17:** 143–155.
30. ASHBURNER, J. & K.J. FRISTON. 1999. Nonlinear spatial normalization using basis functions. Hum. Brain Map. **7:** 254–266.
31. ALEXANDER, A.L. *et al.* 2000. A geometric analysis of diffusion tensor measurements of the human brain. Magn. Reson. Med. **44:** 283–291.
32. KRIEG, W.J.S. 1963. Connections of the Cerebral Cortex. Brain Books. Evanston, IL.
33. CLARKE, S. & J. MIKLOSSY. 1990. Occipital cortex in man: organization of callosal connections, related myelo- and cytoarchitecture, and putative boundaries of functional visual areas. J. Comp. Neurol. **298:** 188–214.
34. HUANG, H. *et al.* 2005. DTI tractography based parcellation of white matter: application to the mid-sagittal morphology of corpus callosum. Neuroimage **26:** 195–205.
35. WANDELL, B.A., A.A. BREWER & R.F. DOUGHERTY. 2005. Visual field map clusters in human cortex. Philos. Trans. R. Soc. Lond. B Biol. Sci. **360:** 693–707.
36. GRILL-SPECTOR, K. & R. MALACH. 2004. The human visual cortex. Annu. Rev. Neurosci. **27:** 649–677.
37. SCHWARTZMAN, A., R.F. DOUGHERTY & J.E. TAYLOR. 2005. Cross-subject comparison of principal diffusion direction maps. Magn. Reson. Med. **53:** 1423–1431.
38. BEAULIEU, C. *et al.* 2005. Imaging brain connectivity in children with diverse reading ability. Neuroimage **25:** 1266–1271.
39. SALAT, D.H. *et al.* 2005. Age-related alterations in white matter microstructure measured by diffusion tensor imaging. Neurobiol. Aging **26:** 1215–1227.
40. DEJERINE, J. 1892. Contribution a l'étude anatomoclinique et clinique des differentes varietes de cecite verbal. C. R. Hebd. Séances Mém. Soc. Biol. **4:** 61–90.

Multiple-Fiber Reconstruction Algorithms for Diffusion MRI

DANIEL C. ALEXANDER

Department of Computer Science, University College London, London WC1E 6BT, United Kingdom

ABSTRACT: **This chapter reviews multiple-fiber reconstruction algorithms for diffusion magnetic resonance imaging (MRI) and provides some initial comparative results for two such algorithms, q-ball imaging and PASMRI, on data from a typical clinical diffusion MRI acquisition. The chapter highlights the problems with standard approaches, such as diffusion-tensor MRI, to motivate a recent set of alternative approaches. The review concentrates on the software implementation of the new techniques. Results of the preliminary comparison show that PASMRI recovers the principal directions of simple test functions more consistently than q-ball imaging and produces qualitatively better results on the test data set. Further simulations suggest that a moderate increase in data quality allows q-ball, which is much faster to run, to recover directions with consistency comparable to that of PASMRI on the test data.**

KEYWORDS: **diffusion MRI; fiber; reconstruction algorithm; white matter; imaging; q-ball; PASMRI**

INTRODUCTION

Diffusion magnetic resonance imaging (MRI) provides a unique probe into the microstructure of materials. The method observes the displacements of particles that are subject to Brownian motion within a sample material. Specifically, the technique measures the probability density function p of particle displacements $\mathbf{x}$ over a fixed time t. The microstructure of the material determines the mobility of the particles within and thus determines p. Conversely, features of p provide information about the material microstructure.

In biomedical diffusion MRI, the particles of interest are usually water molecules. Water is a major constituent of biological tissue. Water molecules within tissue undergo random motion due to thermal fluctuations. Currently, brain imaging is the most common application of biomedical diffusion MRI. The brain has a complex architecture of gray matter areas connected by white matter fibers. In white matter, particles move further on average along fibers than across them since only mobility along the fiber is unhindered by cell walls. Thus, p tends to have ridges in the fiber

Address for correspondence: Daniel C. Alexander, Department of Computer Science, University College London, Gower Street, London WC1E 6BT, United Kingdom.
d.alexander@cs.ucl.ac.uk

**Ann. N.Y. Acad. Sci. 1064: 113–133 (2005). © 2005 New York Academy of Sciences.
doi: 10.1196/annals.1340.018**

directions or, equivalently, the contours of p have peaks in the fiber directions. Since diffusion MRI measures p, it can provide estimates of microstructural fiber orientations. By following fiber orientation estimates from point to point through an image volume, the technique allows noninvasive mapping of the connectivity of the brain.

The basic measurement A^* in diffusion MRI provides an approximation to the Fourier transform of p at wavenumber $\mathbf{q}$:

$$A(\mathbf{q}) = (A^*(\mathbf{0}))^{-1}A^*(\mathbf{q}) = \int_{\mathbb{R}^3} p(\mathbf{x})\cos(\mathbf{q}\cdot\mathbf{x})d\mathbf{x}, \tag{1}$$

where A is the normalized measurement (see refs. 1 and 2 for derivations). The scanner operator controls $\mathbf{q}$ by setting the strength, orientation, and duration of magnetic-gradient pulses in the measurement sequence.

One of the main goals of modern diffusion MRI is to determine, from a set of measurements $A^*(\mathbf{q}_1)$, ..., $A^*(\mathbf{q}_N)$ in each voxel of an image volume, the dominant fiber orientation(s) in each voxel. Diffusion-tensor MRI[3] was the first method to allow mapping of fiber orientations over an image volume and remains the most common. However, a drawback of diffusion-tensor MRI is that it can only reveal a single fiber orientation in each voxel and fails in voxels containing complex tissue architecture with more than one significant fiber orientation. The diffusion spectrum imaging (DSI) method of Wedeen, Tuch, and coworkers[4,5] can resolve the orientations of crossing fibers. However, DSI requires many more measurements than diffusion-tensor MRI. The long acquisition times required for DSI have prohibited its widespread use.

Recently, a new generation of diffusion MRI technique has emerged. These techniques retain the desirable qualities of both diffusion-tensor MRI and DSI and aim to reveal complex tissue architectures with acquisition requirements similar to diffusion-tensor MRI. The insights that give rise to these new techniques lie mostly in the development of new algorithms to reconstruct p (or features thereof) rather than improved measurement techniques. This chapter reviews the algorithms that underlie these new techniques. Since the theory behind the algorithms is well documented in the literature, we concentrate here on the software implementation of the algorithms, which has received less attention. The review highlights the strong computational similarities between the techniques. The aim is to provide a new perspective on the methods to promote further refinement and the development of still better techniques and algorithms in the future.

The next section reviews traditional reconstruction algorithms in diffusion MRI, specifically diffusion-tensor MRI and DSI. The third section reviews the new generation of reconstruction algorithms. The fourth section describes a preliminary performance comparison of two of the more established new techniques: **q**-ball imaging[6] and PASMRI.[7] The experiments do not provide a comprehensive performance analysis of the two techniques, but simply test and compare them on data from a standard acquisition scheme for diffusion-tensor MRI. The results provide a useful insight into the benefits and weaknesses of the new algorithms. The chapter closes in the fifth section with some discussion of the state of the art and areas for future work.

TRADITIONAL RECONSTRUCTION ALGORITHMS

This section briefly reviews the established acquisition and reconstruction techniques in modern diffusion MRI.

Diffusion-Tensor MRI

Diffusion-tensor MRI[3] assumes that particles move according to a simple anisotropic diffusion process and computes the *apparent diffusion tensor* on the assumption that p is a zero-mean trivariate Gaussian distribution:

$$p(\mathbf{x}) = G(\mathbf{x}; D, t), \tag{2}$$

where

$$G(\mathbf{x}; D, t) = [(4\pi t)^3 \det(D)]^{-1/2} \exp\left(-\frac{\mathbf{x}^T D^{-1} \mathbf{x}}{4t}\right), \tag{3}$$

D is the diffusion tensor, and t is the diffusion time. Substitution of equation 2 into equation 1 gives

$$A(\mathbf{q}) = \exp(-t\mathbf{q}^T D\mathbf{q}). \tag{4}$$

If we take the logarithm of equation 4, we see that each $A(\mathbf{q})$ provides a linear constraint on the elements of D. The Gaussian model has six free parameters, which are the elements of the symmetric three-by-three matrix D. To fit the six free parameters, we need a minimum of six $A(\mathbf{q})$ with independent $\mathbf{q}$, although many more are often acquired. Note that six $A(\mathbf{q})$ requires a minimum of seven $A^*(\mathbf{q})$, including one for normalization. The standard approach is to use the linear least-squares fit of D to the log measurements. However, fitting directly using equation 4 (as in ref. 8) can improve results since the error distribution is closer to normal on $A(\mathbf{q})$ than on $\log[A(\mathbf{q})]$ and allows constraints on the diffusion tensor (see ref. 2).

Diffusion-tensor MRI provides two key insights into the material microstructure that simple diffusion-weighted MRI does not. First, it provides rotationally invariant statistics of the anisotropy of p, such as the fractional anisotropy,[9] which reflect the anisotropy of the microstructure. Second, the principal eigenvector of D provides an estimate of the dominant orientation of microstructural fibers.

Most diffusion-tensor MRI measurement schemes acquire more than the minimum seven measurements to reduce the effects of noise. The standard approach[10] is to acquire M measurements with $\mathbf{q} = \mathbf{0}$ and N measurements with nonzero wavenumbers $\mathbf{q}_i$, $i = 1, \ldots, N$. The $|\mathbf{q}_i|$ are all equal, and the diffusion time t, and hence the "diffusion-weighting factor" $b = t|\mathbf{q}|^2$, is fixed for all the $A(\mathbf{q}_i)$. The gradient directions $\hat{\mathbf{q}}_i$ are unique and distributed uniformly over the sphere. We refer to this kind of measurement scheme as a "spherical acquisition scheme" since the $\mathbf{q}_i$ all lie on a sphere in $\mathbf{q}$-space.

The major drawback of diffusion-tensor MRI is that the Gaussian model is often a poor fit to the data. The Gaussian function has ellipsoidal contours, which can have only a single peak. Thus, diffusion-tensor MRI provides only one fiber orientation estimate in each voxel. In regions where fibers cross within a voxel, the contours of

p have multiple peaks, which the Gaussian model cannot capture. When the Gaussian model is poor, the two major selling points of diffusion-tensor MRI fail. First, indices of anisotropy derived from the diffusion tensor underestimate the true directional variability of p. Second, fiber orientation estimates are incorrect. Reference 2 discusses these issues in detail.

Diffusion Spectrum Imaging (DSI)

DSI[4] reconstructs a discrete representation of p directly from measurements on a regular grid of wavenumbers via a fast Fourier transform. The reconstruction gives values of p on a grid of displacements.

From the discrete representation of p, we can compute the orientation distribution function (ODF):

$$\phi(\hat{\mathbf{x}}) = \int_0^\infty p(\alpha\hat{\mathbf{x}})d\alpha, \tag{5}$$

where $\hat{\mathbf{x}}$ is a unit vector in the direction of $\mathbf{x}$. The ODF is the radial projection of p onto the unit sphere. The ODF has peaks in the directions in which p has most mass, which DSI assumes are the fiber directions. In DSI, ϕ is computed numerically by interpolating the grid representation of p. The function ϕ can have multiple pairs of equal and opposite peaks. Each pair provides a separate fiber orientation estimate, which enables DSI to resolve the orientations of crossing fibers.

Qualitative results from DSI (e.g., in refs. 4 and 5) show ODF peaks in the expected fiber directions at known fiber crossings in human and animal brain data. However, the results also show ODFs with multiple peaks in gray matter regions and it is unclear whether these peaks show genuine anatomic structure or simply arise from measurement noise. DSI has the clear advantage over diffusion-tensor MRI since it can resolve multiple fiber orientations. Despite this advantage, though, DSI is not used as widely as diffusion-tensor MRI. The main drawback of the technique is that acquisition times are long since it requires an order of magnitude more measurements than diffusion-tensor MRI to get sufficient detail in the reconstructed p.

NEW RECONSTRUCTION ALGORITHMS

This section reviews the new generation of multiple-fiber reconstruction algorithms. Since DSI can resolve crossing fibers within a voxel, the reconstruction algorithm in the technique (FFT followed by projection onto the sphere) qualifies as a "multiple-fiber reconstruction algorithm". However, the technique relies on the grid arrangement of the sampled wavenumbers. This measurement scheme is uneconomical if all we require is the angular structure of p, which is the part that reveals fiber orientations, since much of the information in the measurements contributes to only the radial structure of p. The techniques that this section includes can, in theory, use sets of measurements acquired at any set of wavenumbers. In practice, however, the arrangement of the $\mathbf{q}_i$ affects the quality of the output. The standard approach for all the methods is to use the spherical sampling scheme common in diffusion-tensor MRI, although other arrangements may improve results.

Compartment and Fiber Models

A simple generalization of diffusion-tensor MRI replaces the Gaussian model for p with a mixture of Gaussian densities:

$$p(\mathbf{x}) = \sum_{i=1}^{n} a_i G(\mathbf{x}; D_i, t), \tag{6}$$

where each $a_i \in [0, 1]$ and $\Sigma_i a_i = 1$. Particle displacements in media containing n distinct compartments, between which no exchange of particles occurs, follow the distribution in equation 6 if the displacement density in the i-th compartment, which has volume fraction a_i, is $G(\mathbf{x}; D_i, t)$.

We take the Fourier transform of equation 6 and substitute into equation 1 to relate the measurement values to the model parameters (D_i and a_i, $i = 1, \ldots, n$):

$$A(\mathbf{q}) = \sum_{i=1}^{n} a_i \exp(-t\mathbf{q}^T D_i \mathbf{q}).$$

The constraint on the model parameters from each measurement is nonlinear, so we must fit the model to the data by nonlinear optimization. The principal eigenvector of each D_i provides a separate fiber orientation estimate. The multicompartment model assumes that the number, n, of distinct fiber populations is known. Practical considerations, such as the number of measurements and the measurement noise level, limit the number of orientations that the method can resolve reliably. Most work to date uses a maximum n of 2.

Two problems accompany the use of multicompartment models. First, the choice of n presents a model-selection problem: in voxels with only one fiber orientation, we lose accuracy by fitting a model with $n \geq 2$. For best results, we need to use the correct n in each voxel. Second, the nonlinear fitting procedure is unstable and starting-point-dependent because of local minima in the objective function. Parker and Alexander[11] and Blyth *et al.*[12] use the spherical-harmonic voxel-classification algorithm proposed in reference 13 to solve the model-selection problem. This method does not extend naturally above $n = 2$, though.[2] Tuch *et al.*[14] threshold the correlation of the measurements with their predictions from the $n = 1$ model in each voxel separately to decide whether to use $n = 1$ or $n = 2$. Constraints on the diffusion tensors in the multicompartment model can help stabilize the fitting procedure. For example, we can enforce positive definiteness, using the Cholesky decomposition,[15] or cylindrical symmetry (see ref. 2), or specific eigenvalues (as in ref. 14), on the component diffusion tensors. Spatial regularization techniques also help overcome the fitting problem by ensuring voxel-to-voxel coherence (see refs. 15 and 16).

A similar model-based approach[17] assumes that particles belong to one of two populations: a restricted population within or around microstructural fibers and a free population unaffected by microstructural barriers. With negligible exchange between the populations, $p = ap_f + (1 - a)p_r$, where p_f is the displacement density for the free population, p_r is that for the restricted population, and a is the fraction of particles in the free population. Behrens *et al.*[17] use an isotropic Gaussian model for p_f. They use a Gaussian model for p_f in which the diffusion tensor has only one nonzero eigenvalue so that particle displacements are restricted to a line. Assaf *et al.*[18] describe a similar approach. They model p_r with Neuman's model for restricted diffusion in a cylinder.[19] The fitted p_r provides the fiber orientation estimate. For p_f, which they call the "hindered compartment", they use an anisotropic Gaussian model. The approaches in references 17 and 18 extend naturally to the multiple-fiber case by including multiple restricted populations in the model. Both methods have the same model-selection and fitting problems as the mixture-of-Gaussian models discussed above.

Deconvolution Methods

Deconvolution methods[17,20,21] generalize the methods in the previous section by assuming a distribution, rather than a discrete number, of fiber orientations. The methods assume that the diffusion MRI signal is the convolution of the *fiber orientation distribution* (FOD) f, which (like the ODF) is a real-valued function of the unit sphere, with the signal $A_f(\cdot;\ \hat{\mathbf{x}})$ from a single fiber with orientation $\hat{\mathbf{x}}$:

$$A(\mathbf{q}) = \int A_f(\mathbf{q};\ \hat{\mathbf{x}})f(\hat{\mathbf{x}})d\hat{\mathbf{x}}. \tag{7}$$

Note that equation 7 assumes that $A_f(\cdot;\ \hat{\mathbf{x}})$ has rotational symmetry about $\hat{\mathbf{x}}$. The methods aim to deconvolve the signal, using a model for $A_f(\cdot;\ \hat{\mathbf{x}})$, to obtain f. Like the ODF, ϕ, f can have multiple pairs of equal and opposite peaks and each pair provides a separate fiber orientation estimate. The methods do not require that the number of peaks or fibers is known and thus do not have the model-selection problem associated with compartment and fiber-model methods.

To implement the method, we can represent f using a linear basis:

$$f(\hat{\mathbf{x}}) = \sum_{k=1}^{K} \beta_k\theta_k(\hat{\mathbf{x}}).$$

We substitute for f in equation 7 and reverse the order of the integral and sum to obtain

$$A(\mathbf{q}) = \sum_{k=1}^{K} (\beta_k \int A_f(\mathbf{q};\hat{\mathbf{x}})\theta_k(\hat{\mathbf{x}})d\hat{\mathbf{x}}). \tag{8}$$

For a set of measurements with wavenumbers $\mathbf{q}_i$, $i = 1, \ldots, N$, we can summarize the set of equations from equation 8 as $\mathbf{A} = \mathbf{XB}$, where $\mathbf{A} = [A(\mathbf{q}_1), \ldots, A(\mathbf{q}_N)]^T$ is the vector of normalized measurements, $\mathbf{B} = (\beta_1, \ldots, \beta_K)^T$ is the vector of basis-function weights, and $\mathbf{X}$ is the matrix with ik-th entry

$$X_{ik} = \int A_f(\mathbf{q}_i;\ \hat{\mathbf{x}})\theta_k(\hat{\mathbf{x}})d\hat{\mathbf{x}}.$$

We solve the matrix equation to obtain the set of basis-function weights that define f via a linear transformation of the measurements: $\mathbf{B} = \mathbf{X'A}$, where $\mathbf{X'} = (\mathbf{X}^T\mathbf{X})^{-1}\mathbf{X}^T$ is the pseudoinverse of $\mathbf{X}$. Since the set of $\mathbf{q}_i$ is identical in each voxel of a typical image volume, we need to compute $\mathbf{X'}$ only once. The computational burden of the method is therefore light (a single matrix multiplication in each voxel) and comparable to that of diffusion-tensor MRI.

References 20 and 21 use the spherical harmonics as the basis for f. References 17 and 20 use Gaussian fiber models to obtain A_f. Tournier *et al.*[21] derive A_f directly from the data by taking an average signal from the most anisotropic voxels; they represent A_f using the spherical-harmonic basis.

q-*Ball Imaging*

Tuch's **q**-ball imaging method[5,6] (henceforth referred to as just "**q**-ball") approximates the ODF by the Funk transform[22] of the diffusion MRI signal at a fixed radius in **q**-space. The Funk transform is a mapping between functions of the sphere. The value of the Funk transform of a spherical function at a point $\hat{\mathbf{x}}$ is the integral of the function over the great circle $C(\hat{\mathbf{x}})$ perpendicular to $\hat{\mathbf{x}}$. Thus, **q**-ball makes the approximation

$$\phi(\hat{\mathbf{x}}) = \int_{C(\hat{\mathbf{x}})} A(Q\hat{\mathbf{q}})d\hat{\mathbf{q}}, \tag{9}$$

where $Q = |\mathbf{q}_i|$ if all the $|\mathbf{q}_i|$ are equal, or Q is the mean or some typical $|\mathbf{q}_i|$ otherwise. To implement the transform, we first interpolate the measurements using a linear basis so that $A(\mathbf{q}) = \Sigma^J_{j=1}\xi_j\psi_j(\mathbf{q})$. We estimate the weights ξ_j, $j = 1, \ldots, J$, of the basis functions ψ_j from the normalized measurements $A(\mathbf{q}_i)$, $i = 1, \ldots, N$. We can write $\mathbf{A} = \mathbf{Y\Xi}$, where $\mathbf{A}$ is the vector of normalized measurements (as in the previous subsection), $\mathbf{\Xi} = (\xi_1, \ldots, \xi_J)^T$, and $Y_{ij} = \psi_j(\mathbf{q}_i)$. Thus, $\mathbf{\Xi} = \mathbf{Y'A}$.

Substituting the linear basis representation of A in equation 9 and reversing the order of the sum and integral, we see that

$$\phi(\hat{\mathbf{x}}) = \sum_{j=1}^{J}\left(\xi_j \int_{C(\hat{\mathbf{x}})} \psi_j(Q\hat{\mathbf{q}})d\hat{\mathbf{q}}\right).$$

We evaluate ϕ at each of a set of unique $\hat{\mathbf{x}}_l$, $l = 1, \ldots, L$, and, representing ϕ using another linear basis so that

$$\phi(\hat{\mathbf{x}}) = \sum_{k=1}^{K} \beta_k\theta_k(\hat{\mathbf{x}}),$$

write the resulting set of equations in matrix form:

$$\mathbf{\Theta C} = \mathbf{\Psi\Xi} = \mathbf{\Psi Y'A}, \tag{10}$$

where $\Theta_{lk} = \theta_k(\hat{\mathbf{x}}_l)$, $\mathbf{C} = (\beta_1, \ldots, \beta_K)^T$, and

$$\Psi_{kj} = \int_{C(\hat{\mathbf{x}}_k)} \psi_j(Q\hat{\mathbf{q}})d\hat{\mathbf{q}}.$$

We solve equation 10 for $\mathbf{C} = \mathbf{\Theta'\Psi Y'A}$, which defines ϕ. The parameters of the ODF are thus a linear transformation of the measurements and we need compute the matrix $\mathbf{\Theta}'\mathbf{\Psi}\mathbf{Y}'$ only once. The computational burden of the method is therefore light and similar to the deconvolution algorithm outlined in the previous section.

Tuch[5,6] uses radial basis functions for the linear bases. Specifically,

$$\theta_k(\hat{\mathbf{x}}) = \exp[-(\sigma^{-1}\cos^{-1}(|\hat{\mathbf{x}} \cdot \hat{\mathbf{y}}_k|))^2], \tag{11}$$

where σ is a constant scaling parameter and the $\hat{\mathbf{y}}_k$, $k = 1, \ldots, K$, are unit vectors evenly distributed on the unit sphere. The ψ_i have similar form, but the scaling parameter σ can have a different value to that used for the θ_k.

In reference 5, Tuch shows analytically that, in the absence of noise, the approximation to ϕ from the Funk transform becomes closer as Q increases. Qualitative results in references 5 and 6 show good agreement between **q**-ball and DSI in a fiber-crossing region in the human brain. However, these results come from high-quality test data from a spherical acquisition scheme with $N = 492$ and with $|\mathbf{q}| = 3.6 \times 10^5$ m^{-1} ($b = 4.0 \times 10^9$ s/m^2) and $|\mathbf{q}| = 5.4 \times 10^5$ m^{-1} ($b = 12.0 \times 10^9$ s/m^2), requiring similar acquisition time to DSI.

Other ODF Estimators

Lin *et al.*[23] propose a similar algorithm to **q**-ball independently. They test their algorithm on data acquired from a phantom containing water-filled capillaries in two orientations, which simulates crossing white matter fibers. The algorithm recovers the orientation of the capillaries consistently.

Ozarslan *et al.*[24] fit higher-order tensor models to measurements from a spherical acquisition scheme. They assume that $A(\mathbf{q})$ decays exponentially with increasing $|\mathbf{q}|$ and fixed $\hat{\mathbf{q}}$. This assumption allows them to estimate the measurements on a regular grid of wavenumbers, which they use as input to the DSI reconstruction process. The method recovers the principal directions of simple test functions. Qualitative results on rat brain data, from a spherical acquisition scheme with $N = 81$ and $b = 1.5 \times 10^9$ s/m^2, are promising. With the exponential radial decay model, however, we can evaluate the integrals required for ϕ more directly, which may improve the results.

PASMRI

Jansons and Alexander's PASMRI algorithm[7] computes another feature of p called the persistent angular structure (PAS). The PAS is the function $\tilde{p}$ of the sphere that, when embedded in three-dimensional space on a sphere of radius r, has a Fourier transform that best fits the normalized measurements, $A(\mathbf{q}_i)$. We make the substitution $p(\mathbf{x}) = \tilde{p}(\hat{\mathbf{x}})r^{-2}\delta(|\mathbf{x}| - r)$ so that the Fourier integral in equation 1 reduces to an integral over the sphere, and seek $\tilde{p}$ for which

$$A(\mathbf{q}_i) = r^{-2} \int \tilde{p}(\hat{\mathbf{x}}) \cos(r\mathbf{q}_i \cdot \hat{\mathbf{x}}) d\hat{\mathbf{x}}. \tag{12}$$

Jansons and Alexander use a maximum-entropy parametrization of $\tilde{p}$,

$$\tilde{p}(\hat{\mathbf{x}}) = \exp\left(\lambda_0 + \sum_{j=1}^{N} \lambda_j \cos(r\mathbf{q}_j \cdot \hat{\mathbf{x}})\right), \tag{13}$$

and minimize

$$\sum_{i=1}^{N} (A(\mathbf{q}_i) - \int \tilde{p}(\hat{\mathbf{x}}) \cos(r\mathbf{q}_i \cdot \hat{\mathbf{x}}) d\hat{\mathbf{x}})^2 \tag{14}$$

with respect to the N+1 parameters, λ_j, $j = 0, \ldots, N$, of $\tilde{p}$ using a Levenberg-Marquardt algorithm. They use numerical approximations of the integrals in equation 14 and run the optimization from a single starting point, which does not guarantee that the global minimum is found. The function $\tilde{p}$ can have any number of pairs of equal-and-opposite peaks and each pair provides a fiber orientation estimate. The parameter r controls the smoothness of $\tilde{p}$.

The nonlinear optimization and numerical integration make the PASMRI algorithm much slower than deconvolution and **q**-ball as implemented above. Note, however, that if we replace the maximum-entropy parametrization of $\tilde{p}$ with a linear basis, we can estimate $\tilde{p}$ via a linear transformation of the data in a similar way to f and ϕ. We substitute

$$\tilde{p}(\hat{\mathbf{x}}) = r^2 \sum_{k=1}^{K} \beta_k \theta_k(\hat{\mathbf{x}})$$

into equation 12 and change the order of the sum and integral to give

$$A(\mathbf{q}) = \sum_{k=1}^{K} (\beta_k \int \theta_k(\hat{\mathbf{x}}) \cos(r\mathbf{q} \cdot \hat{\mathbf{x}}) d\hat{\mathbf{x}}).$$

The set of measurements **A** is then a linear transform of the coefficients $\mathbf{B} = (\beta_1, \ldots, \beta_K)^T$ of $\tilde{p}$, which we can invert to obtain $\mathbf{B} = \boldsymbol{\Theta}'\mathbf{A}$, where

$$\Theta_{ik} = \int \theta_k(\hat{\mathbf{x}}) \cos(r\mathbf{q}_i \cdot \hat{\mathbf{x}}) d\hat{\mathbf{x}}.$$

We need compute $\boldsymbol{\Theta}'$ only once so that the computational burden is similar to the implementations of spherical deconvolution and **q**-ball given above. Conversely, we could implement spherical deconvolution or **q**-ball using a nonlinear representation for f or ϕ. Note the striking similarity between this linearized version of PASMRI and the spherical deconvolution algorithm. If we set $A_f(\mathbf{q}; \hat{\mathbf{x}}) = \cos(r\mathbf{q}; \hat{\mathbf{x}})$, the two algorithms are identical.

EXPERIMENTS AND COMPARISONS

This section shows some comparative results for two of the methods discussed in the previous section, PASMRI and **q**-ball. The results are not intended as a complete performance evaluation of the two methods, but serve to highlight the benefits and problems of these kinds of algorithms.

Brain Data

We begin by showing the output of the two algorithms on a standard clinical diffusion MRI data set originally acquired for diffusion-tensor MRI using a spherical acquisition scheme. The subject was a healthy adult male, who gave informed written consent. The data set has 60 slices. Each slice contains 128×128 voxels reconstructed from a 62×96 measurement array. For these data, $M = 6$, $N = 54$, and each measurement comes from a pulsed-gradient spin-echo sequence with EPI readout, echo time TE = 0.095 s, gradient-pulse duration $\delta = 0.034$ s, separation $\Delta = 0.040$ s, and strength $|\mathbf{g}| = 0.022$ T/m, which gives $|\mathbf{q}_i| = 2.0 \times 10^5$ m^{-1} and $b = 1.15 \times 10^9$ s/m^2. This scheme is typical for whole-brain clinical diffusion-tensor MRI and requires around 20 min of scan time on a 1.5-T GE Signa scanner. In white matter regions, the signal to noise ratio at $\mathbf{q} = \mathbf{0}$, which we shall call S, is around 16 on average.

After eddy-current-distortion correction using slice-by-slice 2D affine registration to the first $\mathbf{q} = \mathbf{0}$ measurement,[25] we compute the PAS and ODF (using **q**-ball) in each voxel. We use the simulations discussed in the next section to choose parameter settings for both algorithms. We set $r = 1.4$ in the PAS computation. In the **q**-ball algorithm, we set the scaling parameter, σ, in equation 11 to $\pi/30$ for the θ_k and to $7\pi/60$ for the ψ_i; we compute the great circle integrals by summing over 48 equally spaced points around the circle and set the number, K, of θ_k to 755. FIGURES 1 and 2 show the PAS and ODF in each voxel of a coronal slice through the data set. The peaks of the ODF are less pronounced than those of the PAS. To emphasize the shape of the ODF, FIGURE 2 in fact shows ϕ^5. The region of interest in FIGURES 1 and 2 contains part of the corpus callosum (top) and a fiber-crossing region in the pons (bottom). In the

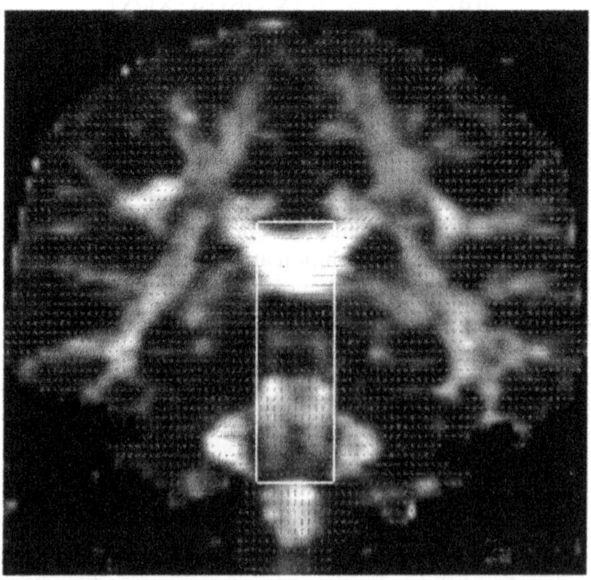

FIGURE 1. The PAS (*in red*) in brain voxels of a coronal slice from the test data set superimposed on the fractional anisotropy map.

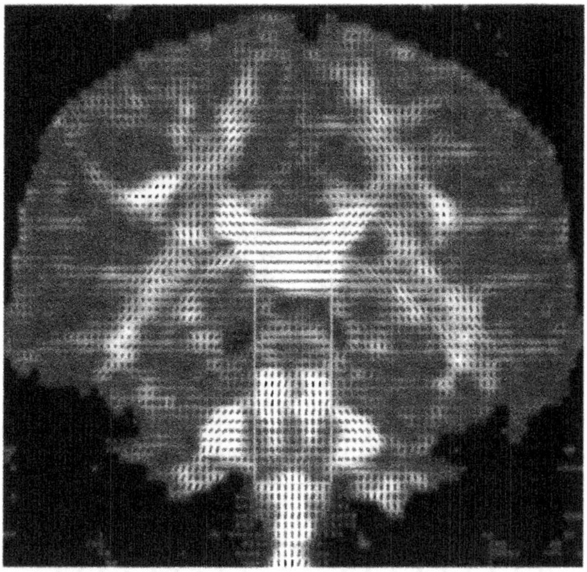

FIGURE 2. The ODF (*in red*) approximated using **q**-ball in brain voxels of the coronal slice in FIGURE 1 superimposed on the fractional anisotropy map.

corpus callosum, we expect a single left-right fiber orientation. In the fiber-crossing region, we expect two fiber orientations: one left-right and one superior-inferior.

Both the PAS and ODF clearly show the expected fiber orientation in the corpus callosum. The PAS shows two pairs of peaks consistently in the crossing fibers in the pons, but the crossing fiber orientations are less clear in the ODF plot. FIGURE 3 shows the peak directions extracted from both the PAS and the ODF in the region of interest marked in FIGURES 1 and 2. While the PAS consistently has two pairs of equal and opposite peaks with the expected directions in the fiber-crossing region, the ODF shows crossing fibers in only about half of those voxels.

To determine the peak directions of the ODF or PAS, we sample the function at each vertex of 1000 random rotations of a regular icosahedron. We find the list of sampled points that are local maxima in the sense that the function is larger at that point than any other sampled location within a search radius ρ, which we set to 0.4. Finally, we refine the locations of the peaks from these local maxima using Powell's local optimization algorithm.[26]

Simulations

The results in the previous section show that the more computationally intensive PASMRI algorithm has some advantage over the more efficient **q**-ball. The PAS appears to resolve the crossing fiber orientations more consistently for these data. In this section, we investigate this apparent advantage further using synthetic data.

We synthesize data by emulating the brain imaging scheme. Given a model for p, we sample its Fourier transform F at $\mathbf{q} = \mathbf{0}$ (M times) and each $\mathbf{q}_i$, $i = 1, \ldots, N$. To each sample, we add a random complex number with independent real and

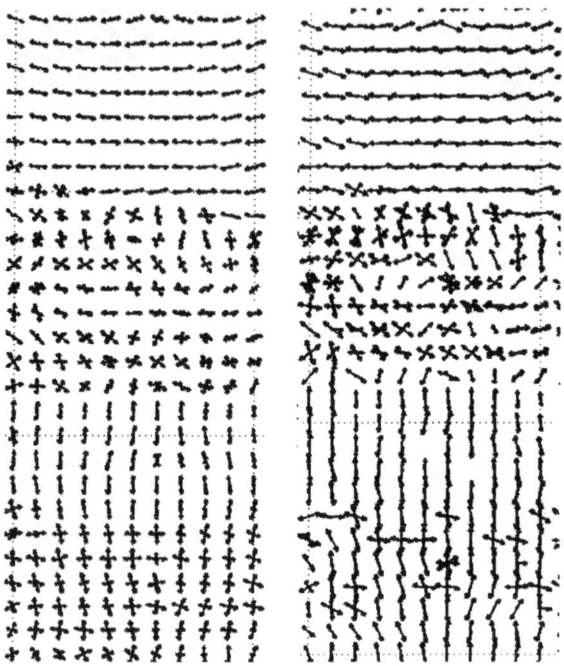

(a) (b)

FIGURE 3. The fiber orientation estimates extracted from **(a)** the PAS and **(b)** the ODF in the region of interest highlighted in FIGURES 1 and 2.

imaginary parts, each with distribution $N(0, \sigma^2)$, where $\sigma = F(\mathbf{0})/S$ and S is the signal to noise at $\mathbf{q} = \mathbf{0}$, as defined in the previous subsection. The modulus of the noisy sample is the synthetic measurement.

We use variations of five basic test functions:

$$p_0(\mathbf{x}) = G(\mathbf{x}; D_0, t)$$
$$p_1(\mathbf{x}) = G(\mathbf{x}; D_1, t)$$
$$p_2(\mathbf{x}) = G(\mathbf{x}; D_4, t)$$
$$p_3(\mathbf{x}) = aG(\mathbf{x}; D_1, t) + (1 - a)G(\mathbf{x}; D_2, t)$$
$$p_4(\mathbf{x}) = [G(\mathbf{x}; D_1, t) + G(\mathbf{x}; D_2, t) + G(\mathbf{x}; D_3, t)]/3$$

where $a \in [0, 1]$ is the mixing parameter for p_3, $G(\mathbf{x}; D, t)$ is defined in equation 3, and the diffusion tensors are

$$D_0 = \mathrm{diag}(\lambda_0, \lambda_0, \lambda_0)$$
$$D_1 = \mathrm{diag}(\lambda_1, \lambda_2, \lambda_2)$$
$$D_2 = \mathrm{diag}(\lambda_2, \lambda_1, \lambda_2)$$
$$D_3 = \mathrm{diag}(\lambda_2, \lambda_2, \lambda_1)$$
$$D_4 = \mathrm{diag}[(\lambda_1 + \lambda_2)/2, (\lambda_1 + \lambda_2)/2, \lambda_2].$$

By default, $a = 0.5$, $\lambda_1 = 1.7 \times 10^{-9}$ m^2/s, and $\mathrm{Tr}(D_i) = \lambda_1 + 2\lambda_2 = 3\lambda_0 = 2.1 \times 10^{-9}$ m^2/s. Thus, p_0 is isotropic, p_1 is anisotropic with prolate ellipsoidal contours, p_2 is anisotropic with oblate ellipsoidal contours, the contours of p_3 have two orthogonal peaks, and those of p_4 have three orthogonal peaks.

To measure the ability of a method to recover principal directions from the test functions, we compute a performance index called the *consistency fraction*, C. We use PASMRI and **q**-ball to estimate the principal directions of p from noisy synthetic data. The result is *consistent* if the number of estimated directions equals the number of principal directions (i.e., peaks of the contours) of p and the estimated directions match the principal directions of p to within a small angular tolerance, which we set to $\cos^{-1}(0.95)$. The consistency fraction is the fraction of 256 independent trials in which the result is consistent. Since the consistency fraction requires the test function to have distinct principal directions, we consider C for only p_1, p_3, and p_4.

Both PASMRI and **q**-ball contain parameters that we can tune to alter performance. For a fair comparison, we must optimize these parameters to maximize the performance of both algorithms. The PASMRI algorithm has the regularization parameter r and the search radius in the peak-finding algorithm. The **q**-ball algorithm has two radial basis function scaling parameters, σ in equation 11, one for the ψ_i and one for the θ_k, as well as a search radius in the peak-finding algorithm. To optimize the settings, we maximize the sum of the consistency fractions for p_1, p_3, and p_4 with the imaging parameters of the brain data over values of r and ρ that are multiples of 0.1 and values of σ that are multiples of $\pi/180$. This procedure gives rise to the settings of r, σ, and ρ used to process the brain data in the previous section. We use the same settings in all the experiments that follow in this section. In **q**-ball, the great circle integral increases in accuracy as the number of summation points increases. We choose 48 because increasing the number further does not affect the sum of the consistency fractions in the parameter-setting optimization. We use $K = 755$ in the **q**-ball algorithm throughout. In the experiments that follow, we use default settings $N = 54$, $M = 6$, $|\mathbf{q}_i| = 2.0 \times 10^5$ m^{-1}, $t = 0.04$ s, $S = 16$, $\lambda_1 = 1.7 \times 10^{-9}$ m^2/s, and $a = 0.5$, unless explicitly stated otherwise in the text.

FIGURE 4 shows how the consistency fractions of both **q**-ball and PASMRI for the anisotropic test functions vary with S. The consistency fraction is generally higher

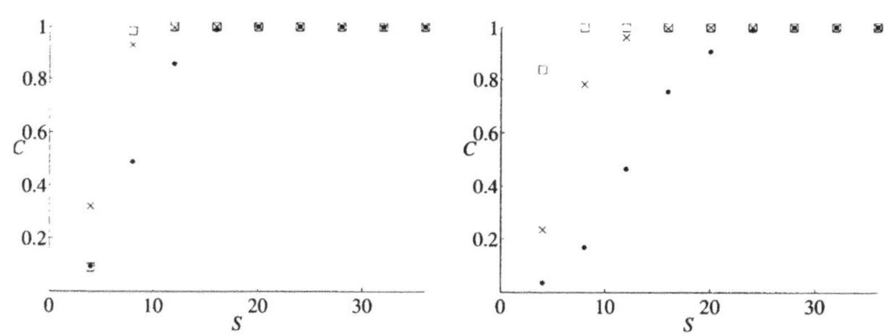

(a) (b)

FIGURE 4. Plots of the consistency fraction, C, of **(a)** the PAS and **(b)** the ODF for p_1 (□), p_3 (×), and p_4 (●) against the signal-to-noise ratio, S, of $A^*(\mathbf{0})$.

for PASMRI than **q**-ball at fixed S. For PASMRI, the consistency fraction is close to 100% for all three functions with $S \geq 16$, but for **q**-ball we need $S \geq 24$ for close to 100% consistency. FIGURES 5 and 6 show examples of $\tilde{p}$ and ϕ, respectively, reconstructed from the synthetic data with various values of S. From FIGURES 5 and 6, we can see that both the PAS and the ODF reflect the structure of the anisotropic test functions p_1, p_3, and p_4 at $S = 16$, which is approximately the signal level in the brain data. However, the isotropic test functions p_0 and p_2 also produce PAS and ODF functions with strong angular structure. In isolation, examples of the PAS or ODF from p_2 are almost indistinguishable from those from p_3. Similarly, the PAS or ODF functions from p_4 look similar to those from p_0.

FIGURE 7 plots the consistency fractions as we vary N. As we expect, C increases with N. For PASMRI, C gets close to 100% at lower N than for **q**-ball. For p_4, for example, PASMRI requires $N = 60$ to achieve near 100% consistency; in contrast, for **q**-ball, $C < 100\%$ even at $N = 120$. The PASMRI algorithm requires $N \approx 20$ to reconstruct the two directions in p_3 consistently, while **q**-ball requires $N = 40$. Note that the results in FIGURE 7 are less reliable at lower N, where C can have significant dependence on the orientation of the test function.

FIGURE 8 shows the dependence of C on the anisotropy of the test functions. Specifically, we vary λ_2 and adjust λ_1 to keep $\mathrm{Tr}(D_i)$, $i = 0, \ldots, 4$, fixed; when $\lambda_2/\mathrm{Tr}(D) = 0$, the diffusion tensor is perfectly anisotropic; when $\lambda_2/\mathrm{Tr}(D) = 1/3$, the diffusion tensor is perfectly isotropic. For p_3 and p_4, PASMRI has higher consistency at each anisotropy than **q**-ball; however, for p_1, **q**-ball retains higher consistency than PASMRI as the anisotropy decreases.

FIGURE 9 shows the consistency fractions as we vary b. For this experiment, we idealize the pulse sequence used to acquire the brain data in the way that Alexander and Barker describe,[27] which increases S at $b = 10^9$ s/m^2 to 20.3. We use the method

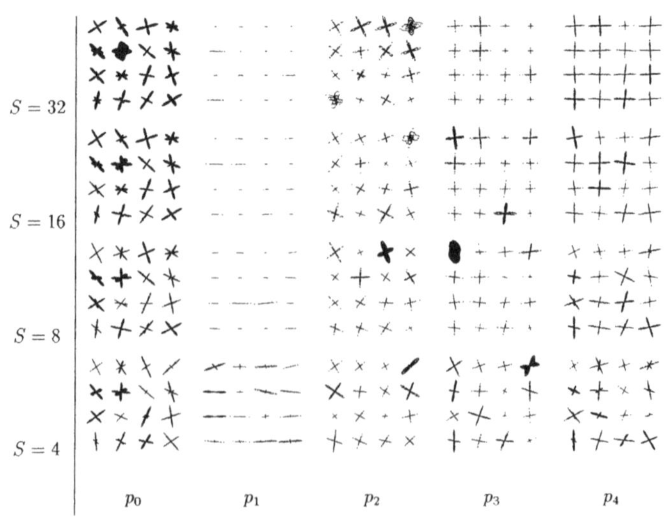

FIGURE 5. Sixteen independent examples of the PAS reconstructed from p_i, $i = 0, \ldots, 4$, for each of various values of the signal-to-noise ratio, S, of $A^*(\mathbf{0})$.

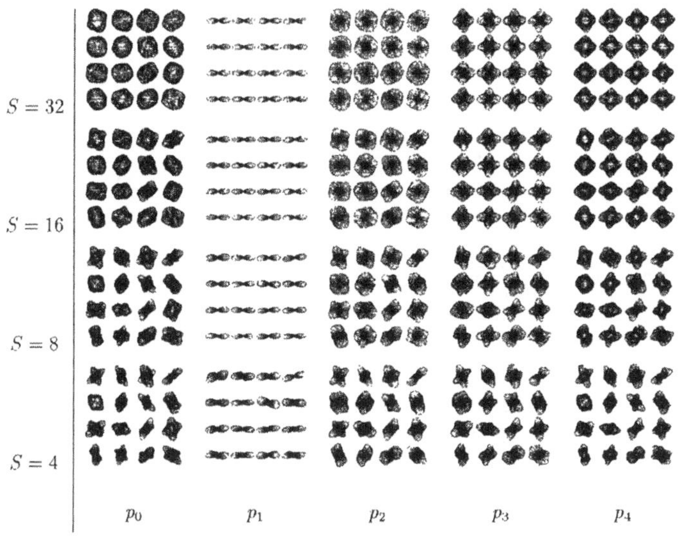

FIGURE 6. Sixteen independent examples of the ODF (in fact, ϕ^5) reconstructed from p_i, $i = 0, \ldots, 4$, for each of various values of the signal-to-noise ratio, S, of $A^*(\mathbf{0})$.

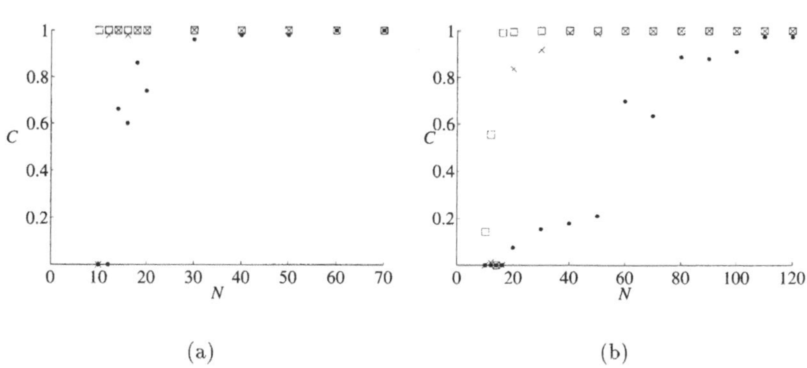

FIGURE 7. Plots of the consistency fraction, C, of **(a)** the PAS and **(b)** the ODF for p_1 (□), p_3 (×), and p_4 (●) against the number of gradient directions, N.

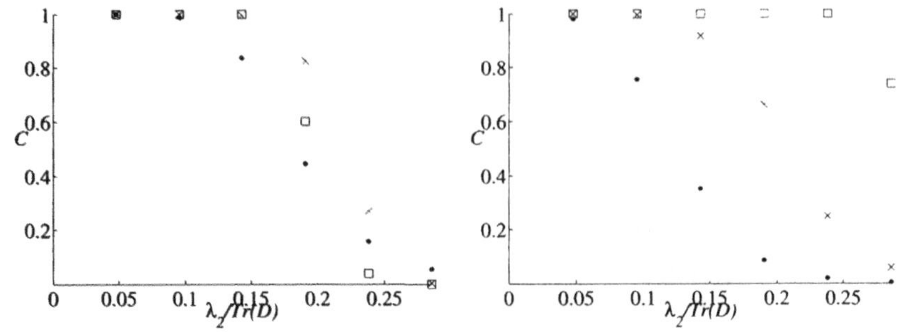

(a) (b)

FIGURE 8. Plots of the consistency fraction, C, of **(a)** the PAS and **(b)** the ODF for p_1 (□), p_3 (×), and p_4 (●) against the anisotropy of the test function characterized by the ratio of the second eigenvalue, λ_2, and the trace of the component diffusion tensors.

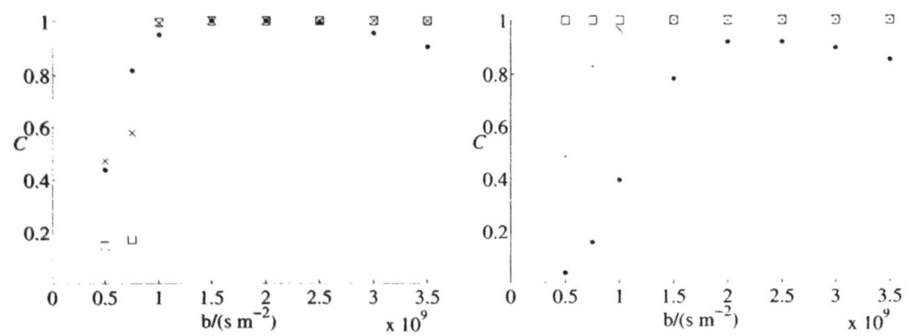

(a) (b)

FIGURE 9. Plots of the consistency fraction, C, of **(a)** the PAS and **(b)** the ODF for p_1 (□), p_3 (×), and p_4 (●) against the b value of the diffusion-weighted measurements in the spherical acquisition scheme.

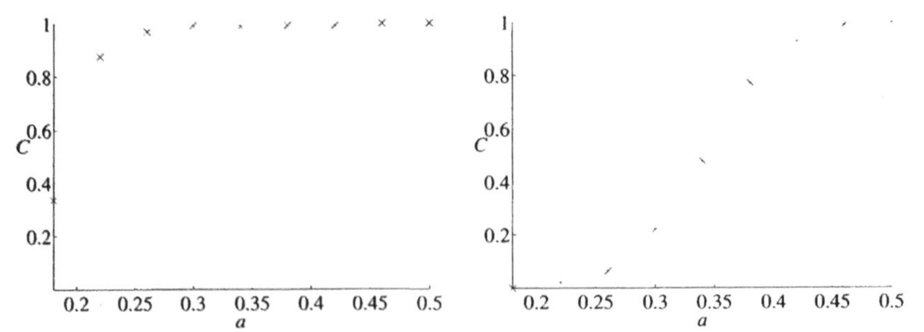

(a) (b)

FIGURE 10. Plots of the consistency fraction, C, of **(a)** the PAS and **(b)** the ODF for p_3 as the mixing parameter, a, varies.

in reference 27 to estimate S at each b, taking T_2 effects into account. The plots show that both algorithms give good consistency for p_1 and p_3 across a wide range of b. For p_4, the results suggest that C peaks at around $b = 2.0 \times 10^9$ s/m^2 for both algorithms. Results in reference 27 suggest that the optimal b is likely to be similar for p_3 as for p_4, but somewhat lower for p_1.

FIGURE 10 shows how the consistency fraction for p_3 depends on the mixing parameter, a. The PASMRI algorithm has $C > 90\%$ for $a \geq 0.26$, while **q**-ball requires $a \geq 0.42$ for $C > 90\%$. FIGURE 11 shows examples of ϕ and $\tilde{p}$ reconstructed from the synthetic data with various values of a. FIGURE 12 shows how the consistency fraction varies with the angle α of rotation of D_2 about the z-axis with D_1 held fixed. For PASMRI, $C > 90\%$ with $\alpha \leq 0.4$. For **q**-ball, C falls below 90% with $\alpha > 0.1$. FIGURE 13 shows examples of ϕ and $\tilde{p}$ reconstructed from the synthetic data with various values of α.

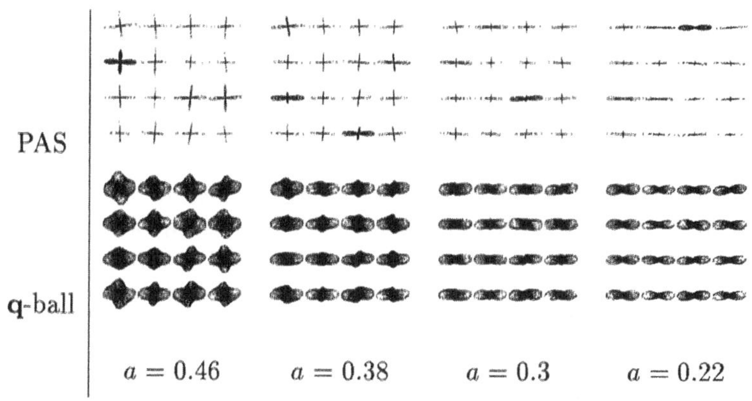

FIGURE 11. Sixteen independent examples of the PAS (*top*) and ODF (*bottom*) reconstructed from p_3 for each of various values of a.

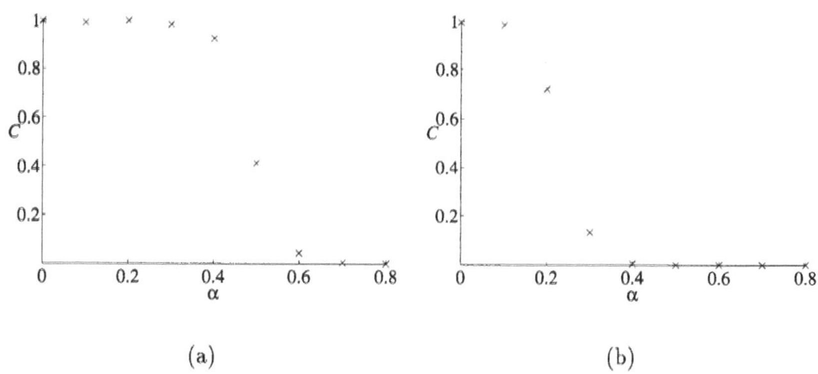

FIGURE 12. Plots of the consistency fraction, C, of **(a)** the PAS and **(b)** the ODF for p_3 as the angle, α, of rotation of D_2 about the z-axis varies.

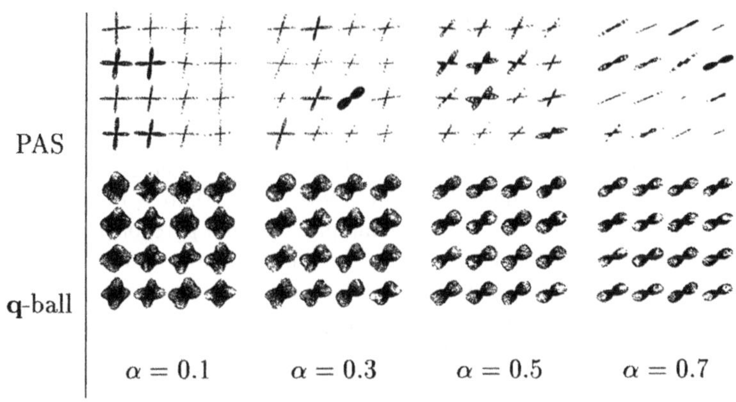

FIGURE 13. Sixteen independent examples of the PAS (*top*) and ODF (*bottom*) reconstructed from p_3 for each of various values of the angle, α, of rotation of D_2 about the z-axis.

DISCUSSION

This chapter has reviewed reconstruction algorithms in diffusion MRI and, in particular, the new generation of multiple-fiber reconstruction algorithms. These algorithms seek to recover the fiber architecture in each voxel of an image volume. A recent set of algorithms, including deconvolution methods, **q**-ball, and PASMRI, estimate features of the particle displacement density that are spherical functions with peaks that provide fiber orientation estimates. The review concentrates on implementation of the methods and highlights their similarity when using a linear basis to represent the spherical functions. Previous results in the literature (e.g., refs. 5 and 21) show that these new multiple-fiber reconstruction algorithms can reveal complex white matter architecture within single voxels. However, most of the testing in the literature has used higher quality data than we can currently acquire routinely over the whole brain on standard hardware. This chapter shows results comparing two of the recent algorithms, PASMRI and **q**-ball, on data from a whole-brain diffusion MRI acquisition sequence that is in routine clinical use. Both methods reveal the expected fiber orientations in regions with a single dominant fiber orientation. The PASMRI algorithm resolves the expected fiber orientations consistently in the fiber crossing at the pons. The **q**-ball algorithm finds the two directions in about half of the voxels in that region.

Simulations show that both PASMRI and **q**-ball pick out genuine angular structure in test functions when the measurements have sufficient signal to noise. The simulations also show that both methods generate spurious angular structure from noisy data synthesized from isotropic test functions. FIGURES 1 and 2 show strong angular structure even in many gray matter and CSF voxels. However, the simulation results show that this structure is likely to be spurious and not to reflect genuine anatomic structure. We can only believe the structure suggested by the PAS or ODF in areas of the brain where we see consistent angular structure over an extended region of neighboring voxels, such as we see from PASMRI in the pons region of the test data. In diffusion-tensor MRI, we can reject fiber orientation estimates from the principal eigenvector by putting a threshold on the anisotropy of the diffusion tensor. However, the simple generalization of this technique for multiple-fiber reconstructions fails. At $S = 16$, the spherical variance of the ODF reconstructed from p_3 using **q**-ball has a mean of 0.109 with a standard deviation of 0.009; for p_2, the mean is 0.153 and the standard deviation is 0.010. For p_4 and p_0, the means and standard deviations are (0.030, 0.005) and (0.027, 0.007), respectively. Reliable methods to distinguish sets of measurements from isotropic and genuinely anisotropic p are an active area of research that will prove instrumental in the future success of multiple-fiber reconstruction algorithms. The voxel-classification algorithm in reference 13 is one technique that can distinguish isotropic and anisotropic voxels; entropy statistics[28] and high-order moments[29] are other possibilities.

The PASMRI algorithm appears to reconstruct directions more consistently than **q**-ball. The simulation results described above show that **q**-ball requires a moderate improvement in data quality to match the performance of PASMRI at reconstructing principal directions: we either increase S by 50% or double the number of measurements, N. However, the simulations suggest that PASMRI will always resolve some borderline cases that **q**-ball cannot. Preliminary experiments with deconvolution methods (not shown) suggest that the basic implementation described in reference 21 produces more false-positive directions than both PASMRI and **q**-ball. Further experiments will compare the techniques more completely.

The PASMRI and **q**-ball algorithms have two fundamental differences that may account for the differences in performance: the two algorithms determine different features of p and they use different kinds of representations for the reconstructed function. Preliminary experiments with the linear version of PASMRI, outlined above, suggest that performance suffers if we use a linear basis in place of the maximum-entropy representation, which can express peaks more precisely. This suggests further that we may improve deconvolution methods and methods that determine the ODF by using an appropriate nonlinear basis.

We have concentrated on extracting a discrete number of fiber orientation estimates from the PAS, ODF, or FOD. However, these features of p potentially reveal more information about the tissue architecture. The theoretical models that underlie the current techniques are a gross simplification of the true particle displacement processes that take place in brain tissue, so any supposed relationship between these features and the true distribution of white matter fibers would require very careful validation. However, we can reasonably expect to derive simple features of the architecture beyond the dominant fiber directions, such as the spread of fiber orientations about the estimates and the fractions of fibers in each dominant direction. So far, we have only scratched the surface of what we can learn from these reconstructions. Parker and Alexander,[30] for example, show that the Hessian of the PAS at the peaks is a reliable predictor of the uncertainty in the peak directions. Further work is required to investigate which features of the microstructure we can determine reliably.

ACKNOWLEDGMENTS

This work has been supported by the MIAS IRC, EPSRC GR/N14248/01 and MRC D2025/31, and EPSRC GR/T22858/01. I also thank Claudia Wheeler-Kingshott from the Institute of Neurology, University College London, who provided the MRI data used to generate the images in this chapter. Dave Tuch, from the Athinoula A. Martinos Imaging Center for Biomedical Imaging, Massachusetts General Hospital, provided the code for the **q**-ball algorithm.

REFERENCES

1. CALLAGHAN, P.T. 1991. Principles of Magnetic Resonance Microscopy. Oxford Sci. Pub. Oxford.
2. ALEXANDER, D.C. 2005. An introduction to computational diffusion MRI: the diffusion tensor and beyond. *In* Visualization and Image Processing of Tensor Fields. Springer-Verlag. Berlin/New York.
3. BASSER, P.J., J. MATIELLO & D. LE BIHAN. 1994. MRI diffusion tensor spectroscopy and imaging. Biophys. J. **66:** 259–267.
4. WEDEEN, V.J., T.G. REESE, D.S. TUCH *et al.* 1999. Mapping fiber orientation spectra in cerebral white matter with Fourier-transform diffusion MRI. *In* Proceedings of the Seventh Annual Meeting of the ISMRM (Philadelphia), p. 321. ISMRM. Berkeley, CA.
5. TUCH, D.S. 2002. Diffusion MRI of complex tissue structure. Ph.D. in Biomedical Imaging at MIT.
6. TUCH, D.S., T.G. REESE, M.R. WIEGELL & V.J. WEDEEN. 2003. Diffusion MRI of complex neural architecture. Neuron **40:** 885–895.
7. JANSONS, K.M. & D.C. ALEXANDER. 2003. Persistent angular structure: new insights from diffusion MRI data. Inverse Probl. **19:** 1031–1046.
8. WANG, Z., B.C. VEMURI, Y. CHEN & T. MARECI. 2003. A constrained variational principle for direct estimation and smoothing of the tensor diffusion field from DWI. *In* Proceedings of the Eighteenth International Conference on Information Processing in Medical Imaging, pp. 660–671. Springer-Verlag. Berlin/New York.
9. BASSER, P.J. & C. PIERPAOLI. 1996. Microstructural and physiological features of tissues elucidated by quantitative diffusion tensor MRI. J. Magn. Reson. **B111:** 209–219.
10. JONES, D.K., M.A. HORSFIELD & A. SIMMONS. 1999. Optimal strategies for measuring diffusion in anisotropic systems by magnetic resonance imaging. Magn. Reson. Med. **42:** 515–525.
11. PARKER, G.J.M., S. LUZZI, D.C. ALEXANDER *et al.* 2005. Non-invasive structural mapping of two auditory-language pathways in the human brain. NeuroImage **24:** 656–666.
12. BLYTH, R., P.A. COOK & D.C. ALEXANDER. 2003. Tractography with multiple fibre orientations. *In* Proceedings of the Eleventh Annual Meeting of the ISMRM (Toronto), p. 240. ISMRM. Berkeley, CA.
13. ALEXANDER, D.C., G.J. BARKER & S.R. ARRIDGE. 2002. Detection and modeling of non-Gaussian apparent diffusion coefficient profiles in human brain data. Magn. Reson. Med. **48:** 331–340.
14. TUCH, D.S., T.G. REESE, M.R. WIEGELL *et al.* 2002. High angular resolution diffusion imaging reveals intravoxel white matter fiber heterogeneity. Magn. Reson. Med. **48:** 577–582.
15. CHEN, Y., W. GUO, Q. ZENG *et al.* 2004. Recovery of intra-voxel structure from HARD DWI. *In* Proceedings of the IEEE International Symposium on Biomedical Imaging (Arlington). IEEE.
16. PASTERNAK, O., N. SOCHEN & Y. ASSAF. 2005. PDE based estimation and regularization of multiple diffusion tensor fields. *In* Visualization and Image Processing of Tensor Fields. Springer-Verlag. Berlin/New York.
17. BEHRENS, T.E.J., M.W. WOLLRICH, M. JENKINSON *et al.* 2003. Characterization and propagation of uncertainty in diffusion-weighted MR imaging. Magn. Reson. Med. **50:** 1077–1088.
18. ASSAF, Y., R.Z. FREIDLIN, G.K. ROHDE & P.J. BASSER. 2004. New modelling and experimental framework to characterize hindered and restricted water diffusion in brain white matter. Magn. Reson. Med. **52:** 965–978.
19. NEUMAN, C.H. 1974. Spin echo of spins diffusion in a bounded medium. J. Chem. Phys. **60:** 4508–4511.

20. ANDERSON, A. & Z. DING. 2002. Sub-voxel measurement of fiber orientation using high angular resolution diffusion tensor imaging. *In* Proceedings of the Tenth Annual Meeting of the ISMRM (Honolulu), p. 440. ISMRM. Berkeley, CA.
21. TOURNIER, J-D., F. CALAMANTE, D.G. GADIAN & A. CONNELLY. 2004. Direct estimation of the fiber orientation density function from diffusion-weighted MRI data using spherical deconvolution. NeuroImage **23:** 1176–1185.
22. HELGASON, S. 1999. The Radon Transform. Birkhäuser. Basel.
23. LIN, C.P., W.Y.I. TSENG, L. KUO *et al.* 2003. Mapping orientation distribution function with spherical encoding. *In* Proceedings of the Eleventh Annual Meeting of the ISMRM (Toronto), p. 2120. ISMRM. Berkeley, CA.
24. OZARSLAN, E., B.C. VEMURI & T. MARECI. 2004. Fiber orientation mapping using generalized diffusion tensor imaging. *In* Proceedings of the IEEE International Symposium on Biomedical Imaging (Arlington). IEEE.
25. SYMMS, M.R., G.J. BARKER, F. FRANCONI & C.A. CLARK. 1997. Correction of eddy-current distortions in diffusion-weighted echo-planar images with a two-dimensional registration technique. *In* Proceedings of the Fifth Annual Meeting of the ISMRM (Vancouver), p. 1723. ISMRM. Berkeley, CA.
26. PRESS, W.H., S.A. TEUKOLSKY, W.T. VETTERING & B.P. FLANNERY. 1988. Numerical Recipes in C. Press Syndicate of the University of Cambridge. New York.
27. ALEXANDER, D.C. & G.J. BARKER. 2005. Optimal imaging parameters for fibre-orientation estimation in diffusion MRI. NeuroImage **27:** 357–367.
28. OZARSLAN, E. & T.H. MARECI. 2003. Anisotropy as a certainty measure in terms of entropy. *In* Proceedings of the Eleventh Annual Meeting of the ISMRM (Toronto), p. 249. ISMRM. Berkeley, CA.
29. CHABERT, S., C.C. MECA & D. LE BIHAN. 2004. Relevance of information about the diffusion distribution *in vivo* given by kurtosis in *q*-space imaging. *In* Proceedings of the Twelfth Annual Meeting of the ISMRM (Kyoto), p. 1238. ISMRM. Berkeley, CA.
30. PARKER, G.J.M. & D.C. ALEXANDER. 2005. Probabilistic anatomic connectivity derived from the microscopic persistent angular structure of cerebral tisue. Philos. Trans. R. Soc. Lond. **360:** 893–902.

The Application of DTI to Investigate White Matter Abnormalities in Schizophrenia

MAREK KUBICKI,[a,b] CARL-FREDRIK WESTIN,[b,c] ROBERT W. McCARLEY,[a] AND MARTHA E. SHENTON[a,b]

[a]*Clinical Neuroscience Division, Laboratory of Neuroscience, Boston VA Healthcare System–Brockton Division, Department of Psychiatry, Harvard Medical School, Brockton, Massachusetts, USA*

[b]*Surgical Planning Laboratory, MRI Division, Department of Radiology, Brigham and Women's Hospital, Harvard Medical School, Boston, Massachusetts, USA*

[c]*Laboratory of Mathematics in Imaging, MRI Division, Department of Radiology, Brigham and Women's Hospital, Harvard Medical School, Boston, Massachusetts, USA*

ABSTRACT: Schizophrenia is a serious and disabling mental disorder that affects approximately 1% of the general population, with often devastating effects on the psychological and financial resources of the patient, family, and larger community. The etiology of schizophrenia is not known, although it likely involves several interacting biological and environmental factors that predispose an individual to schizophrenia. However, although the underlying pathology remains unknown, it has been believed that brain abnormalities would ultimately be linked to the etiology of schizophrenia. This theory was rekindled in the 1970s, when the first computer-assisted tomography (CT) study showed enlarged lateral ventricles in schizophrenia. Since that time, there have been many improvements in MR acquisition and image processing, including the introduction of positron emission tomography (PET), followed by functional MR (fMRI), and diffusion tensor imaging (DTI). These advances have led to an appreciation of the critical role that brain abnormalities play in schizophrenia. While structural MRI has proven to be useful in investigating and detecting gray matter abnormalities in schizophrenia, the investigation of white matter has proven to be more challenging as white matter appears homogeneous on conventional MRI and the fibers connecting different brain regions cannot be appreciated. With the development of DTI, we are now able to investigate white matter abnormalities in schizophrenia.

KEYWORDS: schizophrenia; diffusion tensor imaging (DTI); frontal-temporal connections; white matter

Address for correspondence: Dr. Marek Kubicki, Department of Psychiatry–116A, Boston VA Healthcare System–Brockton Division, Harvard Medical School, 940 Belmont Street, Brockton, MA 02301. Voice: 508-583-4500, ext. 1371; fax: 508-580-0059.
kubicki@bwh.harvard.edu

Ann. N.Y. Acad. Sci. 1064: 134–148 (2005). © 2005 New York Academy of Sciences.
doi: 10.1196/annals.1340.024

INTRODUCTION

Schizophrenia is a serious and disabling mental disorder that affects approximately 1% of the general population, with often devastating effects on the psychological and financial resources of the patient, family, and larger community. It generally afflicts individuals in early adulthood, at a time when they are on the threshold of entering the most productive and formative years of life. The overt, or positive, symptoms of the disorder include auditory hallucinations, disordered thinking, and delusions, while the negative symptoms include avolition, anhedonia, blunted affect, and apathy. Additionally, broad areas of functioning are frequently disturbed including attention, memory, emotion, motivation, thought and language processes, social functioning, and mood regulation.

The etiology of schizophrenia is not known, although it likely involves several interacting biological (e.g., genetic and neurodevelopmental) and environmental factors (e.g., viral infection, fetal insult, drug abuse) that predispose an individual to schizophrenia. Of note, however, although the underlying pathology remains unknown, both Kraepelin[1] and Bleuler[2] (pioneers of schizophrenia) believed that brain abnormalities would ultimately be linked to the etiology of schizophrenia. This theory was rekindled in the 1970s, when the first computer-assisted tomography (CT) study showed enlarged lateral ventricles in schizophrenia.[3] With the introduction of magnetic resonance imaging (MRI) studies, the first MRI study of schizophrenia was conducted in 1984.[4] Since that time, there have been many improvements in MR acquisition and image processing, including the introduction of positron emission tomography (PET), followed by functional MR (fMRI) and diffusion tensor imaging (DTI), all of which have enabled us to exploit more fully information contained in MR and other medical images. These advances have led to an appreciation of the critical role that brain abnormalities play in schizophrenia.

MRI findings in schizophrenia include lateral ventricle enlargement, medial temporal lobe volume reduction (including amygdala-hippocampal complex and parahippocampal gyrus), neocortical superior temporal gyrus volume reduction, frontal and parietal lobe abnormalities, and subcortical abnormalities affecting the cavum septi pellucidi, basal ganglia, corpus callosum, thalamus, and cerebellum (see review in ref. 5). These findings further suggest the involvement of a large number of functionally related brain regions. Of note here, Wernicke[6] and Kraepelin[1] both suggested that schizophrenia might be a disease of insufficient or ineffective communication between these brain regions. This hypothesis has been refueled by recent functional imaging studies, which have demonstrated functional connectivity abnormalities between temporolimbic and prefrontal regions (e.g., refs. 7–11). These findings, as well as recent postmortem and genetic studies that demonstrate myelin-related abnormalities in schizophrenia,[12–14] suggest that not only functional, but also anatomical disconnection between brain regions may be involved in schizophrenia. This latter speculation has led to an interest in investigating white matter fiber tract abnormalities in schizophrenia. Here, the focus is on long, association fiber tracts, consisting of heavily myelinated axons, which interconnect distant brain regions. Of particular interest, and based on earlier speculations, are white matter fiber tracts connecting the frontal and temporal lobes.

Finally, while structural MRI has proven to be useful in investigating and detecting gray matter abnormalities in schizophrenia, the investigation of white matter has proven to be more challenging as white matter appears homogeneous on conventional MRI and the fibers connecting different brain regions cannot be appreciated. With the development of diffusion tensor imaging (DTI), we are now able to investigate white matter abnormalities in schizophrenia.

DTI AND SCHIZOPHRENIA

DTI is the first imaging tool that makes it possible to evaluate white matter fiber tracts in the brain. Moreover, the ability to visualize and to quantify white matter fiber tracts makes DTI a particularly attractive research tool. The first DTI study of schizophrenia was published in 1998,[15] only 4 years after DTI was introduced.[16] Since then, over 20 DTI studies of schizophrenia have been conducted.

Early applications of DTI to schizophrenia investigations focused on the quantification and group comparison of large regions of white matter and thus did not focus on specific anatomically defined fiber tracts (e.g., refs. 17–19). With more recent advances in DT image processing, the field is clearly moving towards a focus on the quantification and group comparison of specific fiber tracts in the brain.

This work has developed along two main directions: (1) a voxel by voxel evaluation of whole brain white matter and (2) measurements of specific fiber bundles based on their anatomical definition. The first type of analysis involves the coregistration of each subject's scan to a common template and can be performed without a priori hypotheses. This method is very popular in gray matter volumetric studies (termed "voxel-based morphometry analyses") and it has also been applied to FA maps in schizophrenia. Findings from these studies show decreased FA within the cingulate bundle,[20,21] uncinate fasciculus,[20] arcuate fasciculus,[21] and corpus callosum[22] in schizophrenia compared to control subjects. Results, however, are not consistent, with some negative findings.[23] Low resolution of the images, heavy partial volume effects, and imperfect registration of the fiber tracts likely contribute to such inconsistent findings. Of note, this method is still in the development phase and, currently, intensity-based registration of FA maps to a common template is being replaced by more precise, higher-order multicomponent registration (e.g., ref. 24).

The second type of analysis involves focusing on testing specific hypotheses about particular connections that may be disrupted in schizophrenia. These studies use small regions of interest (ROIs), defined either manually or automatically, and based on anatomical definitions of the fiber tract. These ROIs have been used to investigate FA within the cingulum bundle, uncinate fasciculus, corpus callosum, and cerebellar peduncles. Because the anatomical definition of ROIs is difficult, particularly given that early DTI images had low resolution, and the measurements were generally taken from one or a small number of slices, results tended to also be equivocal (for review, see ref. 25). This type of analysis, however, is rapidly developing and now takes advantage of new image processing tools such as fiber tracking, where specific fiber bundles can be parcellated based on their anatomical connections (see FIG. 1 and FUTURE DIRECTIONS, below, for further details). Note that, in FIGURE 1, fiber tracts have been delineated using in-house software (www.slicer.org). Here, seed points have been placed manually on a single coronal slice, within three fiber tracts (cingulate bundle, fornix, and uncinate fasciculus), and fiber tracking has been performed using the Runge-Kutta algorithm.

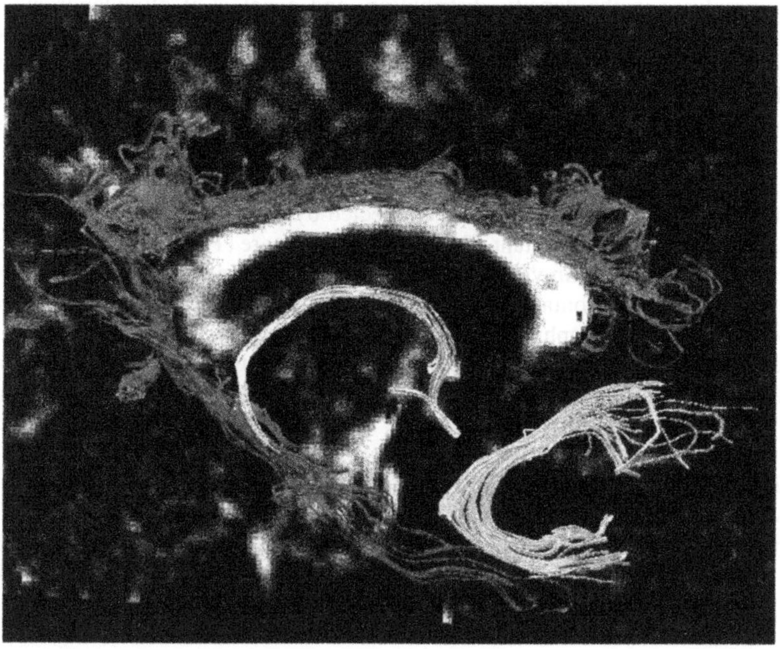

FIGURE 1. Three large frontal-temporal fiber tracts are visualized: the cingulum bundle in red, the fornix in yellow, and the uncinate fasciculus in green.

FRONTOTEMPORAL CONNECTIONS IN SCHIZOPHRENIA

Evidence from the Literature

Our laboratory has used DTI techniques to investigate frontotemporal connections in schizophrenia. The focus of this work has evolved from evidence in the literature, reviewed above, which suggests the importance of temporal and frontal lobe abnormalities in MRI studies of schizophrenia (see also review in ref. 5). While most of these studies highlight gray matter abnormalities in schizophrenia, several studies have reported prefrontal white matter volume reduction,[26,27] temporal white matter volume reduction,[28–30] or a relationship between prefrontal white matter volume reduction and temporal gray matter volume reduction (e.g., STG, amygdala-hippocampal complex, and parahippocampal gyrus).[26,31]

Of further note, Weinberger and coworkers[8] were among the first to report an association between functional deficits observed in the prefrontal cortex in schizophrenic patients and structural abnormalities observed in the medial temporal lobe (i.e., reduced volume of the hippocampus). Since then, several functional studies have reported a disruption of functional connectivity between frontal and temporal lobes in schizophrenia (e.g., refs. 32–35), again emphasizing the likely importance of frontotemporal circuitry abnormalities in the pathophysiology of schizophrenia.

Postmortem investigations of schizophrenia have also reported abnormalities in white matter neuronal distribution in both prefrontal and temporal regions, as well as abnormalities in the density, distribution, and genetics of oligodendrocytes in schizophrenia, further suggesting a disruption in frontal-temporal connections. Moreover, oligodendrocyte abnormalities, involved in the formation of a protective sheath around the axons, as well as an abnormal distribution of interstitial neurons, responsible for neuronal migration guidance, suggest that perhaps not only functional, but also structural disconnectivity may be involved in the pathophysiology of schizophrenia.

The frontal and temporal lobes are connected by multiple long, association fiber tracts. Among the most important of these connections are (1) the uncinate fasciculus, a fiber tract connecting amygdala, uncus, and temporal pole with the subcallosal region and the orbitofrontal gyrus,[36–38] and involved in decision-making, social behavior, and autobiographical and episodic memory—all functions that are disturbed in schizophrenia; (2) the cingulum bundle, a fiber tract interconnecting limbic structures (e.g., dorsolateral prefrontal cortex, cingulate gyrus, parahippocampal gyrus) and involved in attention, emotions, spatial orientation, and memory—functions that are also disrupted in schizophrenia; (3) the fornix, a fiber tract connecting the hippocampus with other brain regions (e.g., thalamus, prefrontal cortex) and involved in spatial learning and memory—functions disrupted in schizophrenia; and finally (4) the arcuate fasciculus, a fiber tract that connects Broca's and Wernicke's areas and is involved in language processing—this fiber tract may also be involved in language and thought disturbances observed in schizophrenia.

Findings from Our Laboratory

Thus far, in our laboratory, we have used DTI to examine the uncinate fasciculus and the cingulate bundle in schizophrenia, and in an additional study we have demonstrated arcuate fasciculus abnormalities in schizophrenia using a voxel-based analysis of DTI. Below, we review these findings.

The uncinate fasciculus (UF), one of the largest connections between the frontal and temporal lobes, may play an important role in emotions, decision-making, and episodic memory, and in pathophysiology of schizophrenia. The UF pathology in schizophrenia has been described in two DTI investigations.[20,39] The ROI study from our laboratory showed a lack of left greater than right fractional anisotropy (FA) asymmetry in schizophrenia that was present in control subjects.[39] In addition, a voxel-based morphometry analysis of FA maps, in controls and schizophrenics, showed FA abnormalities in the UF in schizophrenia.[20] FIGURE 2 shows an example of both an FA and a tensor map showing the UF in a normal control subject.

The cingulum bundle (CB) is a fiber tract interconnecting limbic structures involved in attention, emotions, spatial orientation, and memory, and has been implicated in numerous studies of schizophrenia. Noteworthy are structural MR studies in schizophrenia that show cingulate gyri volume decrease,[28,40,41] histopathological studies that show changes in neuronal organization within the cingulate gyrus,[42–44] and functional studies that show abnormal activation of the cingulate gyri in numerous tasks.[45–47] In addition, functional connectivity, as measured by fMRI, tends to be abnormally modulated by the cingulate gyrus,[48,49] thus furnishing a possible basis for hallucinations. CB integrity abnormalities have thus far also been described in three DTI ROI studies.[50–52] Our laboratory used automatic ROI definition and measured CB on 8 coronal slices covering the anterior part of the fiber bundle. We found

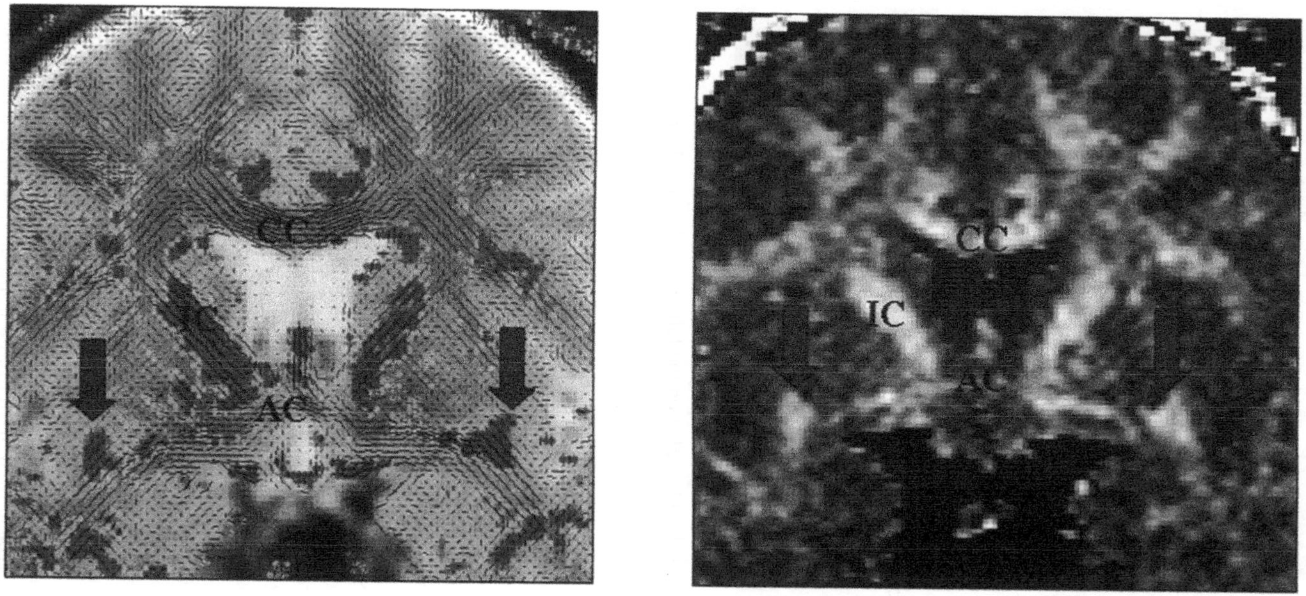

FIGURE 2. The *left panel* shows a tensor map with the largest in-plane component of the diffusion in blue and the out-of-plane component in orange. The *arrows* point to the uncinate fasciculus. The *right panel* shows an FA map: CC, corpus callosum; IC, internal capsule; AC, anterior commissure.

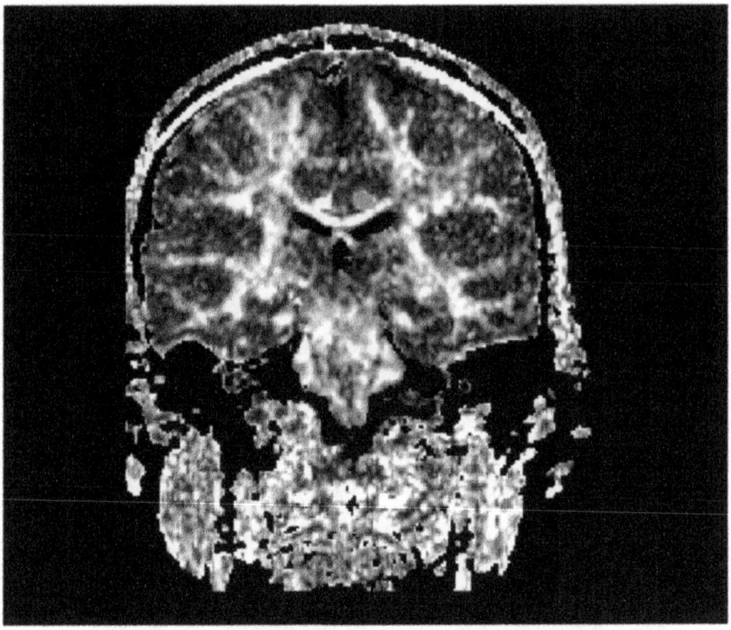

FIGURE 3. An FA map is displayed of a normal control subject, with the ROI for the cingulum bundle displayed as blue on the left and yellow on the right.

decreased mean FA as well as decreased area of this bundle in schizophrenia, bilaterally. In addition, left > right asymmetry was present in both groups. Moreover, decreased diffusion in CB was associated with poorer performance on the Wisconsin Card Sorting Test, which is heavily dependent upon intact communication between prefrontal and anterior cingulate gyri. FIGURE 3 provides an example of a CB ROI drawn in a control subject superimposed over the FA map.

The arcuate fasciculus connects Wernicke's and Broca's areas and is the major language processing tract in the brain. Patients diagnosed with schizophrenia often suffer from deficits in verbal memory, auditory hallucinations, delusions, and thought disorder, with the latter likely related to language deficits observed in behavioral studies, although verbal memory and auditory hallucinations may also be linked to areas of the brain involving auditory perception and language. Functional studies show abnormal activation within Broca's and Wernicke's areas in schizophrenia during various verbal tasks.[35,53–56] Of note here, abnormalities within the arcuate fasciculus have been reported in schizophrenia in two separate voxel-based FA analyses.[20,21] The Hubl *et al.* study[21] also showed an association between arcuate fasciculus abnormalities and auditory hallucinations. Studies from our laboratory also indicate arcuate fasciculus abnormalities in schizophrenia. More specifically, a VBM study, using a higher-order tensorial coregistration method, showed a left arcuate fasciculus FA decrease in schizophrenia compared with control subjects. FIGURE 4 shows results of a voxel-based comparison between controls and schizophrenics,[57] where FA decrease is present within the arcuate fasciculus in schizophrenics.

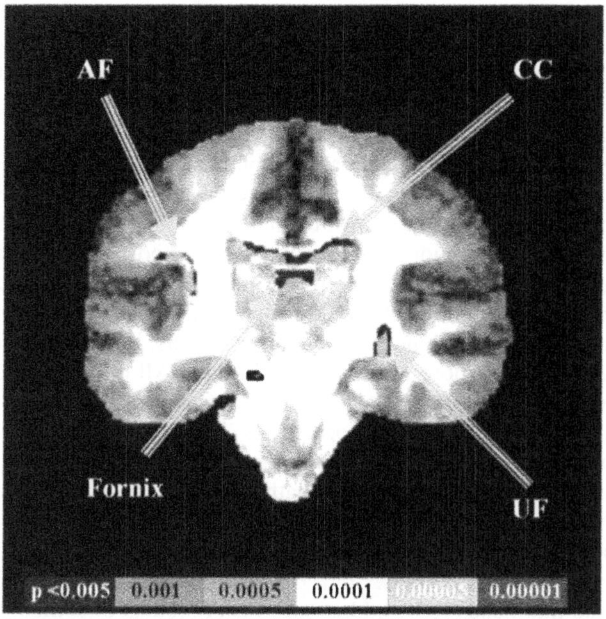

FIGURE 4. Results of a voxel-based analysis show the FA differences between controls and schizophrenics, located primarily in the arcuate fasciculus (AF), corpus callosum (CC), fornix, and uncinate fasciculus (UF).

We have also begun to evaluate the fornix in patients diagnosed with schizophrenia compared with normal controls, where we have shown a decrease in FA in the patient group.[58]

Thus, in summary, empirical data have amassed that clearly support the notion that frontal-temporal connectivity in the brain is likely disrupted in schizophrenia. As DTI provides us with the opportunity to directly visualize and measure these connections, it has become important for studying white matter fiber tracts that connect frontal and temporal lobes. To date, all our investigations have used a small ROI approach, measuring only a small segment of the fiber connection of interest. As this is just the beginning of such investigations, it is important now to develop tools that will enable us to improve the methods used in investigating white matter abnormalities in schizophrenia as well as in other neuropsychiatric disorders (see below under FUTURE DIRECTIONS FOR DTI STUDIES OF SCHIZOPHRENIA).

LIMITATIONS OF DTI IN SCHIZOPHRENIA STUDIES

To date, over 20 studies have investigated white matter abnormalities in schizophrenia using DTI. Results of these studies are frequently inconsistent as there are no accepted gold standards for data acquisition, processing, and analysis. In addition, small sample size, different ROI definitions, low S/N ratio, different number of acquisition angles, and different scanner gradient performance and field strength, as well as partial volume effect due to the low image resolution, all affect the final results. DTI findings in schizophrenia must therefore be understood within the context of the acquisition parameters used as well as in the context of the different methodological strategies adopted by different investigators.

Positive DTI findings in schizophrenia obtained to date are interesting, but methodological limitations, discussed here as well as elsewhere, limit our ability to conclude, with certainty, the nature of white matter pathology in schizophrenia. There is no question that DTI provides a new window of opportunity to evaluate white matter in a manner not possible with conventional MRI. However, we do not as yet know whether the abnormalities observed reflect a decrease in number of axons, a decrease in axonal diameter, thinner myelination sheaths, less coherent fibers, more fiber crossings, or simply more noise in the DTI data. All of these possibilities are supported to some extent by empirical data that show abnormal neuronal number and density,[59,60] as well as myelin abnormalities in both postmortem anatomical studies[12,61] and genetic studies.[13,62,63] Moreover, DTI is not sufficiently specific to differentiate among these aforementioned pathologic processes, and thus we need to consider additional imaging techniques such as magnetization transfer imaging (MTI), MR spectroscopy, and relaxation time measurements in order to increase the specificity of white matter findings.

Another issue that needs to be addressed in DTI studies of schizophrenia is the effect of medication. For example, several DTI studies report a correlation between medication and FA in some regions of the brain,[64,65] while other studies report no such association.[15,19,52,66] Further clarification regarding medication effects is thus needed in order to understand further white matter pathology of schizophrenia. This suggests that unmedicated first-episode patients, a group that has not been studied with DTI, should be investigated.

Finally, there is evidence to suggest that some brain abnormalities progress over time in schizophrenia.[67] In order to better understand such changes over time, and in particular in order to better understand changes observed in DTI, we need to look more closely at the effects of aging on DTI in normal control subjects and adjust for these effects when searching for neurodegenerative signs in schizophrenia.

FUTURE DIRECTIONS FOR DTI STUDIES OF SCHIZOPHRENIA

Over the last 5 years, DTI has become the most important imaging technique to investigate white matter changes in schizophrenia. This technique, however, is not free of limitations, the major two being the relatively low resolution and image distortions. Use of high magnet fields, and the development of fast, low distortion techniques that produce high resolution and high SNR diffusion images, will thus be important in furthering our understanding of white matter pathology in schizophrenia.

Fiber tracking is one of the most promising image processing techniques in terms of both the visualization of fiber pathways in the brain as well as in the quantitative analysis of specific fiber bundles. Future studies will likely be heavily dependent upon the further development and validation of these methods. Their application to schizophrenia will also likely reveal disruptions in connections between brain regions that heretofore could not be evaluated quantitatively. Such studies will also

FIGURE 5. Three-dimensional model of the uncinate fasciculus (UF) is created using fiber tracking. This fiber tracking was created using two ROIs (red and yellow circles on the picture). The UF is colored according to the FA values (green: higher FA values; blue: lower FA values).

likely benefit from the acquisition of small, isotropic voxels, with diffusion encoded in multiple directions. With reliable fiber tracking, it will become easier to define and to measure whole fibers of interest.

FIGURE 5 provides an example of work that will be critical for evaluating fiber bundles in schizophrenia. Here, an automated fiber tracking procedure (described in detail in ref. 24) was used to create all of the major white matter fiber tracts in the brain. Two ROIs were then manually drawn in order to define and to extract the entire uncinate fasciculus fiber tract, which previously was evaluated using only one coronal slice.[39]

Another example of the way fiber tracking will be useful in future studies is displayed in FIGURE 6. Here, manual seeding within the splenium of the corpus callosum reveals fiber traces that are located in close proximity to each other within the corpus callosum, but which nonetheless belong to fiber bundles that connect anatomically distinct brain areas. This work, however, will need to be validated, perhaps in anatomical postmortem studies.

The examples in FIGURES 5 and 6 show the potential strength of using tractography to define fiber bundles of interest. The regions defined by these "fibers of interests", or FOIs, can then further be analyzed using fiber analysis methods such as performing statistics on diffusion-related values along the fibers. If the spatial extent of the FOI region is projected back to the voxel space, that will define the collection of voxels that is traversed with at least one fiber from the FOI, and this set of voxels can then be analyzed using traditional voxel-based methods.

FIGURE 6. Fibers traveling through the splenium of the corpus callosum are shown, where fibers connecting the left and right occipital lobe are displayed in green, fibers interconnecting lateral temporal regions are displayed in red, and fibers connecting medial temporal regions are displayed in blue. Additionally, the genu of the corpus callosum is labeled as "G", the body as "B", the isthmus as "I", and the splenium as "S".

Future DTI analyses will also likely be performed in some fiber-related space, utilizing both fiber tract atlases and precise registration strategies, which will further increase the anatomical accuracy of such studies and allow a direct comparison of the whole fiber bundles. There is, however, still an open question as to how best to construct fiber spaces and atlases. Concepts from traditional voxel-based approaches do not directly translate to fibers. For example, averaging of information in voxel space normally produces an output that is also voxel-based (i.e., average values of a set of aligned MRI scans will produce a new smoother MRI looking scan). In contrast, the average of a set of fibers does not straightforwardly produce a set of smooth fibers. New methods that will be able to infer local coordinate systems defining anatomically relevant positions along the fibers are thus a direction of research that will likely produce interesting results. Automated methods for fiber clustering are an important direction for this type of work[68,69] (FIG. 7).

Future studies investigating white matter abnormalities in schizophrenia should also be combined not only with other structural imaging techniques, but also with functional MRI and PET imaging in order to characterize and to understand more fully the relationship between functional and structural abnormalities in schizophrenia. Using multiple images to evaluate brain abnormalities in schizophrenia will provide a wealth of information that will likely lead to an increased understanding of the neuropathology of schizophrenia, and perhaps even targeted treatments, as we begin to understand brain circuitry better with the use of multiple imaging techniques.

Finally, DTI was introduced to clinical imaging in 1995 and has thus only begun to be used to explore white matter abnormalities in schizophrenia. Nevertheless, it has fast become one of the most popular imaging techniques used in schizophrenia

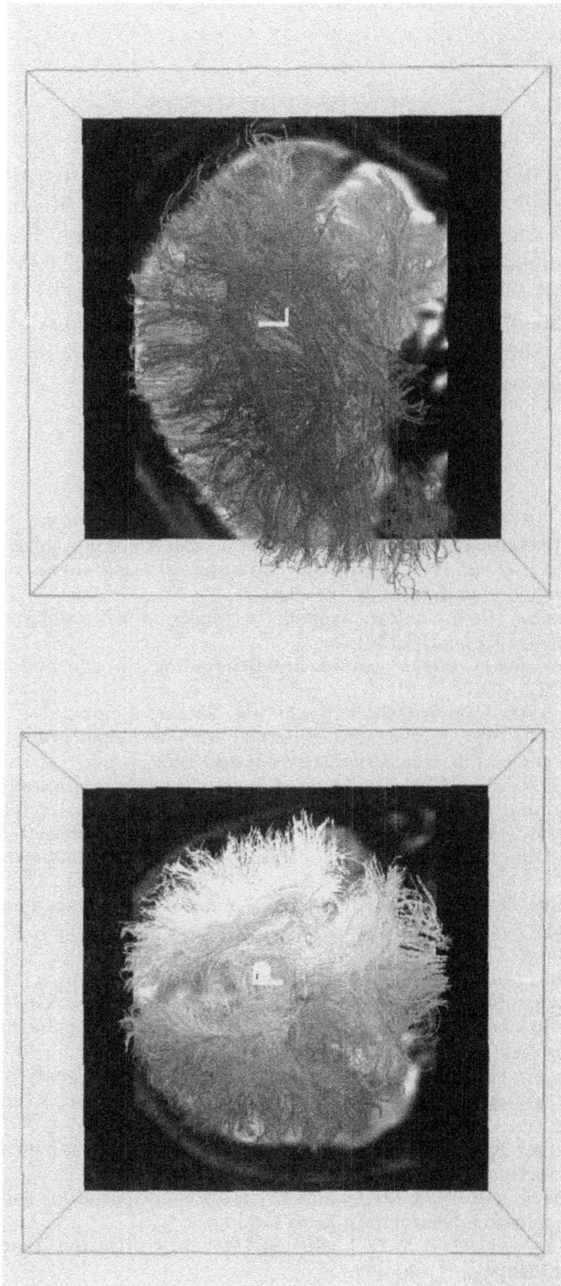

FIGURE 7. Fiber traces from a human brain are colored such that fibers with similar endpoints are assigned similar colors. Slices from a T2-weighted volume add additional understanding of the anatomy. The view of the fiber traces on the left is coronal, while the view on the right is sagittal. [Courtesy of Anders Brun.]

research. With still newer technological advances, we will likely learn even more about white matter abnormalities in this devastating disorder.

ACKNOWLEDGMENTS

We would like to thank Marie Fairbanks for her administrative assistance. Additionally, we gratefully acknowledge the support of the National Alliance for Research on Schizophrenia and Depression (to M. Kubicki), The Wodecroft Foundation (to M. Kubicki), the National Institutes of Health (R03 MH 068464-02 to M. Kubicki, R01 NS 39335 and R01 MH 40799 to R. W. McCarley, and K02 MH 01110 and R01 MH 50747 to M. E. Shenton), and the Department of Veterans Affairs Merit and REAP Awards (to M. E. Shenton and R. W. McCarley). In addition, we thank Anders Brun for his figure showing fiber traces (FIG. 7).

REFERENCES

1. KRAEPELIN, E. 1919/1971. Dementia Praecox. Churchill Livingstone. New York.
2. BLEULER, M. 1911/1954. The concept of schizophrenia. Am. J. Psychiatry **111:** 382–383.
3. JOHNSTONE, E.C. *et al.* 1976. Cerebral ventricular size and cognitive impairment in chronic schizophrenia. Lancet **2:** 924–926.
4. SMITH, R.C. *et al.* 1984. Nuclear magnetic resonance in schizophrenia: a preliminary study. Psychiatry Res. **12:** 137–147.
5. SHENTON, M.E. *et al.* 2001. A review of MRI findings in schizophrenia. Schizophr. Res. **49:** 1–52.
6. WERNICKE, C. 1906. Grundrisse der Psychiatrie. Thieme. Leipzig.
7. WEINBERGER, D.R. 1987. Implications of normal brain development for the pathogenesis of schizophrenia. Arch. Gen. Psychiatry **44:** 660–669.
8. WEINBERGER, D.R. *et al.* 1992. Evidence of dysfunction of a prefrontal-limbic network in schizophrenia: a magnetic resonance imaging and regional cerebral blood flow study of discordant monozygotic twins. Am. J. Psychiatry **149:** 890–897.
9. WEINBERGER, D.R. 1995. From neuropathology to neurodevelopment. Lancet **346:** 552–557.
10. YURGELUN-TODD, D.A. *et al.* 1996. Functional magnetic resonance imaging of schizophrenic patients and comparison subjects during word production. Am. J. Psychiatry **153:** 200–205.
11. SHERGILL, S.S. *et al.* 2000. Mapping auditory hallucinations in schizophrenia using functional magnetic resonance imaging. Arch. Gen. Psychiatry **57:** 1033–1038.
12. URANOVA, N. *et al.* 2001. Electron microscopy of oligodendroglia in severe mental illness. Brain Res. Bull. **55:** 597–610.
13. HAKAK, Y. *et al.* 2001. Genome-wide expression analysis reveals dysregulation of myelination-related genes in chronic schizophrenia. Proc. Natl. Acad. Sci. USA **98:** 4746–4751.
14. DAVIS, K.L. *et al.* 2003. White matter changes in schizophrenia: evidence for myelin-related dysfunction. Arch. Gen. Psychiatry **60:** 443–456.
15. BUCHSBAUM, M.S. *et al.* 1998. MRI white matter diffusion anisotropy and PET metabolic rate in schizophrenia. Neuroreport **9:** 425–430.
16. BASSER, P.J., J. MATTIELLO & D. LEBIHAN. 1994. MR diffusion tensor spectroscopy and imaging. Biophys. J. **66:** 259–267.
17. LIM, K.O. *et al.* 1999. Compromised white matter tract integrity in schizophrenia inferred from diffusion tensor imaging. Arch. Gen. Psychiatry **56:** 367–374.
18. HOPTMAN, M.J. *et al.* 2002. Frontal white matter microstructure, aggression, and impulsivity in men with schizophrenia: a preliminary study. Biol. Psychiatry **52:** 9–14.

19. STEEL, R.M. *et al.* 2001. Diffusion tensor imaging (DTI) and proton magnetic resonance spectroscopy (H MRS) in schizophrenic subjects and normal controls. Psychiatry Res. Neuroimaging Sect. **106:** 161–170.
20. BURNS, J. *et al.* 2003. Structural disconnectivity in schizophrenia: a diffusion tensor magnetic resonance imaging study. Br. J. Psychiatry **182:** 439–443.
21. HUBL, D. *et al.* 2004. Pathways that make voices: white matter changes in auditory hallucinations. Arch. Gen. Psychiatry **61:** 658–668.
22. ARDEKANI, B.A. *et al.* 2003. MRI study of white matter diffusion anisotropy in schizophrenia. Neuroreport **14:** 2025–2029.
23. FOONG, J. *et al.* 2002. Investigating regional white matter in schizophrenia using diffusion tensor imaging. Neuroreport **13:** 333–336.
24. PARK, H.J. *et al.* 2003. Spatial normalization of diffusion tensor MRI using multiple channels. Neuroimage **20:** 1995–2009.
25. KUBICKI, M. *et al.* 2002. Diffusion tensor imaging and its application to neuropsychiatric disorders. Harv. Rev. Psychiatry **10:** 324–336.
26. BREIER, A. *et al.* 1992. Brain morphology and schizophrenia: a magnetic resonance imaging study of limbic, prefrontal cortex, and caudate structures [see comments]. Arch. Gen. Psychiatry **49:** 921–926.
27. PAILLERE-MARTINOT, M. *et al.* 2001. Cerebral gray and white matter reductions and clinical correlates in patients with early onset schizophrenia. Schizophr. Res. **50:** 19–26.
28. SIGMUDSSON, T. *et al.* 2001. Structural abnormalities in frontal, temporal, and limbic regions and interconnecting white matter tracts in schizophrenic patients with prominent negative symptoms. Am. J. Psychiatry **158:** 234–243.
29. SPALLETTA, G. *et al.* 2003. Chronic schizophrenia as a brain misconnection syndrome: a white matter voxel-based morphometry study. Schizophr. Res. **64:** 15–23.
30. MITELMAN, S.A. *et al.* 2003. MRI assessment of gray and white matter distribution in Brodmann's areas of the cortex in patients with schizophrenia with good and poor outcomes. Am. J. Psychiatry **160:** 2154–2168.
31. WIBLE, C.G. *et al.* 1995. Prefrontal cortex and schizophrenia: a quantitative magnetic resonance imaging study. Arch. Gen. Psychiatry **52:** 279–288.
32. FRISTON, K.J. & C.D. FRITH. 1995. Schizophrenia: a disconnection syndrome? Clin. Neurosci. **3:** 89–97.
33. MCGUIRE, P.K. & C.D. FRITH. 1996. Disordered functional connectivity in schizophrenia [editorial]. Psychol. Med. **26:** 663–667.
34. FLETCHER, P. *et al.* 1999. Abnormal cingulate modulation of fronto-temporal connectivity in schizophrenia. Neuroimage **9:** 337–342.
35. KUBICKI, M. *et al.* 2003. An fMRI study of semantic processing in men with schizophrenia. Neuroimage **20:** 1923–1933.
36. KLINGLER, J. & P. GLOOR. 1960. The connections of the amygdala and of the anterior temporal cortex in the human brain. J. Comp. Neurol. **115:** 333–369.
37. EBELING, U. & D. VON CRAMON. 1992. Topography of the uncinate fascicle and adjacent temporal fiber tracts. Acta Neurochir. (Wien) **115:** 143–148.
38. KIER, E.L. *et al.* 2004. MR imaging of the temporal stem: anatomic dissection tractography of the uncinate fasciculus, inferior occipitofrontal fasciculus, and Meyer's loop of the optic radiation. AJNR Am. J. Neuroradiol. **25:** 677–691.
39. KUBICKI, M. *et al.* 2002. Uncinate fasciculus findings in schizophrenia: a magnetic resonance diffusion tensor imaging study. Am. J. Psychiatry **159:** 813–820.
40. CRESPO-FACORRO, B. *et al.* 2000. Insular cortex abnormalities in schizophrenia: a structural magnetic resonance imaging study of first-episode patients. Schizophr. Res. **46:** 35–43.
41. GOLDSTEIN, J.M. *et al.* 1999. Cortical abnormalities in schizophrenia identified by structural magnetic resonance imaging. Arch. Gen. Psychiatry **56:** 537–547.
42. BENES, F.M. *et al.* 1992. Increased GABAA receptor binding in superficial layers of cingulate cortex in schizophrenics. J. Neurosci. **12:** 924–929.
43. BENES, F.M., S.L. VINCENT & M. TODTENKOPF. 2001. The density of pyramidal and nonpyramidal neurons in anterior cingulate cortex of schizophrenic and bipolar subjects. Biol. Psychiatry **50:** 395–406.

44. BOURAS, C. *et al.* 2001. Anterior cingulate cortex pathology in schizophrenia and bipolar disorder. Acta Neuropathol. (Berl.) **102:** 373–379.
45. KIEHL, K.A., P.F. LIDDLE & J.B. HOPFINGER. 2000. Error processing and the rostral anterior cingulate: an event-related fMRI study. Psychophysiology **37:** 216–223.
46. CARTER, C.S. *et al.* 1997. Anterior cingulate gyrus dysfunction and selective attention deficits in schizophrenia: $[^{15}O]H_2O$ PET study during single-trial Stroop task performance. Am. J. Psychiatry **154:** 1670–1675.
47. NORDAHL, T.E. *et al.* 2001. Anterior cingulate metabolism correlates with Stroop errors in paranoid schizophrenia patients. Neuropsychopharmacology **25:** 139–148.
48. FRITH, C.D. *et al.* 1995. Regional brain activity in chronic schizophrenic patients during the performance of a verbal fluency task. Br. J. Psychiatry **167:** 343–349.
49. FORD, J.M. *et al.* 2001. Neurophysiological evidence of corollary discharge dysfunction in schizophrenia. Am. J. Psychiatry **158:** 2069–2071.
50. WANG, F. *et al.* 2004. Anterior cingulum abnormalities in male patients with schizophrenia determined through diffusion tensor imaging. Am. J. Psychiatry **161:** 573–575.
51. SUN, Z. *et al.* 2003. Abnormal anterior cingulum in patients with schizophrenia: a diffusion tensor imaging study. Neuroreport **14:** 1833–1836.
52. KUBICKI, M. *et al.* 2003. Cingulate fasciculus integrity disruption in schizophrenia: a magnetic resonance diffusion tensor imaging study. Biol. Psychiatry **54:** 1171–1180.
53. NIZNIKIEWICZ, M.A. *et al.* 1997. ERP assessment of visual and auditory language processing in schizophrenia. J. Abnorm. Psychol. **106:** 85–94.
54. GUILLEM, F. *et al.* 2001. Memory impairment in schizophrenia: a study using event-related potentials in implicit and explicit tasks. Psychiatry Res. **104:** 157–173.
55. HAZLETT, E.A. *et al.* 2000. Hypofrontality in unmedicated schizophrenia patients studied with PET during performance of a serial verbal learning task. Schizophr. Res. **43:** 33–46.
56. FLETCHER, P.C. *et al.* 1998. Brain activations in schizophrenia during a graded memory task studied with functional neuroimaging. Arch. Gen. Psychiatry. **55:** 1001–1008.
57. KUBICKI, M. *et al.* 2004. DTI and MTR abnormalities in schizophrenia: analysis of white matter integrity. Neuropsychopharmacology **29:** 167.
58. KUROKI, N. *et al.* 2004. Fornix integrity and hippocampal volume in schizophrenia. Biol. Psychiatry **55:** 72S.
59. EASTWOOD, S.L. & P.J. HARRISON. 2003. Interstitial white matter neurons express less reelin and are abnormally distributed in schizophrenia: towards an integration of molecular and morphologic aspects of the neurodevelopmental hypothesis. Mol. Psychiatry **8:** 821–831.
60. KIRKPATRICK, B. *et al.* 2003. Interstitial cells of the white matter in the dorsolateral prefrontal cortex in deficit and nondeficit schizophrenia. J. Nerv. Ment. Dis. **191:** 563–567.
61. HOF, P.R. *et al.* 2003. Loss and altered spatial distribution of oligodendrocytes in the superior frontal gyrus in schizophrenia. Biol. Psychiatry **53:** 1075–1085.
62. TKACHEV, D. *et al.* 2003. Oligodendrocyte dysfunction in schizophrenia and bipolar disorder. Lancet **362:** 798–805.
63. ASTON, C., L. JIANG & B.P. SOKOLOV. 2004. Microarray analysis of postmortem temporal cortex from patients with schizophrenia. J. Neurosci. Res. **77:** 858–866.
64. MINAMI, T. *et al.* 2003. Diffusion tensor magnetic resonance imaging of disruption of regional white matter in schizophrenia. Neuropsychobiology **47:** 141–145.
65. OKUGAWA, G. *et al.* 2004. Subtle disruption of the middle cerebellar peduncles in patients with schizophrenia. Neuropsychobiology **50:** 119–123.
66. FOONG, J. *et al.* 2000. Neuropathological abnormalities of the corpus callosum in schizophrenia: a diffusion tensor imaging study. J. Neurol. Neurosurg. Psychiatry **68:** 242–244.
67. KASAI, K. *et al.* 2003. Progressive decrease of left superior temporal gyrus gray matter volume in patients with first-episode schizophrenia. Am. J. Psychiatry **160:** 156–164.
68. BRUN, A. *et al.* 2003. Coloring of DT-MRI fiber traces using Laplacian eigenmaps: computer aided systems theory (EUROCAST'03). Lect. Notes Comput. Sci. **2809:** 564–572.
69. BRUN, A. *et al.* 2004. Clustering fiber tracts using normalized cuts. *In* Seventh International Conference on Medical Image Computing and Computer-Assisted Intervention. Springer-Verlag. Berlin/New York.

Brain/Language Relationships Identified with Diffusion and Perfusion MRI

Clinical Applications in Neurology and Neurosurgery

ARGYE E. HILLIS

Department of Neurology, Johns Hopkins University School of Medicine, Baltimore, Maryland, USA

ABSTRACT: Diffusion and perfusion MRI have contributed to stroke management by identifying patients with tissue "at risk" for further damage in acute stroke. However, the potential usefulness of these imaging modalities, along with diffusion tensor imaging, can be expanded by using these imaging techniques with concurrent assessment of language and other cognitive skills to identify the specific cognitive deficits that are associated with diffusion and perfusion abnormalities in particular brain regions. This paper illustrates how this combined behavioral and imaging methodology can yield information that is useful for predicting specific positive effects of intervention to restore blood flow in hypoperfused regions of brain identified with perfusion MRI, and for predicting negative effects of resection of particular brain regions or fiber bundles. Such data allow decisions about neurological and neurosurgical interventions to be based on specific risks and benefits in terms of functional consequences.

KEYWORDS: MRI; diffusion; perfusion; brain; language; stroke; DWI; PWI

What sorts of information provided by diffusion and perfusion MRI can be useful in planning treatment of patients with stroke, brain tumor, epilepsy, or other neurological disease? To answer this question, it makes sense to first ask what sorts of information do neurologists and neurosurgeons need to know to guide their interventions. In treatment of stroke, the neurologist needs to know what areas of the brain are already infarcted and what areas are receiving too little blood flow to function or even survive for long periods. Furthermore, it would be useful to be able to predict the consequences of restoring blood flow to a region of hypoperfusion and the consequences of not restoring blood flow (allowing that area to progress to infarction). Only with all of this information can the physician and the patient (and/or the family) make informed decisions regarding whether or not to proceed with an intervention, by considering the potential risks and benefits. Likewise, when a neurosurgeon is considering removal of a brain tumor or removal of an area of brain that is causing intractable epilepsy, he or she needs to know the likelihood that resection of that area of brain will cause particular deficits in order to weigh the risks and benefits. The neurosurgeon

Address for correspondence: Argye E. Hillis, M.D., M.A., Associate Professor of Neurology, Johns Hopkins University School of Medicine, Johns Hopkins Hospital, Phipps 126, 600 North Wolfe Street, Baltimore, MD 21287. Voice: 410-614-2381; fax: 410-614-9807.
argye@jhmi.edu

Ann. N.Y. Acad. Sci. 1064: 149–161 (2005). © 2005 New York Academy of Sciences.
doi: 10.1196/annals.1340.027

also needs to know where critical white matter tracts are located with respect to the lesion and what would be the functional consequences of resecting a particular white matter tract. Armed with this information, the neurosurgeon cannot only determine the potential risks and benefits of surgery, but can also guide the approach to surgery that will minimize the chances of important negative functional consequences. This chapter will focus on studies that can provide this sort of information.

CLINICAL APPLICATIONS IN STROKE

Identifying the Ischemic Penumbra

Diffusion-weighted imaging (DWI) and perfusion-weighted imaging (PWI) were originally introduced as methods for defining areas of acute infarction and areas at risk for infarction in acute stroke.[1–4] The uses and limitations for defining both areas of infarction and areas of potentially salvageable—but at-risk—tissue have been recently reviewed.[5] To briefly summarize, there is evidence that areas that are bright on DWI and dark on apparent diffusion coefficient (ADC) maps correspond to densely ischemic tissue (areas of cytotoxic edema). However, at least a portion of this area may recover if rapidly reperfused (although, in some cases, rapid reperfusion may nevertheless result in damage to the area due to "reperfusion injury").[6,7] Several studies have evaluated the threshold of ADC that is most strongly associated with progression to infarct, but the results have not yet converged on a single answer.[8,9] Likewise, PWI data can yield a number of different hemodynamic maps, including time-to-peak (TTP) maps that reveal the time to peak arrival of a bolus of contrast in each voxel and mean-time-to-transit (MTT) maps that reveal the time between arrival and clearance of the contrast in each voxel. Both measures are closely related to blood flow, but do not provide absolute measures of blood flow. Relative delays in TTP and MTT, compared to corresponding areas in the normal hemisphere, have been used to estimate regions of hypoperfusion. Recent studies have attempted to define the delay (number of seconds delay in TTP or MTT relative to the homologous region in the normal hemisphere) associated with progression to infarct.[10–12] These studies together indicate that the area of hypoperfusion visible on MTT or TTP maps likely includes a core of hypoperfused tissue that may be at risk for infarction (areas with >6-s delay),[13,14] surrounded by an area of hypoperfusion that may not be at risk. These conclusions call into question the simplest hypothesis about DWI and PWI in acute stroke—that the entire PWI abnormality minus the DWI abnormality, or diffusion-perfusion mismatch, represents the "ischemic penumbra", or potentially salvageable tissue. However, another definition of ischemic penumbra is defined as an area getting enough blood to survive, but not enough to function (even if it is not at any immediate risk for progression to infarct). A few studies have attempted to determine the degree of delay in TTP or MTT associated with dysfunction or functional outcome.[15,16] Some such studies indicate that regions with >2.5-s delay in TTP are likely to have somewhat impaired function, but more severe delay is associated with more severe functional consequences.[17,18] Thus, it is plausible that the area of abnormality seen on TTP or MTT maps may be divided into a central area that is likely to infarct; a surrounding area that is dysfunctional, but may not be at risk for infarct; as well as an outer area of benign oligemia or minimally delayed blood flow.

Other studies have calculated CBF maps based on the cerebral blood volume (CBV) in each voxel, divided by MTT. The relative CBF calculated in this way (ratio of calculated CBF in ischemic area to that in the homologous region in the unaffected hemisphere) is strongly correlated with quantitative CBF measured by ^{15}O-PET.[19] Other investigators have produced quantitative CBF maps with PWI based on a single arterial input function measured at the level of the ICA or MCA.[20,21] A more precise method entails identifying the arterial input function separately for each major vasculature territory. However, because these methods have not yet been put into routine clinical use, this chapter will illustrate the clinical usefulness of PWI using relative TTP maps. Based on the available evidence, it will be assumed that a relative delay in TTP (compared to the homologous region in the unaffected hemisphere) of >2.5 s corresponds to dysfunctional tissue, and a relative delay of >6 s corresponds to immediate risk of infarction in the absence of reperfusion.

Predicting Functional Outcome of Intervention

It is not enough to be able to predict whether or not a brain region has neuronal dysfunction or is at risk for infarction. It is also necessary to predict the functional consequences of restoring blood flow to that area versus functional consequences of allowing that area to infarct. For example, suppose PWI shows isolated hypoperfusion (say, mean delayed TTP of 6 s) of left striate cortex, and DWI shows no abnormality in left striate cortex. This hypoperfusion would account for right homonymous hemianopia, and reperfusion would restore vision in the right visual field. This potential benefit would need to be weighed against the risk of intervention to restore blood flow. For instance, intravenous administration of rtPA (thrombolysis) carries a 6.4% risk of symptomatic intracerebral hemorrhage, even if administered within 3 h of symptom onset in patients who meet other criteria for rtPA.[22] Even if the patient has other deficits, these are not likely to improve with reperfusion if the only area of diffusion-perfusion mismatch is the left striate cortex. Therefore, most patients would not accept the risks of intravenous rtPA for the possibility of recovering from right homonymous hemianopia since this deficit is not severely disabling (except perhaps to professional drivers, pilots, and others who depend heavily on full vision). On the other hand, suppose the diffusion-perfusion mismatch involves the entire left posterior cerebral artery territory (including striate cortex, ventral temporo-occipital cortex, and splenium of the corpus callosum; regions that are critical for reading[23–29]). In this case, reperfusion could result in recovery of reading ability, and failure to reperfuse could result in permanent alexia.

To illustrate, a young woman had severe hypoperfusion of bilateral ventral occipital and temporal cortex, including fusiform gyri (FIG. 1, left), associated with visual agnosia (impaired recognition of visually presented objects or pictures) and severe alexia. Her marked hypoperfusion, with large diffusion-perfusion mismatch, was caused by severe intracranial large vessel stenosis due to Moya-Moya disease. Since visual agnosia typically occurs only with bilateral fusiform damage/dysfunction, it was predicted that (1) her small, unilateral acute infarct could not account for her functional deficits and (2) reperfusion of the fusiform gyrus on either side would improve visual recognition. Furthermore, since left[28] or either right or left[29] fusiform gyrus is critical for reading (as discussed more later), it was predicted that restoring blood flow to the left fusiform gyrus would improve her reading as well as

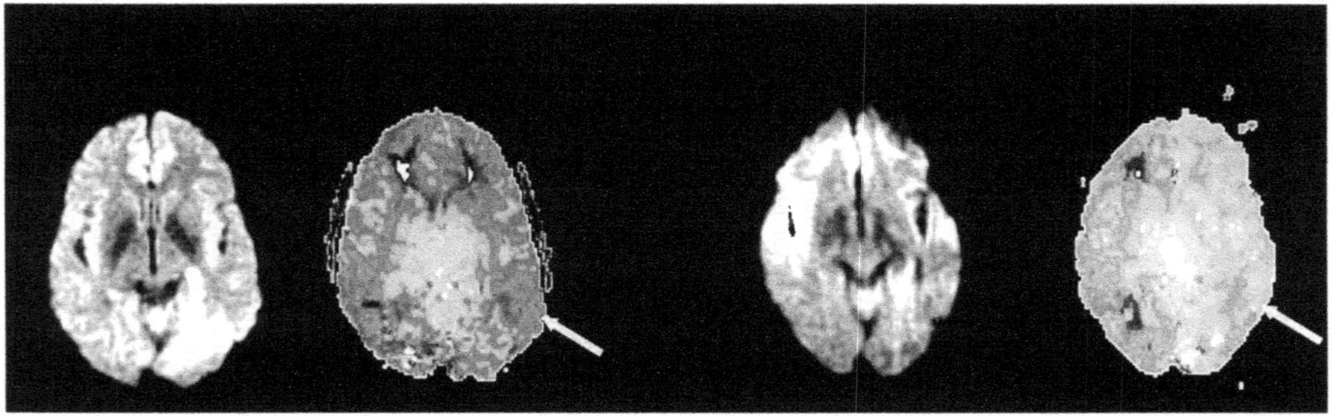

FIGURE 1. Reperfusion of left midfusiform gyrus and surrounding cortex resulted in recovery of naming and reading. **(Two Left Panels)** DWI (*left*) and PWI (*right*) before intervention to improve blood flow. **(Two Right Panels)** DWI and PWI after intervention. In these and all PWI figures, TTP maps are shown; dark areas are areas of delayed TTP ("hypoperfusion"); light areas are areas of relatively normal TTP. The *arrow* points to left midfusiform gyrus.

her visual recognition of objects. Therefore, she underwent a surgical intervention to restore blood flow to left temporal, occipital, and parietal cortex (see ref. 30 for details). Postintervention scans showed markedly improved perfusion of the posterior left hemisphere (FIG. 1, right). As predicted, her scores improved substantially on visual object recognition/naming (from 0% to 76% correct) and comprehension of written words (from 0% to 71% correct) after blood flow was restored to left temporal and occipital cortex, including fusiform gyrus.

Identifying Brain/Behavioral Relationships

The above case demonstrates how DWI and PWI can be useful in predicting outcome when the function of a particular region of brain that is hypoperfused is known. However, the cognitive functions of most cortical regions are not well established. Studies of brain/behavioral relationships using functional imaging or lesion/deficit associations have often yielded conflicting results. Variability across functional imaging studies is often due to differences in methods for evaluating cognitive functions; even rate of presentation can strongly influence results.[31] Variability across lesion studies results in part from the fact that most investigators have studied patients long after stroke, after substantial reorganization of structure/function relationships has occurred. Two patients with identical lesions may have very different deficits many months after stroke due to differences in degree of recovery, reorganization, or focus of rehabilitation. Moreover, lesion studies have traditionally included only patients with chronic impairment in some particular function. Therefore, even if they can identify the probability that a chronic deficit is associated with damage to a particular area, they cannot identify the probability that damage to that area will result in the deficit. Likewise, functional imaging studies (PET and fMRI) show regions of activation—areas that may be engaged in a particular task—but do not provide evidence that those regions are essential to the task.

To identify the probability that a lesion in a given region will cause a particular deficit, my colleagues and I have taken a different approach to lesion/deficit association studies. We study consecutive patients (with and without imaging abnormalities in a given region of interest, with and without the deficit of interest) within 24 h of onset of stroke symptoms, before the opportunity for recovery, substantial reorganization, or rehabilitation. We determine the frequency of the deficit of interest in patients with and without damage or dysfunction of various brain regions. Since patients are studied in the setting of acute stroke, both DWI and PWI are needed to identify the entire region of ischemia and/or hypoperfusion responsible for the deficit. The following example illustrates how DWI, PWI, and cognitive testing carried out in hyperacute stroke can reveal brain/behavior relationships that can help to predict the functional consequences of ischemia or intervention in acute stroke.

Consider two hypotheses about the role of left midfusiform gyrus, derived from functional imaging studies of language. One hypothesis, based on the observation of consistent activation of left midfusiform gyrus during reading of words and pseudowords, is that this region is responsible for accessing a visual word form in the reading process.[32–34] An alternative hypothesis, based on evidence that this region is activated in a variety of lexical tasks with different input and output modalities and on evidence that lesions in this region cause naming as well as reading impairments, is that left midfusiform gyrus is engaged in modality-independent lexical process-

TABLE 1. Percent of patients with deficits in each task, among patients with and without hypoperfusion/infarct of left midfusiform gyrus

	Written word comprehension	Oral picture naming	Oral tactile naming	Written naming
Abnormal midfusiform gyrus	74%	90%	89%	88%
Normal midfusiform gyrus	58%	46%	35%	45%

ing.[35,36] To test these hypotheses, we evaluated performance on written word comprehension, written lexical decision (deciding whether or not a string of letters is a real word), oral reading, oral naming, and written naming of 80 patients with acute left hemisphere stroke within 24 h of onset of symptoms. DWI and PWI were obtained the same day as language testing. TABLE 1 shows that acute dysfunction (hypoperfusion and/or infarct) of left midfusiform gyrus did not reliably result in impaired reading comprehension, but did reliably result in impaired oral naming of pictures and objects presented for tactile identification, and written naming of pictures.[29] These results are consistent with the hypothesis that left midfusiform gyrus is critical for modality-independent lexical processing. This region does not appear to be crucial for written word recognition (reading comprehension and lexical decision), although it may well be consistently engaged in written word recognition (as indicated by functional imaging studies). One proposal that would account for both functional imaging studies and lesion studies is that bilateral midfusiform gyri are activated in computing a prelexical representation of the string of graphemes (letter identities) that is independent of case, font, and location of the stimulus, but only the left or the right is essential for this component of reading.[29]

This information can be of practical use in stroke management. For example, this study indicates that, among patients with acute dysfunction of left midfusiform gyrus, 90% have impaired oral naming and only 74% have impaired reading comprehension. Furthermore, of patients with no imaging abnormality in this region, only 46% have impaired naming and 58% have impaired reading comprehension. Therefore, one would predict that restoring blood flow to hypoperfused left midfusiform gyrus, at least in the absence of ischemia in this region on DWI, is likely to result in recovery of naming (oral and written naming, in response to pictures or objects), but less likely to positively affect reading comprehension. This prediction has been confirmed in several cases. For instance, the patient whose preintervention scans are shown in FIGURE 2 (left) had impaired oral and written naming of pictures and tactile objects at day 1, associated with hypoperfusion of Brodmann's area 37, including left midfusiform gyrus, without infarct in this region on DWI. Increasing his blood pressure resulted in improved blood flow in this region and immediate improvement in naming across modalities. There was no change in reading comprehension or lexical decision when this region was reperfused at day 2 (FIG. 2, right; see refs. 29 and 37 for additional cases).

What about the role of white matter tracts in reading and naming? We have proposed that either left or right midfusiform gyrus is necessary for computing a case-, font-, location-, and orientation-independent representation for reading. A similar, abstract structural description of an object probably must be computed for recognizing

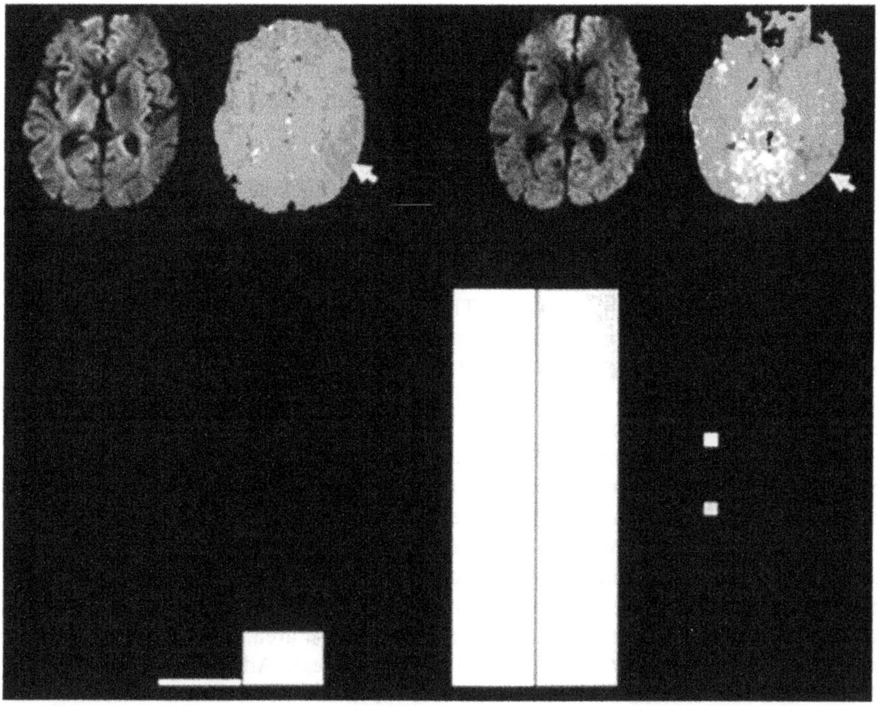

FIGURE 2. Reperfusion of left midfusiform gyrus resulted in improved oral naming of visual and tactile input (modality-independent lexical processing). **(Two Left Panels)** DWI (*left*) and PWI (*right*) before intervention to improve blood flow. **(Two Right Panels)** DWI and PWI after intervention. The *arrow* points to left midfusiform gyrus.

object and pictures. Construction of such a structural description also likely requires either the left or the right fusiform gyrus since only damage to both (bilateral fusiform gyrus) results in visual agnosia. This hypothesis can explain why the patient whose case is illustrated in FIGURE 1 improved in visual recognition of objects and words with reperfusion of left fusiform gyrus (despite persistent hypoperfusion of right midfusiform). However, when the abstract structural description of words and objects is computed in right fusiform, this representation must have access to left hemisphere language areas to name the picture or word. This interhemispheric transfer from fusiform gyrus takes place via white matter tracts in the splenium of the corpus callosum, as demonstrated by tractography.[38] Damage or dysfunction of this tract, along with dysfunction of left fusiform gyrus, would be expected to prevent reading comprehension and naming of visually presented objects. Evidence consistent with this prediction has been reported by Marsh and myself.[39] We described a patient with infarct of left occipital lobe and fusiform gyrus, and hypoperfusion of the splenium, who had severely impaired reading comprehension and naming of pictures (but relatively spared naming to tactile exploration or spoken definition; that is, alexia without agraphia and "optic aphasia"). When the splenium was reperfused, he showed recovery of reading comprehension and improvement in visual naming. He had some residual deficits in modality-independent naming because of the infarct in left mid-fusiform gyrus. It is likely that reperfusion of the splenium allowed the location- and orientation-independent structural description of the picture and word computed in right fusiform gyrus to have access to left hemisphere language areas.

CLINICAL APPLICATIONS IN NEUROSURGERY

In planning resection of a tumor or area thought to be responsible for intractable epilepsy, the surgeon needs to know the risk of disrupting a particular function if a region of brain is surgically removed. This sort of prediction can generally rely on information obtained from the above sort of studies. For example, before removing a tumor involving Wernicke's area (left BA 22), it should be considered that any dysfunction of Wernicke's area carries a high risk of interfering with spoken word comprehension. Among patients studied within 24 h of stroke onset, 95% who had hypoperfusion and/or infarct of Wernicke's area (BA 22) had impaired word comprehension; among patients with spoken word comprehension deficits, 90% had hypoperfusion and/or infarct of Wernicke's area ($P < .0000001$).[17] Furthermore, the worse the dysfunction of Wernicke's area, the worse the impairment of spoken word comprehension. That is, in a study of acute stroke patients, there was a very high correlation between the degree of delay in blood flow, as measured by the mean time to peak arrival of contrast in voxels within Wernicke's area, and the error rate in a word comprehension test[18] (see FIG. 3 for examples). Additional evidence that dysfunction in Wernicke's area is responsible for impairments in word meaning was obtained by showing that reperfusion of Wernicke's area results in recovery of word comprehension.[40]

One caveat in planning surgery is that there is known individual variability in structure/function relationships in the brain. This fact is illustrated in the above cited studies since the risk of a functional impairment with damage in a particular area was never 100%. For example, 5% of patients with acute hypoperfusion/infarct of Wernicke's area had no impairment in word comprehension. This problem of individual variability is likely to be greater in patients for whom surgery is planned since these patients have lesions or chronic epilepsy that may have led to reorganization of structure/function relationships. For example, children with Rasmussen's encephalitis, which results in intractable epilepsy originating from one hemisphere, often show transfer of language functions from the left to the right hemisphere when seizures have a left hemisphere origin; thus, even removal of the entire left hemisphere in such children may result in little or no language impairment.[41,42] It is likely that the extent of such reorganization is dependent on age as well as volume of dysfunctional

---

FIGURE 3. Severity of hypoperfusion in Wernicke's area corresponds to severity of word comprehension deficit. **(Top)** DWI (*left*) and PWI (*right*) of a patient with mild relative delay in TTP (mean: 2.6 s, compared to the homologous region on the right) in Wernicke's area and mild word comprehension impairment at day 1. **(Middle)** DWI and PWI of a patient with moderate relative delay in TTP (mean: 4.7 s) in Wernicke's area and moderate word comprehension impairment at day 1. **(Bottom)** DWI and PWI of a patient with severe relative delay in TTP (mean: 7.1 s) in Wernicke's area and severe word comprehension impairment at day 1. The *arrow* points to Wernicke's area.

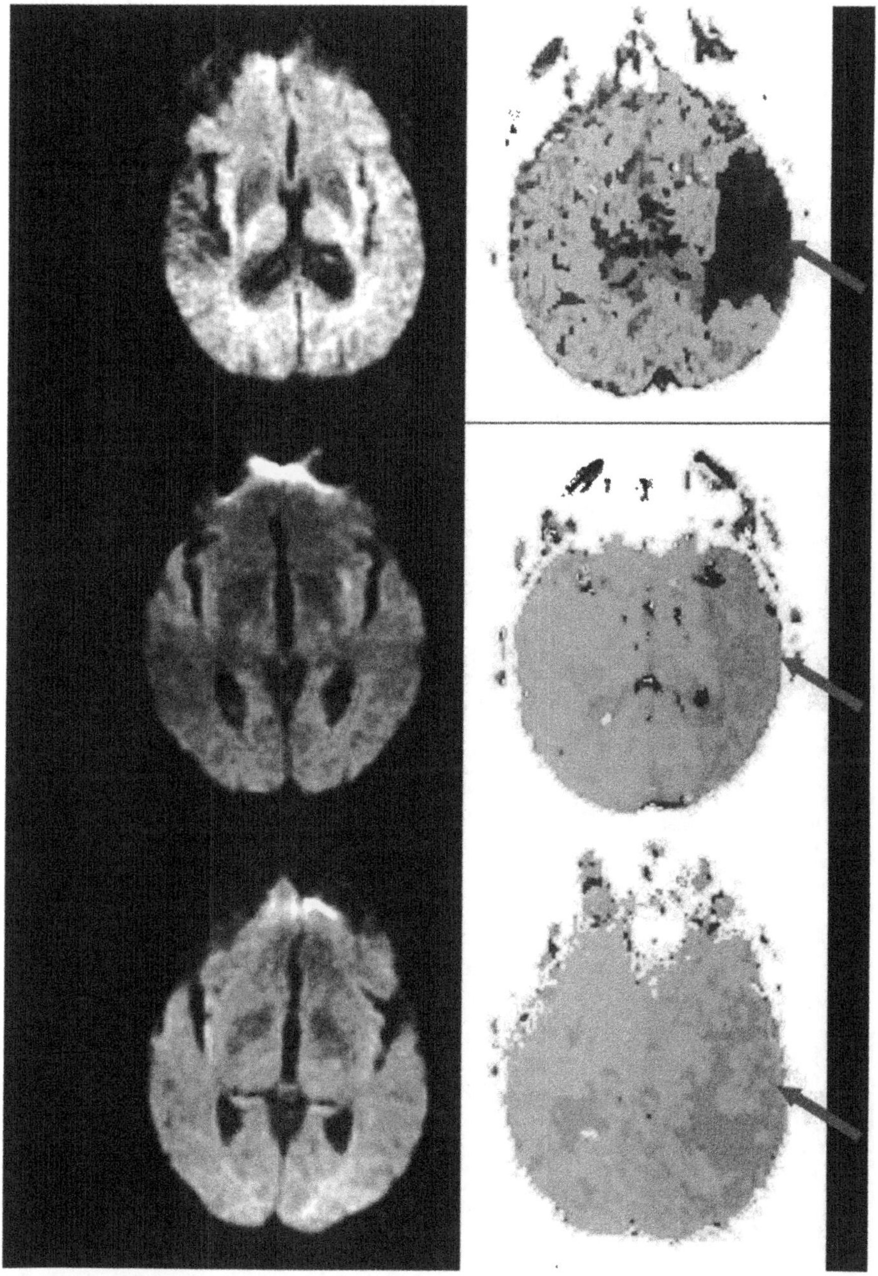

FIGURE 3. *See previous page for legend.*

brain tissue. One way to address this individual variability is to cause "temporary lesions" with cortical stimulation during awake craniotomy[43,44] or via implanted subdural grids[45] to evaluate probable effects of removal prior to surgery in each patient. However, since this sort of testing must be limited (due to risks of seizure caused by stimulation), it is still necessary to have a reasonable hypothesis about the cognitive functions that depend on a given region of brain in order to plan testing.

Surgery can also cut white matter pathways that may have crucial roles in cognitive functions (e.g., by connecting cortical areas involved in a particular function). Until recently, identified functions of particular white matter tracts in cognition were predominantly limited to interhemispheric tracts. For example, the long-standing hypothesis that the splenium of the corpus callosum provides the only connection between the left and right occipital lobes has recently been confirmed with tractography.[38] Combined lesions of the splenium and the left occipital lobe can account for many cases of alexia without agraphia due to inability of visual information (which is processed in the right occipital lobe only due to damage to the left occipital lobe) to be transferred to the left hemisphere language areas, as described above.

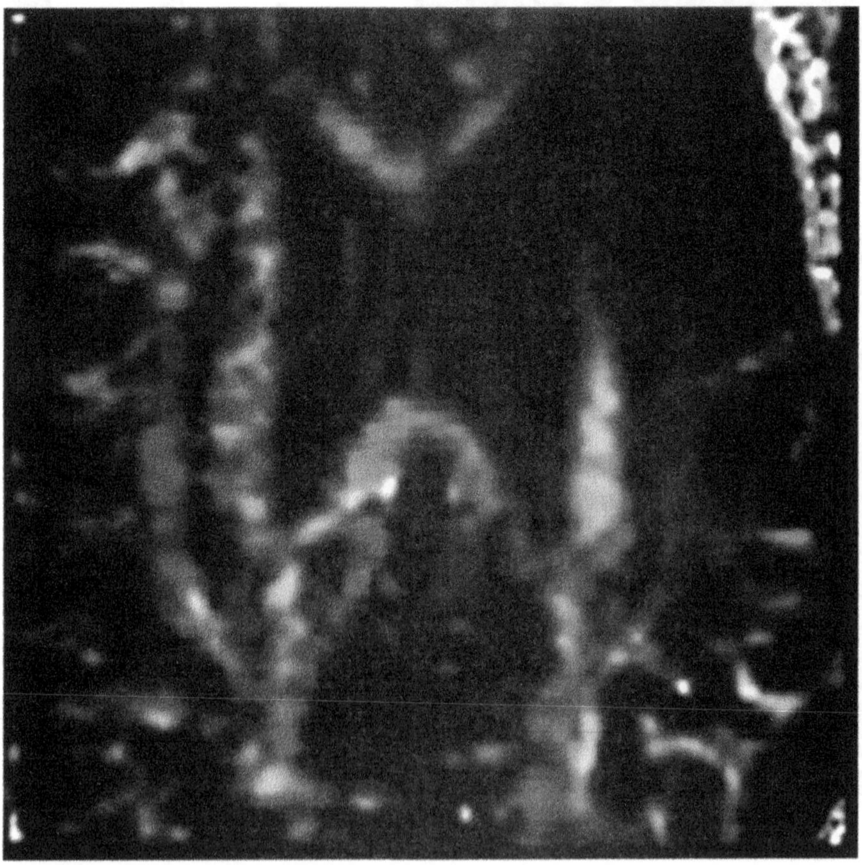

FIGURE 4. Disruption in the arcuate fasciculus associated with disproportionately spared word and sentence repetition.

Another long-standing hypothesis about the role of a particular white matter tract in language has concerned the arcuate fasciculus. The left arcuate fasciculus, which appears to connect Wernicke's area and Broca's area, has been hypothesized to be essential for verbal repetition.[46] Thus, it has been proposed that lesions of the arcuate fasciculus are responsible for "conduction aphasia", characterized by disproportionately impaired repetition, by disconnecting auditory input processing in Wernicke's area from spoken output processing in Broca's area.[47] However, this hypothesis has not been confirmed by cases of focal lesions in the arcuate fasciculus. For example, the patient whose DTI is shown in FIGURE 4 had a focal disruption of the arcuate fasciculus on the left, but had a transcortical aphasia at the onset of stroke, characterized by disproportionately *spared* sentence repetition.[48] Future studies of specific language deficits associated with lesions in particular white matter tracts after stroke or other focal neurological disease are likely to shed light on the necessary roles of these tracts in particular language and other cognitive tasks. In particular, such studies may reveal necessary connections between cortical areas that comprise a neural network of regions supporting particular cognitive tasks. This information, along with preoperative DTI in particular patients in whom surgery is planned, may be important in planning surgery in such a way as to spare those white matter tracts that are essential for language and other critical cognitive skills. Furthermore, longitudinal DTI after stroke is likely to reveal the effects of Wallerian degeneration resulting from focal lesions. For example, lesions in motor cortex or in the pyramid tracts eventually result in degeneration of the distal pyramid tract, quite far from the original lesions, as revealed by subsequent DTI.[38]

SUMMARY AND CONCLUSIONS

DWI has revolutionized acute stroke therapy by revealing the site and extent of densely ischemic tissue within minutes to hours of onset of stroke symptoms. DWI, together with PWI, can estimate the area of dysfunctional, but potentially salvageable neural tissue that may be at risk for infarction unless blood flow is restored. Further, DWI and PWI, along with cognitive testing, in hyperacute stroke can provide information about the probability of impairment in a particular cognitive function if a region of hypoperfusion is to become permanently infarcted, or if a region is to be surgically removed. Such probability data can be useful in prognosis in acute stroke[49] and can be useful in decision-making in neurology (acute stroke management) and neurosurgery (e.g., planning routes and extent of surgery in cases of intractable epilepsy or tumor resection; predicting potential benefits of revascularization). DTI is likely to become increasingly important for identifying the role of white matter tracts in cognitive and motor functions and identifying functional risks associated with damage to white matter tracts.

ACKNOWLEDGMENTS

The research reported in this paper was supported by NIH Grant No. RO1 DC 05375 to A. Hillis.

REFERENCES

1. LUTSEP, H.L. *et al.* 1997. Clinical utility of diffusion-weighted magnetic imaging in the assessment of ischemic stroke. Ann. Neurol. **41:** 574–580.
2. MARKS, M.P. *et al.* 1999. Evaluation of early reperfusion and IV tPA therapy using diffusion- and perfusion-weighted MRI. Neurology **52:** 1792–1798.
3. RORDORF, G. *et al.* 1998. Regional ischemia and ischemic injury in patients with acute middle cerebral artery stroke as defined by early diffusion-weighted and perfusion-weighted MRI. Stroke **29:** 939–943.
4. WARACH, S. *et al.* 1995. Acute human stroke studied by whole brain echo planar diffusion-weighted MRI. Ann. Neurol. **37:** 232–241.
5. KIDWELL, C.S., J.R. ALGER & J.L. SAVER. 2003. Beyond mismatch: evolving paradigms in imaging the ischemic penumbra with multimodal magnetic resonance imaging. Stroke **34:** 2729–2735.
6. KIDWELL, C.S. *et al.* 2000. Thrombolytic reversal of acute human cerebral ischemic injury shown by diffusion/perfusion magnetic resonance imaging. Ann. Neurol. **47:** 462–469.
7. CHALELA, J.A. *et al.* 2004. Early magnetic resonance imaging findings in patients receiving tissue plasminogen activator predict outcome: insights into the pathophysiology of acute stroke in the thrombolysis era. Ann. Neurol. **55:** 105–112.
8. NA, D.G. *et al.* 2004. Diffusion-weighted MR imaging in acute ischemia: value of apparent diffusion coefficient and signal intensity thresholds in predicting tissue at risk and final infarct size. Am. J. Neuroradiol. **25:** 1331–1336.
9. SCHAEFER, P.W. *et al.* 2003. Assessing tissue viability with MR diffusion and perfusion imaging. Am. J. Neuroradiol. **24:** 436–443.
10. NEUMANN-HAEFELIN, T. *et al.* 1999. Diffusion and perfusion-weighted MRI: the DWI/PWI mismatch region in acute stroke. Stroke **30:** 1591–1597.
11. PARSONS, M.W. *et al.* 2000. MR perfusion imaging: acute rCBF is more accurate than rMTT or rCBV in prediction of infarct size. Stroke **31:** 275.
12. SCHLAUG, G. *et al.* 1999. The ischemic penumbra: operationally defined by diffusion and perfusion MRI. Neurology **53:** 1528–1537.
13. THIJS, V.N. *et al.* 2001. Relationship between severity of MR perfusion deficit and DWI lesion evolution. Neurology **57:** 1205–1211.
14. SHIH, L.C. *et al.* 2003. Perfusion-weighted magnetic resonance imaging thresholds identifying core, irreversibly infarcted tissue. Stroke **34:** 1425.
15. BARBER, P.A. *et al.* 1998. Prediction of stroke outcome with echo planar perfusion- and diffusion-weighted MRI. Neurology **51:** 418–426.
16. BEAULIEU, C. *et al.* 1999. Longitudinal magnetic resonance imaging study of perfusion and diffusion in stroke: evolution of volume and correlation with clinical outcome. Ann. Neurol. **46:** 568–578.
17. HILLIS, A.E. *et al.* 2000. MR perfusion imaging reveals regions of hypoperfusion associated with aphasia and neglect. Neurology **55:** 782–788.
18. HILLIS, A.E. *et al.* 2001. Hypoperfusion of Wernicke's area predicts severity of semantic deficit in acute stroke. Ann. Neurol. **50:** 561–566.
19. MIHARA, F. *et al.* 2003. Semi-quantitative CBF and CBF ratios obtained using perfusion-weighted MR imaging. Neuroreport **14:** 725–727.
20. OSTERGAARD, L. *et al.* 1996. High resolution measurement of cerebral blood flow using intravascular tracer bolus passages. Part I: Mathematical approach and statistical analysis. Magn. Reson. Med. **36:** 715–725.
21. OSTERGAARD, L. *et al.* 1996. High resolution measurement of cerebral blood flow using intravascular tracer bolus passages. Part II: Experimental comparison and preliminary results. Magn. Reson. Med. **36:** 726–736.
22. THE NATIONAL INSTITUTE OF NEUROLOGICAL DISORDERS AND STROKE RT-PA STROKE STUDY GROUP. 1995. Tissue plasminogen activator for acute ischemic stroke. N. Engl. J. Med. **333:** 1581–1587.
23. DEJERINE, J. 1891. Sur un cas de cécité verbale avec agraphie, suivi d'autopsie. C. R. Hebd. Séances Mém. Soc. Biol. **Ninth series:** 197–201.

24. DEJERINE, J. 1892. Contribution à l'étude anatomo-pathologique et clinique des différentes variétés de cécité verbale. Mém. Soc. Biol. **4:** 61–90.
25. COHEN, L. *et al.* 2004. The pathophysiology of letter-by-letter reading. Neuropsychologia **42:** 1768–1780.
26. BINDER, J.R. & J.P. MOHR. 1992. The topography of callosal reading pathways: a case-control analysis. Brain **115:** 1807–1826.
27. CHIALANT, D. & A. CARAMAZZA. 1998. Perceptual and lexical factors in a case of letter-by-letter reading. Cogn. Neuropsychol. **15:** 167–201.
28. COHEN, L. *et al.* 2003. Visual word recognition in the left and right hemispheres: anatomical and functional correlates of peripheral alexias. Cereb. Cortex **13:** 1313–1333.
29. HILLIS, A.E. *et al.* 2005. The roles of the "visual word form area" in reading. Neuroimage **24:** 548–559.
30. WITYK, R. *et al.* 2002. Perfusion weighted MRI in adult moyamoya syndrome: characteristic patterns and change after surgical intervention—case report. Neurosurgery **51:** 1499–1506.
31. GRABOWSKI, T.J. & A.R. DAMASIO. 2000. Investigating language with functional imaging. *In* Brain Mapping: The Systems. Academic Press. San Diego.
32. COHEN, L. *et al.* 2002. Language-specific tuning of visual cortex? Functional properties of the visual word form area. Brain **125:** 1054–1069.
33. COHEN, L. *et al.* 2000. The visual word form area: spatial and temporal characterization of an initial stage of reading in normal subjects and posterior split brain patients. Brain **123:** 291–307.
34. DEHAENE, S. *et al.* 2001. Cerebral mechanisms of word masking and unconscious repetition priming. Nat. Neurosci. **4:** 752–758.
35. PRICE, C.J. & J.T. DEVLIN. 2003. The myth of the visual word form area. Neuroimage **19:** 473–481.
36. PRICE, C.J. *et al.* 2003. Cortical localization of the visual and auditory word form areas: a reconsideration of the evidence. Brain Lang. **86:** 272–286.
37. HILLIS, A.E. *et al.* 2002. Reperfusion of specific brain regions by raising blood pressure restores selective language functions in subacute stroke. Brain Lang. **79:** 495–510.
38. HUANG, H. *et al.* 2005. DTI tractography based parcellation of white matter: application to the midsagittal morphology of corpus callosum. In preparation.
39. MARSH, E.B. & A.E. HILLIS. 2005. Cognitive and neural mechanisms underlying reading and naming: evidence from optic aphasia and letter-by-letter reading. In preparation.
40. HILLIS, A.E. & J. HEIDLER. 2002. Mechanisms of early aphasia recovery: evidence from MR perfusion imaging. Aphasiology **16:** 885–896.
41. BOATMAN, D. *et al.* 1999. Language recovery after left hemispherectomy in children with late-onset seizures. Ann. Neurol. **46:** 579–586.
42. VANLANCKER-SIDTIS, D. 2004. When only the right hemisphere is left: studies in language and communication. Brain Lang. **91:** 199–211.
43. OJEMANN, G.A. 1994. Cortical Stimulation and Recording in Language. Academic Press. New York/London.
44. LESSER, R., M. LUDERS & N. DINNER. 1986. Electrical stimulation of Wernicke's area interferes with comprehension. Neurology **36:** 658–663.
45. HART, J., R. LESSER & B. GORDON. 1992. Selective interference with the representation of size in the human by direct cortical stimulation. J. Cogn. Neurosci. **4:** 337–344.
46. GESCHWIND, N. 1965. Disconnection syndromes in animals and man. Brain **88:** 237–294; 585–644.
47. BENSON, D.F. 1979. Aphasia, Alexia, and Agraphia. Churchill Livingstone. New York.
48. SELNES, O.A. *et al.* 2002. MR diffusion tensor imaging documented arcuate fasciculus lesion in a patient with normal repetition performance. Aphasiology **16:** 897–902.
49. REINECK, L. *et al.* 2005. The "diffusion-clinical mismatch" predicts early language recovery in acute stroke. Neurology **64:** 828–833.

White Matter and Behavioral Neurology

CHRISTOPHER M. FILLEY

University of Colorado School of Medicine and Denver Veterans Affairs Medical Center, Denver, Colorado, USA

ABSTRACT: Although the study of higher brain function has traditionally focused on the cortical gray matter, recent years have witnessed the recognition that white matter also makes an important contribution to cognition and emotion. White matter comprises nearly half the brain volume and plays a key role in development, aging, and many neurologic and psychiatric disorders across the life span. More than 100 disorders exist in which white matter neuropathology is the primary or a prominent feature. A variety of neurobehavioral syndromes may result from these disorders; the concept of white matter dementia has been introduced as characteristic of many patients with white matter involvement, and a wide range of focal neurobehavioral syndromes and psychiatric disorders can also be related to dysfunction of myelinated tracts. Understanding the neurobehavioral aspects of white matter disorders is important for clinical diagnosis, treatment, prognosis, and research on brain-behavior relationships. Central to these investigations is the use of modern neuroimaging techniques, which have already provided substantial information on the characterization of white matter and its disorders, and which promise to advance our knowledge further with continued innovation. Diffusion tensor imaging is an exciting method that will assist with the identification of critical white matter tracts in the brain, and the localization of specific lesions that can be correlated with neurobehavioral syndromes. A behavioral neurology of white matter is thus emerging in which clinical observation combined with sophisticated neuroimaging will enable elucidation of the role of white matter connectivity in the distributed neural networks subserving higher brain function.

KEYWORDS: white matter; behavioral neurology; magnetic resonance imaging; dementia; diffusion tensor imaging

INTRODUCTION

Behavioral neurology can be defined as the subspecialty of neurology devoted to the effects of brain lesions on human behavior. The primary concern of behavioral neurologists is the study of brain-behavior relationships as revealed by the clinical syndromes encountered when individuals develop brain diseases or injuries that disrupt normal cognitive or emotional function. The principal avenue of study is the lesion method, whereby neurobehavioral deficits seen clinically can be associated with specific sites of brain involvement. The rapidly advancing field of neuroradiology

Address for correspondence: Christopher M. Filley, Behavioral Neurology Section, UCHSC, 4200 East Ninth Avenue, Denver, CO 80262. Voice: 303-315-6461; fax: 303-315-5867. christopher.filley@uchsc.edu

**Ann. N.Y. Acad. Sci. 1064: 162–183 (2005). © 2005 New York Academy of Sciences.
doi: 10.1196/annals.1340.028**

has increasingly assumed a major role in identifying these lesions accurately and noninvasively. Insights disclosed by these investigations can in turn yield a better understanding of the normal brain as it performs its myriad of uniquely human operations.

Traditionally, the focus of behavioral neurology—and much of neuroscience generally—has been on the cerebral cortex, the thin sheet of gray matter on the brain surface where many complex cognitive and emotional operations are recognized to occur. Indeed, the term "higher cortical function" is commonly used to describe the province of behavioral neurology. Whereas this emphasis is entirely appropriate and has resulted in major advances in the understanding of human behavior, it is well to recall that the cerebral cortex, on average, is only the outermost three millimeters of the brain and that other regions cannot be overlooked in considering brain-behavior relationships.

White matter, the densely packed collection of myelinated axons coursing between widely dispersed gray matter areas, occupies nearly one-half of the human brain, and its billions of fibers provide the intricate connectivity that links all brain regions into functional ensembles known as distributed neural networks.[1,2] In recent years, a growing awareness of the importance of white matter has emerged in the clinical and basic neurosciences, and information on this often neglected area is rapidly accumulating. Much of this progress can be directly attributed to the introduction of magnetic resonance imaging (MRI) some two decades ago, which enabled the detailed visualization of both normal and abnormal white matter. From a clinical perspective, the white matter is vulnerable to a wide range of neuropathologic insults from disease or injury, and many syndromes resulting from these lesions fall squarely within the scope of behavioral neurology. The evidence gathered from these inquiries convincingly documents that the white matter, like the gray matter, is essential for all realms of human behavior. This review will summarize the understanding of white matter and its disorders from the perspective of behavioral neurology.[1,2]

BACKGROUND

White matter was first distinguished from gray matter by the Renaissance anatomist, Andreas Vesalius, in 1543.[2] In the seventh book of his masterwork, *De Humani Corporis Fabrica*, Vesalius clearly demonstrated that the cerebral white matter could be differentiated grossly from the overlying gray matter (FIG. 1). Later, as the neurosciences developed, the structure of myelin became elucidated, and the coalescence of myelinated fibers into discrete tracts was appreciated.[2] These tracts were then seen to provide structural connections between gray matter areas within the brain, and speculations on the function of white matter tracts began to appear.[2] In the 19th century, Jean Martin Charcot greatly advanced the understanding of white matter from a neurologic perspective with his detailed studies of multiple sclerosis (MS), the most important demyelinative disease of the central nervous system (CNS). In 1965, Norman Geschwind proposed the notion of cerebral disconnection as a mechanism of neurobehavioral dysfunction, thus strongly implicating the importance of white matter lesions in brain-behavior relationships.[3] At the turn of the 21st century, the notion of distributed neural networks has become widely accepted, and the obligate participation of myelinated tracts in these networks further implies that white matter has a key role in the elaboration of human behavior.[4,5] In parallel with this develop-

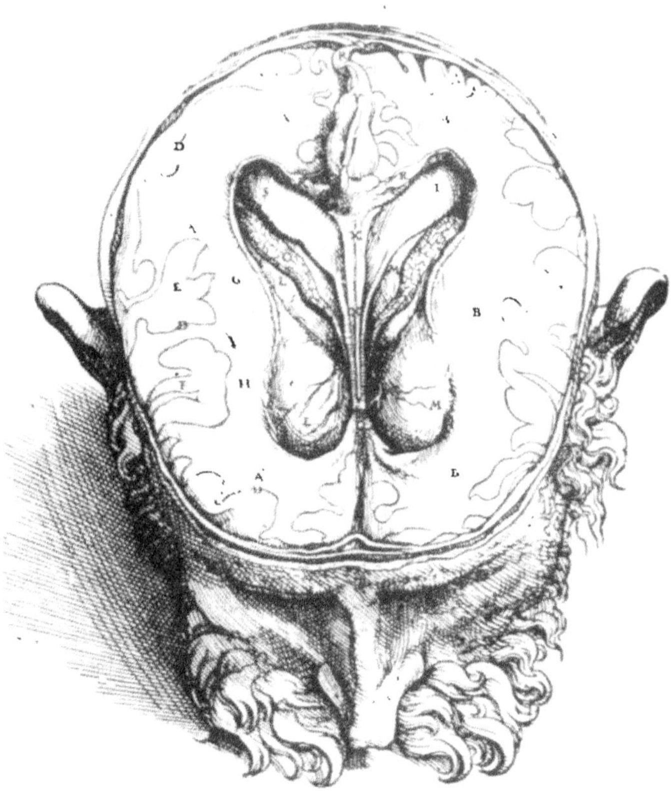

FIGURE 1. Drawing from the seventh book of Vesalius' *De Humani Corporis Fabrica* (1543) in which the white matter of the brain is clearly demarcated from the gray matter. Reprinted with permission from ref. 63.

ment, the impressive growth of modern neuroimaging has vigorously propelled the field forward by enabling the viewing of white matter and its functional specializations in increasingly elegant detail. Today, there is a proliferation of information on the white matter, its role in normal brain function, and its relevance to human illness.

WHITE MATTER ANATOMY AND PHYSIOLOGY

White matter refers to the regions where there is a preponderance of axons coated with myelin, and these areas occupy 40–50% of the adult human brain.[2] At the microscopic level, white matter consists of collections of closely apposed axons wrapped concentrically in myelin, the insulation of nerve cells made up of roughly 70% lipid and 30% protein.[2] Oligodendrocytes, the glial cells responsible for myelinating the axons, are numerous, as are astrocytes, ependymal cells, and blood vessels. Macroscopically, the white matter can be seen to form fiber collections known as tracts, fasciculi, bundles, peduncles, and lemnisci. The three major fiber systems are the

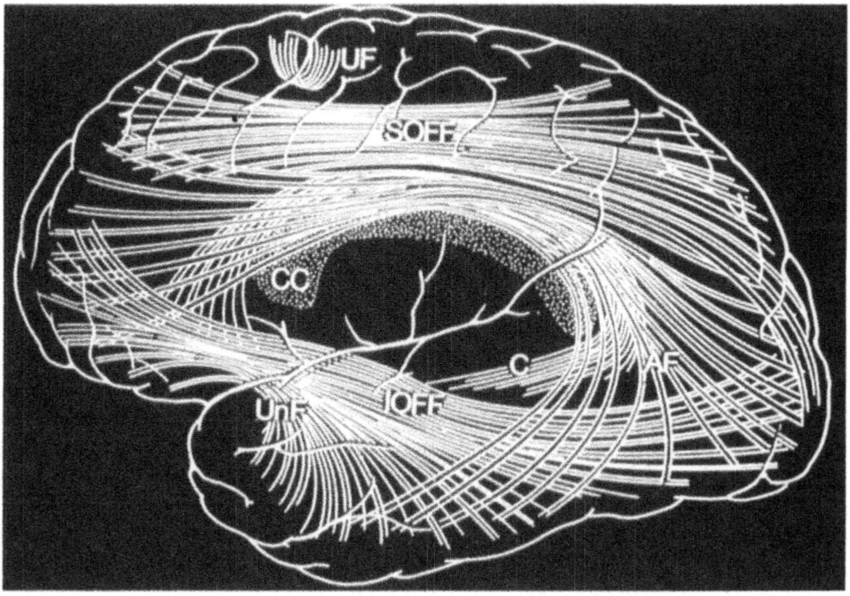

FIGURE 2. Drawing of major white matter tracts in the human brain (CC, corpus callosum; UF, U fiber; SOFF, superior occipitofrontal fasciculus; IOFF, inferior occipitofrontal fasciculus; AF, arcuate fasciculus; C, cingulum; UnF, uncinate fasciculus).

projection fibers, the association fibers, and the commissural fibers. In general, the most critical tracts for the purposes of behavioral neurology are the association tracts that travel within the hemispheres and the commissural tracts that course between them (FIG. 2).[2] Many other smaller tracts, such as the fornix and the median forebrain bundle, also merit attention. However, with the exception of the corpus callosum and some of the association systems, relatively little is known of the neurobehavioral affiliations of these tracts.[2] A better understanding of the function of the many myelinated systems in the brain is a high priority in clinical and basic neuroscience.

The physiology of white matter is organized to maximize the rapid and efficient transfer of information within the nervous system. Myelin forms a sheath that encircles the axon in such as a way that small regions, known as nodes of Ranvier, are left unwrapped; these regions permit the phenomenon of saltatory conduction, whereby the propagation of the action potential along the axon is dramatically accelerated. This feature contributes to the greatly increased conduction velocity of myelinated axons in comparison to those that are unmyelinated. In the normal brain, the white matter ensures that neuronal conduction swiftly occurs, and in all probability contributes to the speed of information processing that typifies normal cognition.[1,2] Another feature of white matter is also likely to impact the organization of higher function; at the macroscopic level, a striking neuroanatomic observation is the abundance of white matter in the frontal regions.[2] The frontal lobes have the highest degree of connectivity of any lobe of the brain and are uniquely positioned to exert an integrative influence via white matter tracts on mental operations performed by other cerebral areas.[5]

DEVELOPMENT AND AGING

The white matter manifests a unique pattern of development throughout the life span. In early life, brain myelination is not complete for many years, probably well beyond the second decade, in contrast to the full complement of brain neurons being present at birth.[2] Conversely, in the later decades of life, a slow, but steady loss of white matter occurs that may in fact be more crucial than the loss of neurons, now thought to be not as pronounced as once believed.[1,2] Although more research on white matter maturation and regression is needed, important clinical implications of this lifelong trajectory of myelinated systems may soon emerge.

In children and adolescents, the myelination of white matter, particularly in the frontal lobes, has been proposed to correlate with the acquisition of mature aspects of personality such as motivation, comportment, and executive function.[1,2] The appearance of these attributes, well known to reflect the function of the frontal lobes, may signify the establishment of connections between the frontal lobes and other brain regions by white matter tracts, thus contributing to the mature behavioral repertoire of adults.[1,2] In addition, a variety of psychiatric disorders in childhood may be influenced by abnormal white matter maturation.[6] Although cortical events including synapse formation, dendritic pruning, and long-term potentiation are clearly important in human maturation, the influence of myelination patterns on human behavioral maturation deserves careful attention.

In contrast, the aging process may involve a progressive diminution of white matter volume. In one autopsy study of normal brains from age 20 to 90, for example, white matter volume loss was just 12%, while neocortical volume loss was 28%.[7] This selective tissue loss has been proposed to account for normal cognitive changes in aging such as slowed speed of information processing, diminished attentional capacity, and forgetfulness.[8] Normal age-related changes in white matter may also contribute to the development of mild cognitive impairment (MCI),[9] a controversial concept recently proposed to represent a transitional stage between normal aging and dementia. Although MCI is most often considered a precursor of Alzheimer's disease (AD),[9] the neuropathologic basis of MCI is unclear and may be explained by white matter loss in some cases.[10] Even if MCI becomes established as a harbinger of AD, age-related changes in white matter cannot be neglected as having a potential impact on cognitive function in older individuals. Normal aging changes affecting myelin must also be considered when the impact of superimposed disease or injury, such as infarction or traumatic brain injury, further damages the white matter.

CONVENTIONAL NEUROIMAGING

Although the appearance of computerized tomography (CT) in the 1970s vastly improved the clinical visualization of the brain, it was the advent of MRI in the 1980s that first permitted the detailed interpretation of the structure of white matter.[2] This major advance permitted a much improved, widespread, and noninvasive depiction of both gray and white matter in living patients without exposure to ionizing radiation. In particular, the capacity of T2-weighted (FIG. 3) and fluid attenuated inversion recovery (FLAIR; FIG. 4) sequences to permit the viewing of myelinated regions quickly led to MRI becoming the preferred method for imaging the white

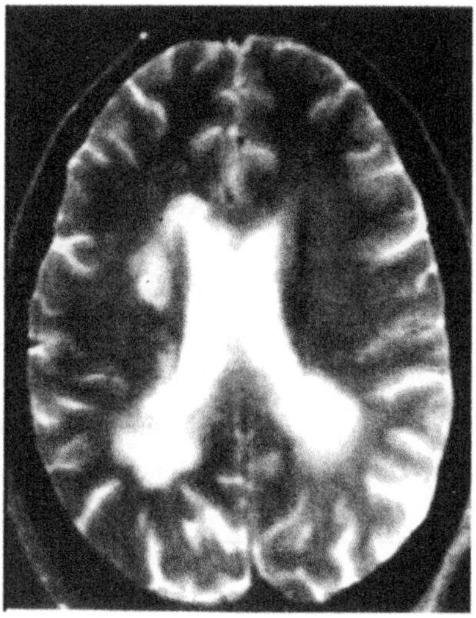

FIGURE 3. T2-weighted MR image of a patient with MS showing multiple demyelinative plaques in the periventricular regions of the brain. Reprinted with permission from ref. 64.

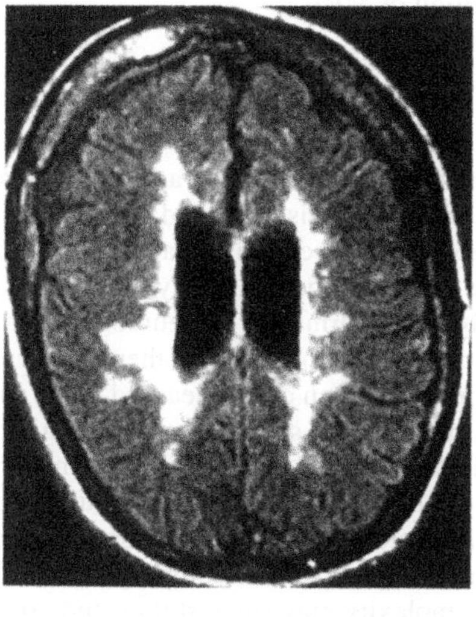

FIGURE 4. Fluid attenuated inversion recovery (FLAIR) MR image of a patient with MS demonstrating multiple demyelinative plaques. This image affords more precise localization of the lesions than the T2-weighted image in FIGURE 3. Reprinted with permission from ref. 65.

TABLE 1. White matter disorders

Category	Example
Genetic	Metachromatic leukodystrophy
Demyelinative	Multiple sclerosis
Infectious	AIDS dementia complex
Inflammatory	Systemic lupus erythematosus
Toxic	Toluene leukoencephalopathy
Metabolic	Cobalamin deficiency
Vascular	Binswanger's disease
Traumatic	Traumatic brain injury
Neoplastic	Gliomatosis cerebri
Hydrocephalic	Normal pressure hydrocephalus

matter and its disorders. The practice of neurology has been changed as much by MRI as it was by its predecessor CT; MRI, for example, has become the most important diagnostic test for MS because of its excellent depiction of demyelinative lesions in the CNS (FIGS. 3 and 4). Old disorders became better understood with MRI, and new ones were soon discovered, leading to a proliferation of clinical research.[2] In addition, detailed observations of white matter at all ages of the human life span were facilitated.[2] One of the major benefits of this research was the opportunity to seek correlations of clinical features with white matter changes on MRI, particularly with regard to higher functions.[2] Although this process is still in its germinal stage, the study of the neurobehavioral aspects of white matter has been enormously facilitated by the application of MRI.

WHITE MATTER DISORDERS

The cerebral white matter is vulnerable to an extraordinarily wide spectrum of neuropathology (TABLE 1). These diseases and injuries vary greatly in terms of etiology, pathogenesis, extent and location of pathology, natural history, prognosis, and treatment.[2] Whereas details of the neurology of these disorders can be found in standard neurologic textbooks, information on neurobehavioral consequences is more limited. The objective here is to develop common themes that occur in these disorders, as from this synthetic approach much can be learned about higher brain function.

One point to keep in mind is that some neuropathologic overlap exists between the white matter disorders and those primarily involving gray matter. In MS, for example, demyelination can be seen in cortical gray matter areas to a variable extent[11] and, conversely, in AD, the prototypical gray matter dementia, damage to white matter tracts occurs from Wallerian degeneration accompanying neuronal loss in the cerebral cortex.[12] Whereas this complexity may suggest that study of white matter disorders as a group may be unproductive, it is clear that much overlap of this kind exists in neurology, psychiatry, and indeed all branches of medicine. There is as much utility in examining details of various white matter disorders as there is in pursuing fine distinctions between degenerative diseases of the cortex such as AD and frontotemporal dementia, or the subtleties of bipolar disorder in contrast to schizophrenia. The focus on white matter serves to introduce a host of testable hypotheses and also to inform the investigation of distributed neural networks that necessarily involve integrated systems of gray and white matter structures working in concert.

The cerebral white matter disorders can be classified as genetic, demyelinative, infectious, inflammatory, toxic, metabolic, vascular, traumatic, neoplastic, and hydrocephalic (TABLE 1).[2] Each category involves a fundamentally distinct pathophysiologic basis, and even the entities within these categories can vary in many clinical and neuropathologic respects. Remarkably, all of these disorders—in excess of 100—can be seen after careful review of available clinical literature to be associated with some form of cognitive or emotional dysfunction.[2] This observation alone validates the emphasis on white matter, but in addition similarities in neurobehavioral dysfunction unify all of these etiologic categories, thus further justifying an emphasis on the white matter as a substrate of higher function in humans.

A comprehensive account of all these disorders is beyond our scope, but a brief survey of selected neurologic entities can highlight the diversity of clinical and neuropathologic phenomena. White matter disorders can involve any age group and may first come to the attention of neurologists, psychiatrists, pediatricians, internists, geriatricians, or psychologists. In infants and children, genetic diseases such as metachromatic leukodystrophy (MLD) illustrate the failure of brain myelin to develop normally, and the subsequent dysmyelination leads to early disability and death, or (in occasional older individuals) a common sequence of psychosis followed by dementia.[13,14] In contrast, young adults are at high risk for the demyelinative diseases such as MS, in which inflammatory destruction of myelin (FIGS. 3 and 4) leads to neurologic and neurobehavioral disability.[15] MS illustrates how a white matter disease formerly regarded as having little significance for behavioral neurology has proven, with the application of modern neuroimaging and clinical assessment, to have major neurobehavioral sequelae.[1,2,15] A number of infectious diseases that produce cognitive decline may affect the white matter, including the acquired immune deficiency syndrome (AIDS) dementia complex[16] and progressive multifocal leukoencephalopathy (PML).[17] Similarly, noninfectious inflammatory diseases such as systemic lupus erythematosus (SLE) can impact the white matter by immune-related processes, and the multiple manifestations of neuropsychiatric lupus are being increasingly recognized.[18] One of the most interesting observations with MRI was in the category of toxic disorders; among the large and growing number of white matter toxins recently identified, the common industrial and household solvent, toluene, has been associated with a disabling leukoencephalopathy in solvent abusers that correlates with the severity of dementia produced (FIG. 5).[19–21] Metabolic disorders of white matter have been noted, among them the commonly sought dementia of vitamin B_{12} (cobalamin) deficiency.[22] One of the most common white matter disorders is the variant of vascular dementia referred to variably as Binswanger's disease (BD) or subcortical ischemic vascular dementia (SIVD).[23,24] Closely related to BD is the genetic disease, cerebral autosomal dominant arteriopathy with subcortical infarcts and leukoencephalopathy (CADASIL).[25,26] Traumatic brain injury (TBI) qualifies as a white matter disorder because of the common white matter lesion in TBI patients known as diffuse axonal injury.[27] A variety of brain neoplasms also primarily affect white matter, and the entity gliomatosis cerebri serves especially well to illustrate the

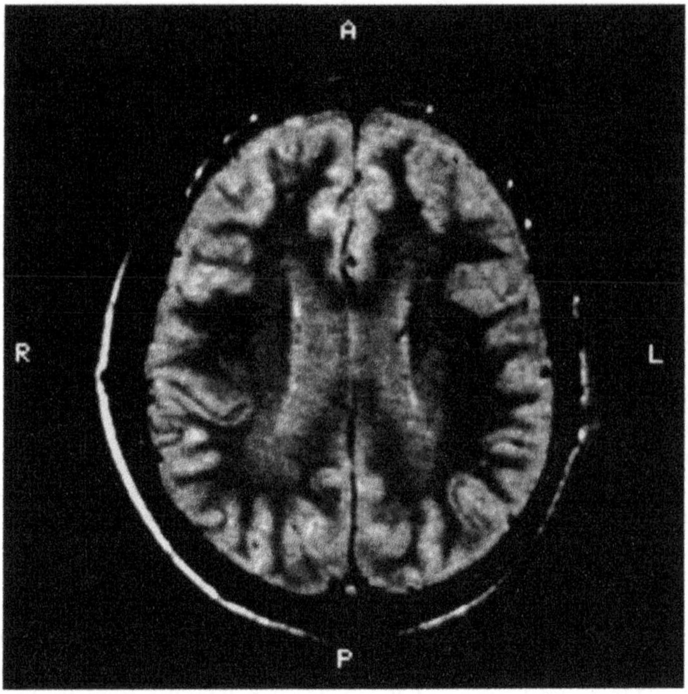

FIGURE 5. T2-weighted MRI scan of a patient with toluene leukoencephalopathy. There is diffuse hyperintensity of the subcortical white matter without a specific area of predilection. Reprinted with permission from ref. 2.

adverse consequences of neoplastic white matter infiltration on neurobehavioral status.[28] Finally, hydrocephalus from any cause exerts its major effect on the periventricular white matter, and examples in children (early hydrocephalus) and adults (normal pressure hydrocephalus) both indicate that white matter damage of this sort can result in major neurobehavioral sequelae.[29,30]

PROGNOSIS AND TREATMENT OF WHITE MATTER DISORDERS

The clinician's ultimate goal in any setting is the best possible patient outcome. Because of the great variety of white matter disorders, however, summary statements about prognosis and treatment can only be given in general terms. While recognizing that specific details of individual disorders may vary considerably, our aim will be to suggest principles pertaining to all of them in an effort to understand the behavioral neurobiology of white matter as a whole.

The prognosis of white matter disorders depends on many factors, including the nature of the neuropathologic insult, the extent and location of white matter damage, the chronicity of the disorder, the age of the patient, and the coexistence of other neurologic and medical disorders. Despite this complex clinical picture, one consistent theme on outcome is that there appears to be a spectrum of severity in white matter disorders, from mild and reversible white matter changes to severe and irreversible necrosis, that generally correlates with prognosis.[2,20] This phenomenon has been most evident in the toxic leukoencephalopathies,[2] but likely applies to other white matter insults as well. For example, there is considerable evidence that, in cases where the neuropathology is severe enough to cause axonal loss, a less favorable outcome can be expected in many white matter disorders, including the leukodystrophies, MS, central pontine myelinolysis and other metabolic encephalopathies, the AIDS dementia complex, BD, and CADASIL.[31]

The observation that white matter pathology can be mild and reversible suggests that, in general, the prognosis of white matter disorders is more optimistic than in gray matter disorders, most notably AD, in which cell bodies, synapses, and receptors are primarily destroyed by the disease process. In many cases, such as subtle toxic leukoencephalopathies characterized by myelin edema alone, full recovery can be expected after withdrawal of the offending agent because the myelin and axons have not been permanently damaged.[20] Even when there is significant myelin damage, substantial recovery can often be predicted if axons are spared.[20]

For clinicians, these observations imply that both prevention and early treatment after diagnosis offer the most potential to improve patient care and outcome. Preventive efforts are of course to be encouraged for many disorders, including infectious, metabolic, toxic, vascular, and traumatic etiologies of white matter damage.[2] Early treatment of individuals who develop many of the white matter disorders can retard, arrest, or reverse the neuropathologic process before axonal loss occurs.[2] As medical and surgical treatments for the various white matter disorders continue to improve, opportunities for effective early intervention should rapidly expand.[2] It has been encouraging, for example, that treatment of MS with immunomodulatory agents such as interferon-beta-1a appears to have a beneficial effect on cognition.[32] Moreover, exciting possibilities are now being considered for many white matter disorders using gene therapy[33] and stem cells.[34] Additional options may include the use of diverse neuropharmacologic agents for target symptoms in these patients, such as CNS stimulants for cognitive slowing[35] and cholinesterase inhibitors for attention and memory loss.[36]

NEUROBEHAVIORAL SYNDROMES ASSOCIATED WITH WHITE MATTER DISORDERS

For those interested in brain-behavior relationships, study of the brain white matter offers an excellent opportunity because most of this component of the brain is devoted to distributed neural networks concerned with higher, not elemental, function.[1,2] The key question is *how* these myelinated systems participate in the networks upon which all higher functions are thought to depend. The role of white matter tracts in human behavior can be directly addressed by the evaluation of individuals with white matter lesions who experience cognitive dysfunction, emotional distress, or both. This process may pose formidable methodological hurdles, but the clinical method of history-taking and physical examination can be joined with neuroimaging to yield a powerful method of correlating the lesion with the behavioral alteration.

In clinical practice, individuals presenting with neurologic disorders can often be best approached by an attempt to classify their symptoms and signs into recognizable constellations known as syndromes. In behavioral neurology, a syndrome diagnosis is a vital component of patient care, as it is crucial that such important syndromes as dementia, amnesia, aphasia, agnosia, visuospatial dysfunction, neglect, executive dysfunction, and the like be distinguished from one another before specific diagnostic tests and therapeutic modalities are recommended. The syndrome approach proves particularly helpful in considering the white matter disorders, in which a surprising number of neurobehavioral syndromes are encountered in addition to well-known elemental neurologic deficits such as hemiparesis, spasticity, visual loss, ataxia, and sphincteric dysfunction. Many patients with white matter disorders, in fact, present with neurobehavioral issues even before other neurologic features come to clinical attention, and the classification of the specific syndrome of impairment has important implications for early diagnosis and treatment. From a theoretical point of view, the recognition of neurobehavioral syndromes in these disorders serves to organize thinking about the role of white matter lesions in brain-behavior relationships generally. A thorough review of the white matter disorders[2] has disclosed that these syndromes can be classified as involving (1) cognitive dysfunction or dementia, (2) focal neurobehavioral disturbances, and (3) neuropsychiatric dysfunction.[2]

WHITE MATTER DEMENTIA

Among the wide array of white matter disorders, cognitive dysfunction is the most common neurobehavioral syndrome that can be related to white matter pathologic burden, and in many cases this disturbance is sufficiently severe to merit the term dementia. Although much clinical evidence favors this view,[1,2] it must be acknowledged that the prevalences of cognitive dysfunction and dementia from white matter disorders as a group are unknown because epidemiologic data of this nature are unavailable. However, the typically diffuse distribution of white matter neuropathology implies that syndromes reflecting widespread white matter dysfunction are far more likely than syndromes reflecting isolated regions of white matter involvement.[1,2] Evidence that focal neurobehavioral syndromes are considerably less common is available from studies of MS, in which cognitive dysfunction or dementia may afflict as many as 65% of patients,[37] whereas aphasia occurs in less than 1%.[38] Similarly, although neuropsychiatric syndromes such as depression are frequent in patients with white matter disorders, the prevalence of these syndromes remains debatable and, in any case, because they may result from many etiologies, there is a less secure association with white matter pathology. Thus, cognitive impairment, especially when it progresses to dementia, is currently the most important source of clinical distress and functional disability that relates directly to damage in the white matter.

White matter dementia is a term introduced to call attention to the morbidity caused by disabling cognitive loss in patients with white matter disorders.[39] Whereas mild cognitive dysfunction is more common than dementia in early stages and may be the presenting feature of the disease,[40] a level of severity justifying the diagnosis of dementia often follows as the disease advances. In MS, for example, a disease that typically involves little or no overt cognitive dysfunction at its onset, it is estimated that 10–20% of patients will develop dementia in the course of the disease.[41] This

TABLE 2. White matter dementia

- Sustained attention deficit
- Executive dysfunction
- Memory retrieval deficit
- Visuospatial impairment
- Psychiatric dysfunction
- Normal language
- Normal extrapyramidal function
- Normal procedural memory

observation implies that effective intervention at an early stage may prevent the development of dementia in many cases. Because understanding the origin and improving the prevention of dementia are such high priorities, the concept of white matter dementia is intended to alert clinicians and researchers to these issues in the context of white matter and its many disorders.

The clinical recognition of early cognitive dysfunction in the white matter disorders, however, may be far from straightforward. Many patients with these disorders present with subtle cognitive symptoms and signs, frequently commingled with other neurologic or medical features of their disease, that challenge the clinician to interpret the relationship of white matter pathologic burden to cognitive status. The range of clinical features heralding the onset of cerebral white matter involvement is impressively broad and may include inattention, executive dysfunction, confusion, memory loss, personality change, depression, somnolence, lassitude, and fatigue. The nonspecific clinical profile of many affected patients often suggests a primary psychiatric disorder, and indeed many patients with white matter dementia have had, in retrospect, early psychiatric dysfunction that antedated measurable cognitive impairment.[2] This interdigitation of cognitive and emotional features may prove particularly challenging in the evaluation of individuals presenting with all varieties of white matter disorder.

Given that the problem of cognitive decline in white matter disorders is important and underappreciated, a deeper understanding of its clinical manifestations is imperative. At this point, only a general portrait of this syndrome is possible as the relevant clinical observations in patient populations have been unsystematic and difficult to compare. Nevertheless, a preliminary profile of deficits and strengths has been developed that may prove useful in diagnosis, counseling, rehabilitation, and research on new therapeutic strategies (TABLE 2).[1,2] From the available information across many lines of inquiry, this profile consists of a sustained attention deficit, executive dysfunction, memory retrieval deficit, visuospatial impairment, and psychiatric dysfunction, and normal language, extrapyramidal function, and procedural memory.[1,2] Pilot data using MS as a prototype white matter disorder suggest that this specific combination appears to differ from that seen with cortical dementias such as AD, in which there are prominent memory encoding deficits (amnesia) and aphasia, but relatively normal attention,[1,2,42] and from subcortical gray matter dementias such as Huntington's disease, in which extrapyramidal function and procedural memory are

TABLE 3. Focal neurobehavioral syndromes

- Amnesia
- Aphasia
  - Broca's
  - Transcortical motor
  - Conduction
  - Wernicke's
  - Global
  - Mixed transcortical
- Apraxia
- Alexia
  - Pure alexia
  - Alexia with agraphia
- Developmental dyslexia
- Gerstmann's syndrome
- Agnosia
  - Visual
  - Auditory
- Neglect
- Visuospatial dysfunction
- Akinetic mutism
- Executive dysfunction
- Callosal disconnection

impaired.[1,2,43] Although much work remains to be done, these data suggest that the white matter may have a unique role in the organization of cognition and emotion.

Sustained attention deficits, executive dysfunction, and memory retrieval deficits may be particularly salient. These problems are typical of patients with white matter disorders and relate to a general slowing of cognition, often called impaired speed of information processing.[2] Neuroanatomically, it is of interest that sustained attention (concentration, vigilance), executive function, and memory retrieval are all closely associated with frontal lobe function, and most white matter disorders show a predilection for the frontal white matter.[2] Even when white matter lesions are situated in more posterior cerebral locations, frontal lobe functions are still affected,[44] probably because of the dense connectivity between frontal and other regions. Visuospatial skills are also affected in white matter disorders,[2] and a wide variety of psychiatric concerns can be implicated as well (see below). In contrast, language is typically normal or mildly affected because of the sparing of language-related cortex,[2] a fact that may lead the clinician to overlook cognitive dysfunction that involves non-linguistic domains. Extrapyramidal function also tends to be intact in white matter disorders, in keeping with the relative sparing of deep gray matter structures.[2] For the same reason, procedural (skill) memory is spared and, although this domain cannot be routinely tested in clinical settings, the theoretical implications of intact procedural memory can be useful in guiding research on brain function.[43,45]

FOCAL NEUROBEHAVIORAL SYNDROMES

The syndrome of white matter dementia vividly serves to illustrate the impact of white matter disorders on cognition, but other neurobehavioral syndromes can also be ascribed to white matter lesions of the brain. In addition to cognitive loss and dementia, a wide range of focal neurobehavioral syndromes have been described (TABLE 3).[2] These disturbances are classic syndromes in behavioral neurology such as aphasia and amnesia; further, although such syndromes are rightly seen as more common with cortical lesions, recent findings have amply confirmed that they may also follow white matter damage. The important point is that focal syndromes are associated with discrete, isolated white matter lesions, in contrast to white matter dementia that results from the more typical widespread pathology of white matter disorders. As examples of this principle, reports have appeared documenting isolated amnesia associated with bilateral mamillothalamic tract infarction[46] and conduction aphasia related to an acute MS plaque in the left arcuate fasciculus.[47] Although uncommon in comparison to the syndromes caused by diffuse white matter damage, the focal neurobehavioral syndromes illustrate the importance of white matter tracts in all domains of mental activity. Further observations of these syndromes and their neuroanatomic correlates can be expected, steadily contributing to the understanding of the many neural networks that subserve the higher functions.

NEUROPSYCHIATRIC SYNDROMES

Abnormalities of white matter have also been associated with a wide spectrum of emotional disturbances, comprising a group that can be referred to as the neuropsychiatric syndromes.[2] This category, however, is less securely defined than the neurobehavioral syndromes discussed above because the correlation of white matter pathology with psychiatric syndromes is less clear. Some suggestion exists that the burden of white matter pathology contributes to emotional dysfunction, but other factors play a role as well and the origin of psychiatric impairment must be considered multifactorial. Nevertheless, much information on the role of white matter in emotional function has recently been assembled, and new insights are illuminating not only the white matter disorders themselves, but many psychiatric diseases as well.

It is useful to consider these neuropsychiatric syndromes in two general groups: (i) the psychiatric problems described in patients with known white matter disorders and (ii) the psychiatric diseases in which white matter abnormalities have been implicated (TABLE 4). In patients with known white matter disorders, numerous reports have documented the presence of depression, mania, psychosis, pathologic affect, euphoria, and fatigue.[2] The correlations between these syndromes and measures of white matter dysfunction are uncertain, however, leaving open the possibilities that psychiatric syndromes may be related only indirectly or not at all to the specific white matter disorder. A complex interplay of many factors including white matter

TABLE 4. Neuropsychiatric syndromes

Psychiatric syndromes in white matter disorders	Psychiatric disorders with white matter abnormalities
Depression	Schizophrenia
Mania	Depression
Psychosis	Bipolar disorder
Pathologic affect	Attention deficit hyperactivity disorder
Euphoria	Autism
Fatigue	Aggression

injury may contribute to emotional distress, and further investigation is clearly needed in this area.[2]

In considering primary psychiatric diseases, typically considered idiopathic and unrelated to known structural brain damage, a number of intriguing reports have recently appeared that exploit MRI techniques to examine the structure of white matter. In schizophrenia, for example, conventional and advanced MRI techniques have detected microstructural white matter abnormalities, and widespread myelin and oligodendrocyte dysfunction leading to altered cerebral connectivity has been implicated.[48] Much evidence also supports an association between white matter changes in geriatric depression,[49] although a firm correlation has yet to be established.[2] MRI white matter hyperintensities are more common in bipolar disorder patients than in the general population.[50] A diminished volume of right frontal white matter has been found to correlate with impaired sustained attention in children with attention deficit hyperactivity disorder.[51] In contrast, an increase in the volume of hemispheric white matter of all lobes has been observed in autism.[52] Finally, diffusion tensor imaging (DTI) studies of schizophrenic men have found a correlation of inferior frontal lobe white matter abnormalities with impulsive aggression.[53] These results indicate that detailed examination of the white matter may offer important new insights into psychiatric disease by concentrating on disruption of the neural networks devoted to emotional function.[2,54]

DIFFUSION TENSOR IMAGING

In recent years, MRI techniques far in advance of those used in routine clinical practice have been developed that further refine the depiction of white matter. DTI is an ideal tool for the study of white matter because it can reveal the course and structural integrity of specific tracts and permit an investigation of their participation in cognitive and emotional operations.[55–57] An abundance of further information on DTI is provided elsewhere in this volume, but the technique is based on the principle of anisotropy, a term referring to the propensity for water in the normal state to diffuse along the direction of white matter tracts. In contrast, damaged white matter is characterized by isotropic diffusion, which is correspondingly less directional and more random. By exploiting these phenomena *in vivo*, DTI can better define the anatomy of normal white matter tracts and the changes they undergo when subjected

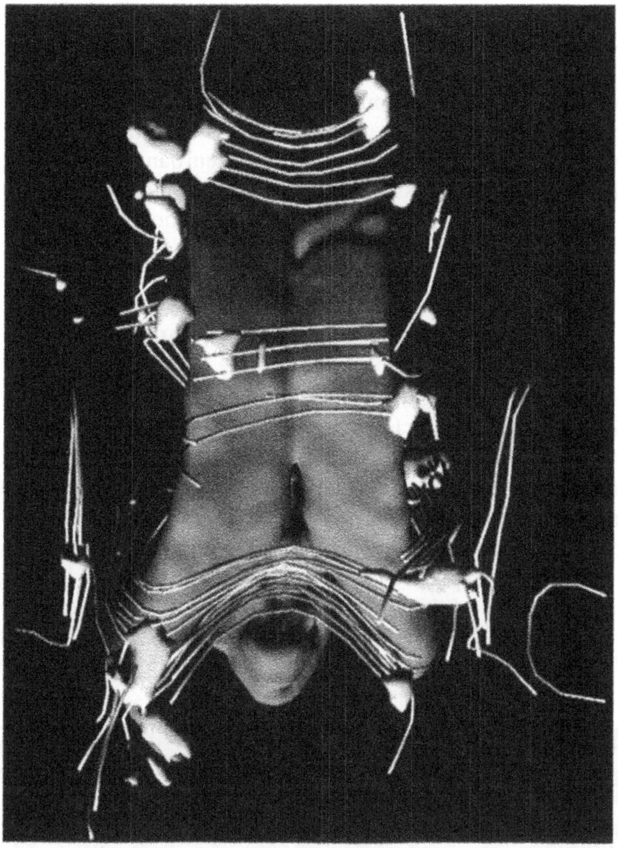

FIGURE 6. DTI study of fiber pathways related to demyelinative lesions in a patient with early MS (blue: corpus callosum; red: white matter fibers; green: demyelinative plaques). [Courtesy of Jack H. Simon, Department of Radiology, University of Colorado Health Sciences Center, Denver, CO.]

to various age-related and neuropathologic processes.[55–57] Particularly intriguing from the perspective of behavioral neurology is the new field of tractography, an application of DTI aimed at delineating the connectivity of both abnormal and normal white matter regions.[55–57] Color-coded maps using DTI are being used to illuminate the effect of white matter disorders such as MS on specific tracts in the brain (FIG. 6). This development presents an entirely new way of imaging sizable and previously obscure regions of the brain, and promises to address the issue of brain connectivity raised decades ago,[3] but still largely unexplored. The application of DTI to both normal and abnormal white matter may offer an enormous step forward for clinicians dealing with ill patients and for researchers attempting to understand the impact of a specific lesion on a given clinical problem. Soon, clinicians and researchers interested in the white matter areas may be able to identify individual tracts with the same specificity as cortical regions and basal ganglia are identified today. By improving the understanding of the origin, course, and destination of white matter tracts, DTI may in fact redefine the neuroanatomy of white matter, which is notably vague and incomplete in today's textbooks; the assessment of white matter in the living brain offers a welcome alternative to the traditional neuroanatomic methods that are obliged to rely on postmortem tissue from humans or other animals. If this promise of DTI technology is fulfilled, the benefits to those concerned with brain-behavior relationships can hardly be overstated.

MAPPING NEURAL NETWORKS

In parallel with advances in structural neuroimaging such as DTI, techniques for the examination of brain function have been equally noteworthy. The cerebral cortex has been the focus of these methods because of its high degree of metabolic activity. The first method to be used for this purpose was positron emission tomography (PET), in which a radiolabeled isotope is injected into the bloodstream and brain metabolic activity is then measured.[58] This technique has enabled impressive depictions of the localization of specific cortical functions (FIG. 7). A less expensive, but also less elegant technique of this type is single photon emission computed tomography (SPECT).[59] Most recently, functional MRI (fMRI) has been added to the armamentarium of functional imaging methods. This method exploits the degree of blood

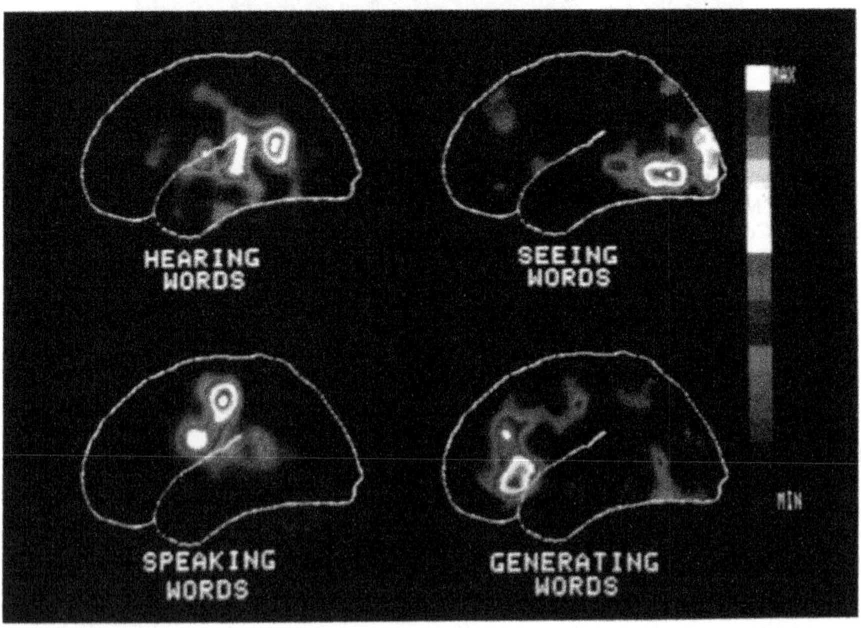

FIGURE 7. PET scan showing the localization of selected cognitive functions in the cerebral cortex. Reprinted with permission from ref. 66.

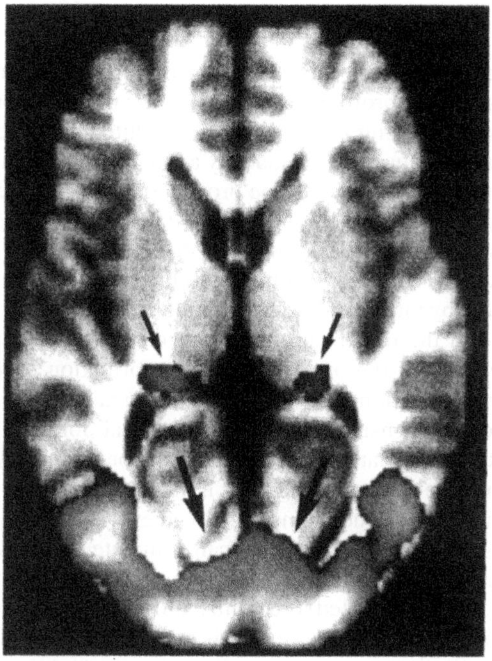

FIGURE 8. fMRI scan demonstrating the localization of cortical and subcortical visual areas (occipital cortex and lateral geniculate nuclei) activated by visual stimuli. Reprinted with permission from ref. 67.

oxygenation in relation to cognitive activity and has more favorable spatial resolution than PET and SPECT. fMRI has been able to demonstrate the gray matter components of neural networks, such as the visual system (FIG. 8), and much effort is being invested in the imaging of cognitive systems as well. Whereas some technical issues remain problematic, fMRI does offer the prospect of more refined analyses of brain-behavior correlations.[60]

The advantages of structural and functional neuroimaging are proving complementary in the pursuit of understanding the architecture of higher function. Gray matter, particularly of the cerebral cortex, is inextricably linked with white matter tracts in the distributed neural networks subserving all domains of mental activity. While PET and fMRI can identify cortical regions involved in cognitive processing, DTI and other related methods can serve to establish the connectivity between these areas. If distributed neural networks consist of both gray and white matter structures linked into coherent arrays, then it should be possible to combine the techniques that can characterize both of these components in formulating a unifying picture.[61] The goal of mapping neural networks is currently motivating a major effort in cognitive neuroscience and related disciplines.

SUMMARY

The white matter is a major constituent of the brain, and the syndromes discussed above serve to illustrate the profound effect that white matter disorders may exert on cognition and emotion. Among these syndromes, white matter dementia appears to be the most important because of its frequency and major clinical impact. Regardless of the specific neurobehavioral syndrome produced, however, the white matter disorders bear directly upon the notion of distributed neural networks. These networks, consisting of widespread ensembles of neurons dedicated to discrete neurobehavioral functions, have recently come to dominate thinking about the higher functions in general. White matter, by virtue of its providing the connectivity in all these networks, plays a pivotal role in the organization of human mental life. In a general sense, whereas the gray matter of the brain subserves information *processing*, the white matter provides for information *transfer*; both are critical for the efficient operations of the neural network responsible for a specific mental domain. In the presence of damaged white matter, information processing occurs only in a slowed and inefficient manner, and there may be no processing at all if the white matter is severely impaired. Considerations such as these lend themselves to an evolving behavioral neurology of white matter, an organizing framework that can stimulate further study and insight.

The study of higher function in humans requires a consideration of all the neural tissues in the brain. Long neglected as a contributor to the organization of cognitive and emotional operations, the white matter is now the subject of substantial effort being devoted to improving our understanding. Among the many approaches that can usefully address this area, the study of individuals with white matter disorders offers a wealth of clinical insights that exploit the time-tested lesion method of behavioral neurology. This process will be complemented by sophisticated neuroimaging techniques such as DTI that increasingly enable detailed visualization of white matter tracts as they participate in the cognitive and emotional operations of distributed neural networks. In practical terms, an appreciation of the neurobehavioral importance of white matter disorders can be of great clinical benefit, especially as it appears that early recognition and treatment can often have an important influence on outcome. In theoretical terms, a focus on the white matter and its disorders promises to expand our knowledge of the brain as an extraordinarily complex organ in which the connectivity provided by white matter is central to cognition, emotion, and consciousness itself.[62] A more complete portrait of the organ of the mind can be anticipated as the details of white matter structure and function become clarified.

REFERENCES

1. Filley, C.M. 1998. The behavioral neurology of cerebral white matter. Neurology **50:** 1535–1540.
2. Filley, C.M. 2001. The Behavioral Neurology of White Matter. Oxford University Press. London/New York.
3. Geschwind, N. 1965. Disconnexion syndromes in animals and man. Brain **88:** 237–294; 585–644.
4. Mesulam, M-M. 1990. Large-scale neurocognitive networks and distributed processing for attention, language, and memory. Ann. Neurol. **28:** 597–613.
5. Mesulam, M-M. 2000. Behavioral neuroanatomy: large-scale neural networks, association cortex, frontal syndromes, the limbic system, and hemispheric specializations.

In Principles of Behavioral and Cognitive Neurology. Second edition, pp. 1–120. Oxford University Press. London/New York.
6. DURSTON, S., H.E. HULSHOFF POL, B.J. CASEY *et al.* 2001. Anatomical MRI of the developing human brain: what have we learned? J. Am. Acad. Child Adolesc. Psychiatry **40:** 1012–1020.
7. PAKKENBERG. B. & H.J.G. GUNDERSEN. 1997. Neocortical neuron number in humans: effect of sex and age. J. Comp. Neurol. **384:** 312–320.
8. FILLEY, C.M. & C.M. CULLUM. 1994. Attention and vigilance functions in normal aging. Appl. Neuropsychol. **1:** 29–32.
9. PETERSEN, R.C., G.E. SMITH, S.C. WARING *et al.* 1996. Mild cognitive impairment: clinical characterization and outcome. Arch. Neurol. **56:** 303–308.
10. MARUYAMA, M., T. MATSUI, H. TANJI *et al.* 2004. Cerebrospinal fluid tau protein and periventricular white matter lesions in patients with mild cognitive impairment: implications for 2 major pathways. Arch. Neurol. **61:** 716–720.
11. AMATO, M.P., M.L. BARTOLOZZI, V. ZIPOLI *et al.* 2004. Neocortical volume decrease in relapsing-remitting MS patients with mild cognitive impairment. Neurology **63:** 89–93.
12. BOZALLI, M., A. FALINI, M. FRANCHESCHI *et al.* 2002. White matter damage in Alzheimer's disease assessed *in vivo* using diffusion tensor imaging magnetic resonance imaging. J. Neurol. Neurosurg. Psychiatry **72:** 742–746.
13. FILLEY, C.M. & K.F. GROSS. 1992. Psychosis with cerebral white matter disease. Neuropsychiatry Neuropsychol. Behav. Neurol. **5:** 119–125.
14. HYDE, T.M., J.C. ZEIGLER & D.R. WEINBERGER. 1992. Psychiatric disturbances in metachromatic leukodystrophy: insights into the neurobiology of psychosis. Arch. Neurol. **49:** 401–406.
15. FEINSTEIN, A. 1999. The Clinical Neuropsychiatry of Multiple Sclerosis. Cambridge University Press. Cambridge/London/New York.
16. BENCHERRIF, B. & D.A. ROTTENBERG. 1998. Neuroimaging of the AIDS dementia complex. AIDS **12:** 233–244.
17. BERGER, J.R. & M. CONCHA. 1995. Progressive multifocal leukoencephalopathy: the evolution of a disease once considered rare. J. Neurovirol. **1:** 5–18.
18. WEST, S. 1994. Neuropsychiatric lupus. Rheum. Dis. Clin. N. Am. **20:** 129–158.
19. FILLEY, C.M., R.K. HEATON & N.L. ROSENBERG. 1990. White matter dementia in chronic toluene abuse. Neurology **40:** 532–534.
20. FILLEY, C.M. & B.K. KLEINSCHMIDT-DEMASTERS. 2001. Toxic leukoencephalopathy. N. Engl. J. Med. **345:** 425–432.
21. FILLEY, C.M., W. HALLIDAY & B.K. KLEINSCHMIDT-DEMASTERS. 2004. The effects of toluene on the central nervous system. J. Neuropathol. Exp. Neurol. **63:** 1–12.
22. KEALEY, S.M. & J.M. PROVENZALE. 2002. Tensor diffusion imaging in B_{12} leukoencephalopathy. J. Comput. Assisted Tomogr. **26:** 952–955.
23. CAPLAN, L.R. 1995. Binswanger's disease—revisited. Neurology **45:** 626–633.
24. KRAMER, J.H., B.R. REED, D. MUNGAS *et al.* 2002. Executive dysfunction in subcortical ischemic vascular disease. J. Neurol. Neurosurg. Psychiatry **72:** 217–220.
25. FILLEY, C.M., L.L. THOMPSON, C-I. SZE *et al.* 1999. White matter dementia in CADASIL. J. Neurol. Sci. **163:** 163–167.
26. HARRIS, J.G. & C.M. FILLEY. 2001. CADASIL: neuropsychological findings in three generations of an affected family. J. Int. Neuropsychol. Soc. **7:** 768–774.
27. HURLEY, R.A., J.C. MCGOWAN, K. ARFANAKIS *et al.* 2004. Traumatic axonal injury: novel insights into evolution and identification. J. Neuropsychiatry Clin. Neurosci. **16:** 1–7.
28. FILLEY, C.M., B.K. KLEINSCHMIDT-DEMASTERS, K.O. LILLEHEI *et al.* 2003. Gliomatosis cerebri: neurobehavioral and neuropathological observations. Cogn. Behav. Neurol. **16:** 149–159.
29. FLETCHER, J.M., T.P. BOHAN, M.E. BRANDT *et al.* 1992. Cerebral white matter and cognition in hydrocephalic children. Arch. Neurol. **49:** 818–824.
30. DEL BIGIO, M.R. 1993. Neuropathological changes caused by hydrocephalus. Acta Neuropathol. **85:** 573–585.
31. MEDANA, I.M. & M.M. ESIRI. 2003. Axonal damage: a key predictor of outcome in human CNS diseases. Brain **126:** 515–530.

32. FISCHER, J.S., R.L. PRIORE, L.D. JACOBS *et al.* 2001. Neuropsychological effects of interferon beta-1a in relapsing multiple sclerosis. Ann. Neurol. **48:** 885–892.
33. HSICH, G., M. SENA-ESTEVES & X.O. BREAKEFIELD. 2002. Critical issues in gene therapy for neurologic disease. Hum. Gene Ther. **13:** 579–604.
34. RICE, C.M., C.A. HALFPENNY & N.J. SCOLDING. 2003. Stem cells for the treatment of neurological disease. Transfus. Med. **13:** 351–361.
35. WEITZNER, M.A., C.A. MEYERS & A.D. VALENTINE. 1995. Methylphenidate in the treatment of neurobehavioral slowing associated with cancer and cancer treatment. J. Neuropsychiatry Clin. Neurosci. **7:** 347–350.
36. ARCINIEGAS, D.B., L.E. ADLER, J. TOPKOFF *et al.* 1999. Attention and memory function after traumatic brain injury: cholinergic mechanisms, sensory gating, and a hypothesis for further investigation. Brain Injury **13:** 1–13.
37. RAO, S.M. 2004. Cognitive function in patients with multiple sclerosis: impairment and treatment. Int. J. MS Care **1:** 9–22.
38. LACOUR, A., J. DE SEZE, E. REVENCO *et al.* 2004. Acute aphasia in multiple sclerosis: a multicenter study of 22 patients. Neurology **62:** 974–977.
39. FILLEY, C.M., G.M. FRANKLIN, R.K. HEATON *et al.* 1988. White matter dementia: clinical disorders and implications. Neuropsychiatry Neuropsychol. Behav. Neurol. **1:** 239–254.
40. FRANKLIN, G.M., L.M. NELSON, C.M. FILLEY *et al.* 1989. Cognitive loss in multiple sclerosis: case reports and review of the literature. Arch. Neurol. **46:** 162–167.
41. RAO, S.M. 1996. White matter disease and dementia. Brain Cogn. **31:** 250–268.
42. FILLEY, C.M., R.K. HEATON, L.M. NELSON *et al.* 1989. A comparison of dementia in Alzheimer's disease and multiple sclerosis. Arch. Neurol. **46:** 157–161.
43. LAFOSSE, J.M., J.R. CORBOY, M.A. LEEHEY *et al.* 2002. Neuropsychological support for the concept of white matter dementia. Neurology **58**(suppl. 3): A355–A356.
44. TULLBERG, M., E. FLETCHER, C. DECARLI *et al.* 2004. White matter lesions impair frontal lobe function regardless of their location. Neurology **63:** 246–253.
45. GABRIELI, J.D., G.T. STEBBINS, J. SINGH *et al.* 1997. Intact mirror-tracing and impaired rotary-pursuit skill learning in patients with Huntington's disease: evidence for dissociable memory systems in skill learning. Neuropsychology **11:** 272–281.
46. YONEOKA, Y., N. TAKEDA, A. INOUE *et al.* 2004. Acute Korsakoff syndrome following mammillothalamic tract infarction. AJNR **25:** 964–968.
47. ARNETT, P.A., S.M. RAO, M. HUSSAIN *et al.* 1996. Conduction aphasia in multiple sclerosis: a case report with MRI findings. Neurology **47:** 576–578.
48. DAVIS, K.L., D.G. STEWART, J.I. FRIEDMAN *et al.* 2003. White matter changes in schizophrenia: evidence for myelin-related dysfunction. Arch. Gen. Psychiatry **60:** 443–456.
49. ALEXOPOULOS, G.S., D.N. KIOSSES, S.J. CHOI *et al.* 2002. Frontal white matter microstructure and treatment response of late-life depression: a preliminary study. Am. J. Psychiatry **159:** 1929–1932.
50. LENOX, R.H., T.D. GOULD & H.K. MANJI. 2002. Endophenotypes in bipolar disorder. Am. J. Med. Genet. **114:** 392–406.
51. SEMRUD-CLIKEMAN, M., R.J. STEINGARD, P. FILIPEK *et al.* 2000. Using MRI to examine brain-behavior relationships in males with attention deficit disorder with hyperactivity. J. Am. Acad. Child Adolesc. Psychiatry **39:** 477–484.
52. HERBERT, M.R., D.A. ZEIGLER, N. MAKRIS *et al.* 2004. Localization of white matter volume increase in autism and developmental language disorder. Ann. Neurol. **55:** 530–540.
53. HOPTMAN, M.J., J. VOLAVKA, G. JOHNSON *et al.* 2002. Frontal white matter microstructure, aggression, and impulsivity in men with schizophrenia: a preliminary study. Biol. Psychiatry **52:** 9–14.
54. FILLEY, C.M. 2001. Neurobehavioral Anatomy. Second edition. University Press of Colorado. Boulder, CO.
55. WAKANA, S., H. JIANG, L.M. NAGAE-POETSCHER *et al.* 2004. Fiber tract–based atlas of human white matter anatomy. Radiology **230:** 77–87.
56. MOSELEY, M., R. BAMMER & J. ILLES. 2002. Diffusion-tensor imaging of cognitive performance. Brain Cogn. **50:** 396–413.
57. TAYLOR, W.D., E. HSU, K. RANGA RAMA KRISHNAN *et al.* 2004. Diffusion tensor imaging: background, potential, and utility in psychiatric research. Biol. Psychiatry **55:** 201–207.

58. CABEZA, R. & L. NYBERG. 1997. Imaging cognition: an empirical review of PET studies with normal subjects. J. Cogn. Neurosci. **9:** 1–26.
59. ALAVI, A. & L.J. HIRSCH. 1991. Studies of central nervous system disorders with single photon emission computed tomography and positron emission tomography. Semin. Nucl. Med. **21:** 58–81.
60. PRICHARD, J.W. & J.L. CUMMINGS. 1997. The insistent call from functional MRI. Neurology **48:** 797–800.
61. WERRING, D.J., C.A. CLARK, G.J. PARKER *et al.* 1999. A direct demonstration of both structure and function in the visual system: combining diffusion tensor imaging with functional magnetic resonance imaging. Neuroimage **9:** 352–361.
62. MESULAM, M-M. 2000. Brain, mind, and the evolution of connectivity. Brain Cogn. **42:** 4–6.
63. SAUNDERS, J.B.D.M. & C.D. O'MALLEY. 1973. The Illustrations from the Works of Andreas Vesalius of Brussels. Dover. New York.
64. ATLAS, S.W., Ed. 1996. Magnetic Resonance Imaging of the Brain and Spine. Second edition. Lippincott-Raven. Philadelphia.
65. MILLER, D.H., J. KESSELRING, W.I. MCDONALD *et al.* 1997. Magnetic Resonance in Multiple Sclerosis. Cambridge University Press. London/New York.
66. PECHURA, C.M. & J.B. MARTIN, Eds. 1991. Mapping the Brain and Its Functions: Integrating Enabling Technologies into Neuroscience Research. National Academy Press. Washington, D.C.
67. KANDEL, E.R., J.H. SCHWARTZ & T.M. JESSELL, Eds. 2000. Principles of Neural Science. Fourth edition. McGraw–Hill. New York.

Adolescents with Disruptive Behavior Disorder Investigated Using an Optimized MR Diffusion Tensor Imaging Protocol

TIE-QIANG LI,[a] VINCENT P. MATHEWS,[a] YANG WANG,[a] DAVID DUNN,[b] AND WILLIAM KRONENBERGER[b]

[a]*Department of Radiology,* [b]*Department of Psychiatry, Indiana University School of Medicine, Indianapolis, Indiana 46202, USA*

ABSTRACT: Adolescents with disruptive behavior disorder (DBD) and controls were investigated using an optimized MR diffusion tensor imaging (DTI) protocol in order to assess any possible structural abnormalities associated with DBD. Thirty-six patients and 40 normal subjects were examined. The extracted diffusion fractional anisotropy (FA) results demonstrate that for the DBD patients there is significantly reduced FA in both the frontal and left temporal regions. The largest brain area with significantly reduced FA is located within the arcuate fasciculus, which has projections extending from the temporal lobe to the frontal lobe along the lateral ventricle, lateral to the tapetum. The reduced FA reflects directly a lower extent of myelination and less coherent fiber track structures in the fasciculus, which in turn may indicate communication weakness among the associated cortical areas. The detected white matter microstructural abnormality, therefore, may be related to the developmental deficits observed in the patient group.

KEYWORDS: adolescents; brain; diffusion tensor imaging (DTI); disruptive behavior disorder (DBD); fractional anisotropy (FA); white matter

INTRODUCTION

Diffusion tensor imaging (DTI) is a promising technique for studying white matter development in the brain and for noninvasive mapping of neuronal connectivity.[1–3] The power of DTI to reveal the details of microstructure in biological tissues relies on the fact that the random, three-dimensional, diffusion-driven molecular motion is sensitive to the tissue structure at a microscopic scale.[4] In the time window of a DTI experiment, which is typically on the order of ~100 ms, the water molecules in the tissue can, on average, move about 10 μm. The mobility of water molecules in tissues may not be the same in all directions since the physical arrangement of tissues or the presence of obstacles limits molecular movement in some directions. This directional dependence of molecular mobility, named as diffusion anisotropy, in a heterogeneous biological system has to be described using at least a second-order symmetric

Address for correspondence: Vincent P. Mathews, Department of Radiology, Indiana University School of Medicine, Indianapolis, IN 46202.
vmathews@iupui.edu

Ann. N.Y. Acad. Sci. 1064: 184–192 (2005).
doi: 10.1196/annals.1340.034

tensor,[5,6] rather than a scalar, as would be appropriate for an isotropic system. The ability for DTI to quantitatively and noninvasively measure the tensor characteristics of water self-diffusion in microscopic biological environments provides a unique opportunity to study myelination maturation, map white matter tracts connecting different brain regions, and reveal the neurological underpinning for normal and abnormal brain functions.

A DTI study entails usually three steps:[7] (1) determination of the six diffusion tensor elements in the laboratory frame with independent diffusion-weighted imaging measurements; (2) evaluation of eigenvalues and eigenvectors of the diffusion tensor; and (3) calculations of diffusion anisotropy and directionality. Because of the requirement for stringent gradient hardware, pulse sequence programming, and post-processing techniques, this procedure, until more recently, was limited to experimental and research settings with MRI physics support. However, the number of DTI capable facilities has recently become more widely available and has been applied in the study of numerous brain pathologies, including demyelinating disorders,[8] epilepsy,[9] brain tumor,[10] and ischemia.[11] In addition to neurological diseases with apparent morphological abnormalities, DTI is also capable of detecting microstructural abnormalities in white matter associated with several neuropsychiatric disorders,[12,13] as well as normal and abnormal brain development.[7,14–16] It is clear that the DTI technique can provide very important insights into the white matter structural changes in the brain that occur during childhood and adolescence.

The objective of the present study was to use DTI for detection of possible white matter microstructural abnormalities associated with disruptive behavior disorder (DBD) in adolescents. DBD is among the most common psychiatric disorders and is associated with substantial short-term and long-term risks.[17,18] Despite improvement in psychotherapeutic and evaluation techniques, a substantial number of adolescents fail to respond to treatment, posing threats to the community and to the adolescents themselves. One factor hindering the treatment of adolescents with DBD is a lack of a known neurobiological etiology. Therefore, a neurobiological investigation is important for improving our understanding of the neurological underpinnings to the aggressiveness of this patient group and may suggest directions for medical treatment of adolescents at risk for dangerous behavior.

MATERIALS AND METHODS

Subjects

Seventy-six adolescents with an average age of 14 ± 1 years were studied. Thirty-six subjects (male/female: 25/11) were diagnosed as patients with DBD, and others were normal controls with no psychiatric diagnosis. For a diagnosis of DBD, all subjects were screened with the DSM-IV criteria for Conduct Disorder or Oppositional-Defiant Disorder, using the K-SADS semistructured diagnostic interview. In order to ensure that the DBD group consisted of individuals with aggressive tendencies, subjects in the DBD group were also required to meet one of the Conduct Disorder symptoms of aggression toward people or animals within the past 6 months. The control subjects (male/female: 21/19) underwent the same battery of cognitive and psychological tests and were required to have no psychiatric diagnosis (screened with clinical interview) and no contact with a mental health professional within the past 3 years.

Protocol

The MR measurements were conducted on a 1.5-T GE Signa whole-body MRI scanner equipped with Echo-Speed gradient system. A standard birdcage CP transmit-receive RF coil was used. All subjects underwent an MRI examination using a protocol that included the following scans: (1) T1-weighted three-plane localizers; (2) high-resolution T1-weighted whole-brain scan using 3D spoiled gradient-recalled echo sequence; and (3) DTI measurements using a custom-made pulse sequence based on single-shot echo-planar imaging acquisition. The main data acquisition parameters were the following: TR = 6 s, spatial resolution = $1.9 \times 1.9 \times 5$ mm^3, matrix size = 128×128, and 22 contiguous slices parallel to the AC-PC line. The distribution of the diffusion-weighting gradients was optimized by minimizing the error propagation from the diffusion-weighted images to the final diffusion tensor image parameters using a downhill simplex algorithm. The design of DTI schemes consisted of 20 distributed diffusion-weighting gradients with $b = 1000$ s/mm^2 and 3 samplings at $b = 0$. The condition number of the transformation matrix for the corresponding DTI scheme was 1.32, which is lowest among the different DTI schemes reported in the literature.[19–21] Each DTI scan lasted 138 s. At least 2 measurements were performed in each subject. To minimize possible motion artifacts in MRI scans, each subject used an individually fitted bite-bar during the entire MRI scans. Each subject was carefully positioned by experienced research technologists so that the AC-PC line was approximately in the transverse image plane. Prior to the MRI scanning, the guardians of the subjects signed consent forms, which were approved by the local institutional review board.

Data Analysis

In order to extract the intersubject comparable diffusion tensor parameters, the MR DTI data from each subject were analyzed using an optimization algorithm that takes into consideration the following three models simultaneously:[22] (1) eddy current artifact correction by estimating the whole-brain-based shearing, scaling, and translation effects; (2) motion artifact correction based on 3D rigid-body motions; and (3) second-order self-diffusion tensor element calculation based on the Stejskal-Tanner equation.[4,23,24] The optimization was performed globally in a least-squares sense for all the voxels inside the brain simultaneously instead of voxel-by-voxel.

After the diffusion tensor calculation, in order to extract intersubject comparable fractional anisotropy (FA) parameter maps, the following postprocessing steps were applied: (1) estimation of eigenvalues and eigenvectors of the diffusion tensor using the single-value decomposition algorithm; (2) calculation of the FA index on the voxel-by-voxel basis according to its definition; and (3) spatial normalization of the FA parameter maps into the Talairach template using SPM2 with the help of the high-resolution 3D anatomical scan. In order to identify the possible brain regions in DBD patients with significantly different FA in comparison with age-matched normal controls, Student's t test statistics were applied to the normalized FA image data. The statistical significance was assessed by using the following two criteria: (1) a pixel-wise t score threshold of 3.2 ($P < 0.001$) and (2) a minimum in-plane cluster size of three interconnected voxels. For each brain region with significant FA deficit, the average region of interest (ROI) FA value for each subject was also determined.

RESULTS

FIGURE 1 shows a typical set of FA parameter maps obtained from a normal control subject. As represented by the contrast differences in the images, the water in white matter tissues exhibits much stronger diffusion anisotropy than that in gray matter. FIGURE 2 shows three cross sections of the normalized average FA data for the normal controls with overlaid t score results, illustrating the brain regions where DBD patients have significantly lowered FA. As depicted by the crossing lines in FIGURE 2, the displayed cross sections reveal the largest ROI where DBD patients have significantly FA reduction in comparison to the normal adolescents. The center of the ROI is located at (–40 mm, –20 mm, 26 mm) on the Talairach coordinate, which was identified as the arcuate fasciculus.

The ROI average FA values from individual subjects are exhibited in FIGURE 3, and the group averages for the controls and DBD patients are also numerically summarized in TABLE 1. As shown, the FA at the arcuate fasciculus for the DBD group is about 13% lower than that for the normal controls on average.

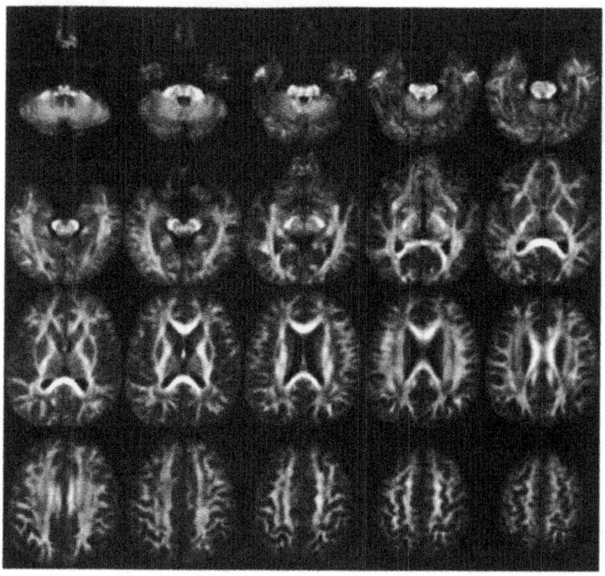

FIGURE 1. A typical set of FA images measured in a normal adolescent using our optimized DTI protocol and postprocessing method. The FA parameter itself valued between 0 and 1 is presented in gray scale.

TABLE 1. The average FA values ± standard deviations for the normal controls and the DBD patient group as measured from the ROI at the arcuate fasciculus

	Control	DBD
n	40	36
FA	0.45 ± 0.02	0.39 ± 0.02

NOTE: On average, the FA at the arcuate fasciculus for the DBD group is about 13% lower than that for the normal controls.

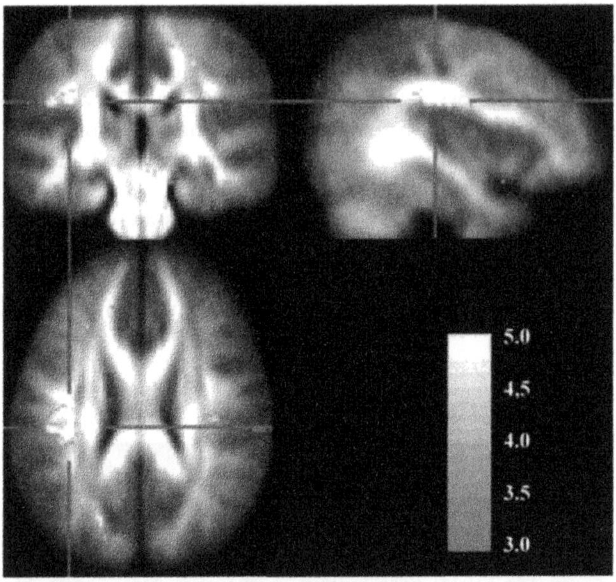

FIGURE 2. Student's t test results overlaid on the spatially normalized average FA data. The cross sections were chosen along the crossing lines, which locate the largest ROI with significant FA deficit in the DBD patient group. The statistical significance of the t test results were assessed by using a t score threshold of ≥3.2 ($P < 0.001$) and a minimum in-plane cluster size of three interconnected voxels. The center of the ROI is located at (–40 mm, –20 mm, 26 mm) on the Talairach coordinate and was identified as the arcuate fasciculus.

In addition to the significant FA deficit in left frontal lobe, the patients with DBD have also less extensive, but significant, diffusion FA deficits in the prefrontal regions. The lower FA implicates a reduced extent of myelination and less coherent fiber tracts. This is further confirmed qualitatively by fiber-tracking results from the ROIs with significantly reduced FA, thus indicating that white matter fiber tract connections in DBD patients may be less extensive and less coherent in these areas.

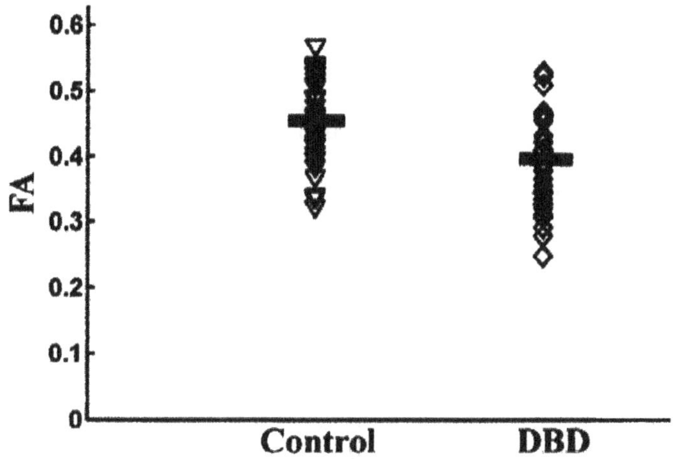

FIGURE 3. The average FA values from individual subjects as evaluated from the ROI at the arcuate fasciculus. The lines indicate the group averages for the normal controls and DBD patients.

DISCUSSION

Because myelination is critical for the functional development of the brain,[25] detailed knowledge on the maturation time course of white matter is not only important to improve our understanding of the normal development of cognitive and emotional functions, but also useful for the detection of individuals at risk for developmental disorders. DTI is becoming a very promising tool for the noninvasive examination of the myelination of axons. Some special challenges, such as motion artifacts and limited available scanning time, are present in using DTI with pediatric populations.[7] Over the last few years, there has been an accelerated interest in this area, as evidenced by the increasing number of published DTI studies.[7,9–11,13–16,26–31] DTI has been used to investigate the anisotropy abnormalities in children with brain injuries[16] and different neurological diseases.[13,16,27,29,30] The results from these studies all indicate that the measurement of diffusion anisotropy with DTI can play an important role in the detection of brain injuries, the assessment of developmental abnormalities, and monitoring of interventions aimed at treating cognitive deficits.

More intriguing findings are the relationships between FA deficiencies in specific brain regions and cognitive performances. For instance, Klingberg *et al.*[13] used DTI to study the relationship between reading disability and microstructure integrity of white matter in the temporoparietal region as indexed by FA and concluded that microstructure integrity of white matter in this area may influence reading ability by determining the communication strength between cortical areas involved in visual, auditory, and language processing. As outlined above, the results from the current study on the DBD population indicate significantly reduced FA in the arcuate fasciculus, which comprises white matter fibers extending from the temporal lobe to the frontal lobe. The involved pathways play a very important role in normal language capability. Preliminary analysis of the neuropsychological test results showed that patients with DBD perform poorer than age-matched controls on reading and vocabulary capability. This would indicate that our result is also in line with the findings from Klingberg's study.[13] If vigorous correlations between FA index in a specific brain area and cognitive deficits associated with a specific neuropsychiatric disorder can be established, this would provide a neurological basis for diagnosis and treatment.

Similar lines of support for the notion above can also be found in DTI studies of normal subjects at different ages.[28,31,32] It seems that both the myelination time course in normal children and demyelination in the elderly are closely related to the changes in cognitive functions. Neil *et al.*[31] studied the mean diffusion coefficient and anisotropy of normal brain in newborns and found mean diffusivity was negatively correlated with gestational age, whereas diffusion anisotropy in centrum semiovale white matter was positively correlated. Klingberg *et al.*[28] reported significant differences in diffusion anisotropy in frontal lobe central white matter of children (mean age: 10 years) when compared with adults (mean age: 27 years). Pfefferbaum *et al.*[32] investigated the age-related decline in brain white matter anisotropy in a group of healthy adults with an age span of five decades (23–76 years) and found significantly reduced FA with aging in many different brain areas. More recent studies[33–35] showed that the age-related decrease in diffusion anisotropy in elders is positively correlated with cognitive decline. This further confirmed the notion that the white matter fiber tract connections between different cortical regions provide very important biological and structural bases for cognitive abilities.[36,37]

As discussed above, DTI has great potential as a tool for noninvasive study of neurological and psychological diseases. It may provide some new insights into the neurological basis for various neuropsychiatric disorders. Regarding technical aspects of this study, the second-order diffusion tensor model seems to provide generally a good fit to the DTI experimental data. The fitting results can reveal the anisotropic nature of white matter and its directionality. One significant limitation of the model is that a voxel contains multiple fibers of different orientations. The voxel FA value simply reflects a population-weighted spatial average over the fibers within the voxel, despite the microscopic nature of the FA parameter itself. To resolve the properties of the individual fibers, either more complicated models[38,39] or high-resolution imaging techniques[36,37] have to be employed, which require lengthening of the data acquisition time.

ACKNOWLEDGMENTS

This work was supported by research grants from the Center for Successful Parenting and the Whitaker Foundation (Grant No. RG-02-0085).

REFERENCES

1. Basser, P.J. *et al.* 2000. *In vivo* fiber tractography using DT-MRI data. Magn. Reson. Med. **44:** 625–632.
2. Conturo, T.E. *et al.* 1999. Tracking neuronal fiber pathways in the living human brain. Proc. Natl. Acad. Sci. USA **96:** 10422–10427.

3. MORI, S. *et al.* 1999. Three-dimensional tracking of axonal projections in the brain by magnetic resonance imaging. Ann. Neurol. **45:** 265–269.
4. BASSER, P.J. & C. PIERPAOLI. 1996. Microstructural and physiological features of tissues elucidated by quantitative-diffusion-tensor MRI. J. Magn. Reson. **B111:** 209–219.
5. ALDROUBI, A. & P.J. BASSER. 1999. Reconstruction of vector and tensor fields from sampled discrete data. Contemp. Math. **247:** 1–15.
6. BEAULIEU, C. 2002. The basis of anisotropic water diffusion in the nervous system—a technical review. NMR Biomed. **15:** 435–455.
7. LI, T.Q. & M.D. NOSEWORTHY. 2002. Mapping the development of white matter tracks with diffusion tensor imaging. Dev. Sci. **5:** 293–300.
8. CASSOL, E. *et al.* 2004. Diffusion tensor imaging in multiple sclerosis: a tool for monitoring changes in normal-appearing white matter. Multiple Sclerosis **10:** 188–196.
9. BRIELLMANN, R.S. *et al.* 2003. Correlation between language organization and diffusion tensor abnormalities in refractory partial epilepsy. Epilepsia **44:** 1541–1545.
10. GAUVAIN, K.M. *et al.* 2001. Evaluating pediatric brain tumor cellularity with diffusion-tensor imaging. AJR Am. J. Roentgenol. **177:** 449–454.
11. SEGHIER, M.L. *et al.* 2004. Combination of event-related fMRI and diffusion tensor imaging in an infant with perinatal stroke. Neuroimage **21:** 463–472.
12. LIM, K.O. & J.A. HELPERN. 2002. Neuropsychiatric applications of DTI—a review. NMR Biomed. **15:** 587–593.
13. KLINGBERG, T. *et al.* 2000. Microstructure of temporo-parietal white matter as a basis for reading ability: evidence from diffusion tensor magnetic resonance imaging. Neuron **25:** 493–500.
14. NEIL, J. *et al.* 2002. Diffusion tensor imaging of normal and injured developing human brain—a technical review. NMR Biomed. **15:** 543–552.
15. HUPPI, P.S. *et al.* 1998. Microstructural development of human newborn cerebral white matter assessed *in vivo* by diffusion tensor magnetic resonance imaging. Pediatr. Res. **44:** 584–590.
16. HUPPI, P.S. *et al.* 2001. Microstructural brain development after perinatal cerebral white matter injury assessed by diffusion tensor magnetic resonance imaging. Pediatrics **107:** 455–460.
17. KRONENBERGER, W.G. & D.W. DUNN. 2003. Learning disorders. Neurol. Clin. **21:** 941–952.
18. DUNN, D.W. & W.G. KRONENBERGER. 2003. Attention-deficit/hyperactivity disorder in children and adolescents. Neurol. Clin. **21:** 933–940.
19. SKARE, S. *et al.* 2000. Condition number as a measure of noise performance of diffusion tensor data acquisition schemes with MRI. J. Magn. Reson. **147:** 340–352.
20. SKARE, S. *et al.* 2000. Noise considerations in the determination of diffusion tensor anisotropy. Magn. Reson. Imaging **18:** 659–669.
21. JONES, D.K., M.A. HORSFIELD & A. SIMMONS. 1999. Optimal strategies for measuring diffusion in anisotropic systems by magnetic resonance imaging. Magn. Reson. Med. **42:** 515–525.
22. ANDERSSON, J.L. & S. SKARE. 2002. A model-based method for retrospective correction of geometric distortions in diffusion-weighted EPI. Neuroimage **16:** 177–199.
23. STEJSKAL, E.O. & J.E. TANNER. 1965. Spin diffusion measurement: spin echoes in the presence of time-dependent field gradient. J. Chem. Phys. **42:** 288–298.
24. TANNER, J.E. 1983. Intracellular diffusion of water. Arch. Biochem. Biophys. **224:** 416–428.
25. PAUS, T. *et al.* 2001. Maturation of white matter in the human brain: a review of magnetic resonance studies. Brain Res. Bull. **54:** 255–266.
26. GUO, A.C. *et al.* 2001. Evaluation of white matter anisotropy in Krabbe disease with diffusion tensor MR imaging: initial experience. Radiology **218:** 809–815.
27. ITO, R. *et al.* 2001. Diffusion tensor brain MR imaging in X-linked cerebral adrenoleukodystrophy. Neurology **56:** 544–547.
28. KLINGBERG, T. *et al.* 1999. Myelination and organization of the frontal white matter in children: a diffusion tensor MRI study. Neuroreport **10:** 2817–2821.
29. NAGY, Z. *et al.* 2003. Preterm children have disturbances of white matter at 11 years of age as shown by diffusion tensor imaging. Pediatr. Res. **54:** 672–679.

30. NAIDU, S. *et al.* 2001. Neuroimaging studies in Rett syndrome. Brain Dev. **23**(suppl. 1): S62–S71.
31. NEIL, J. *et al.* 1998. Normal brain in human newborns: apparent diffusion coefficient and diffusion anisotropy measured by using diffusion tensor MR imaging. Radiology **209:** 57–66.
32. PFEFFERBAUM, A. *et al.* 2000. Age-related decline in brain white matter anisotropy measured with spatially corrected echo-planar diffusion tensor imaging. Magn. Reson. Med. **44:** 259–268.
33. O'SULLIVAN, M. *et al.* 2004. Diffusion tensor imaging of thalamus correlates with cognition in CADASIL without dementia. Neurology **62:** 702–707.
34. O'SULLIVAN, M. *et al.* 2004. Diffusion tensor MRI correlates with executive dysfunction in patients with ischaemic leukoaraiosis. J Neurol. Neurosurg. Psychiatry **75:** 441–447.
35. MOSELEY, M. 2002. Diffusion tensor imaging and aging—a review. NMR Biomed. **15:** 553–560.
36. BUTTS, K. *et al.* 1997. Isotropic diffusion-weighted and spiral-navigated interleaved EPI for routine imaging of acute stroke. Magn. Reson. Med. **38:** 741–749.
37. LI, T-Q., K. KIM & M. MOSELEY. 2005. High-resolution diffusion-weighted imaging with interleaved variable-density spiral acquisitions. J. Magn. Reson. Imaging **21:** 468–475.
38. FRANK, L.R. 2001. Anisotropy in high angular resolution diffusion-weighted MRI. Magn. Reson. Med. **45:** 935–939.
39. TUCH, D.S. *et al.* 2002. High angular resolution diffusion imaging reveals intravoxel white matter fiber heterogeneity. Magn. Reson. Med. **48:** 577–582.

Principal Diffusion Direction in Peritumoral Fiber Tracts

Color Map Patterns and Directional Statistics

AARON S. FIELD,[a,b] YU-CHIEN WU,[c] AND ANDREW L. ALEXANDER[c,d]

Departments of [a]Radiology, [b]Biomedical Engineering, [c]Medical Physics, and [d]Psychiatry, University of Wisconsin–Madison, Madison, Wisconsin, USA

ABSTRACT: The ability of diffusion tensor imaging (DTI) to probe the ultrastructural properties of biological tissues presents new possibilities for DTI-based tissue characterization, with the potential for greater pathologic specificity than conventional imaging methods. This is urgently needed in the diagnosis and treatment of cerebral neoplasms, where clinical decisions depend on the ability to discriminate tumor-involved from uninvolved tissue, a major shortcoming of conventional imaging. Several investigators have attempted to make this determination on the basis of the apparent diffusion coefficient (ADC) or the fractional anisotropy (FA), with mixed results. The directionally encoded color map, with hues reflecting tensor orientation and intensity weighted by FA, provides an aesthetic and informative summary of DTI features throughout the brain in an easily interpreted format. The use of these maps is becoming increasingly common in both basic and clinical research, as well as in purely clinical settings. These examples serve to demonstrate our approach to the quantitation of regional diffusion tensor distributions using directional statistical methods.

KEYWORDS: diffusion tensor imaging (DTI); apparent diffusion coefficient (ADC); fractional anisotropy (FA); color map; direction; edema; glioma

INTRODUCTION

The ability of diffusion tensor imaging (DTI) to probe the ultrastructural properties of biological tissues presents new possibilities for DTI-based tissue characterization, with the potential for greater pathologic specificity than conventional imaging methods. This is urgently needed in the diagnosis and treatment of cerebral neoplasms, where clinical decisions depend on the ability to discriminate tumor-involved from uninvolved tissue, a major shortcoming of conventional imaging. For example, it is frequently impossible to determine whether T2 prolongation in the white matter (WM) surrounding a glioma represents tumor infiltration or bland (tumor-free) edema. Several investigators have attempted to make this determination on the basis of the

Address for correspondence: Aaron S. Field, M.D., Ph.D., Assistant Professor, Department of Radiology, University of Wisconsin, 600 Highland Avenue, CSC E3/311, Madison, WI 53792. Voice: 608-263-7952; fax: 608-265-4152.

as.field@hosp.wisc.edu

Ann. N.Y. Acad. Sci. 1064: 193–201 (2005). © 2005 New York Academy of Sciences.

doi: 10.1196/annals.1340.037

apparent diffusion coefficient (ADC)[1–12] and/or the fractional anisotropy (FA),[5,9–11] with mixed results. The most recent studies employing the most sophisticated DTI methodology have revealed a tendency for tumor infiltration to cause greater FA reductions than edema.[5,10,11] However, these differences have been small and they may prove less significant to decision-making in individual patients than to statistical comparisons of sizable groups. We have observed substantial overlap of FA distributions between infiltrating gliomas and bland edema.[13]

The ADC and FA constitute just a fraction of the information available from the diffusion tensor. The examination of individual eigenvalues, as well as measures of

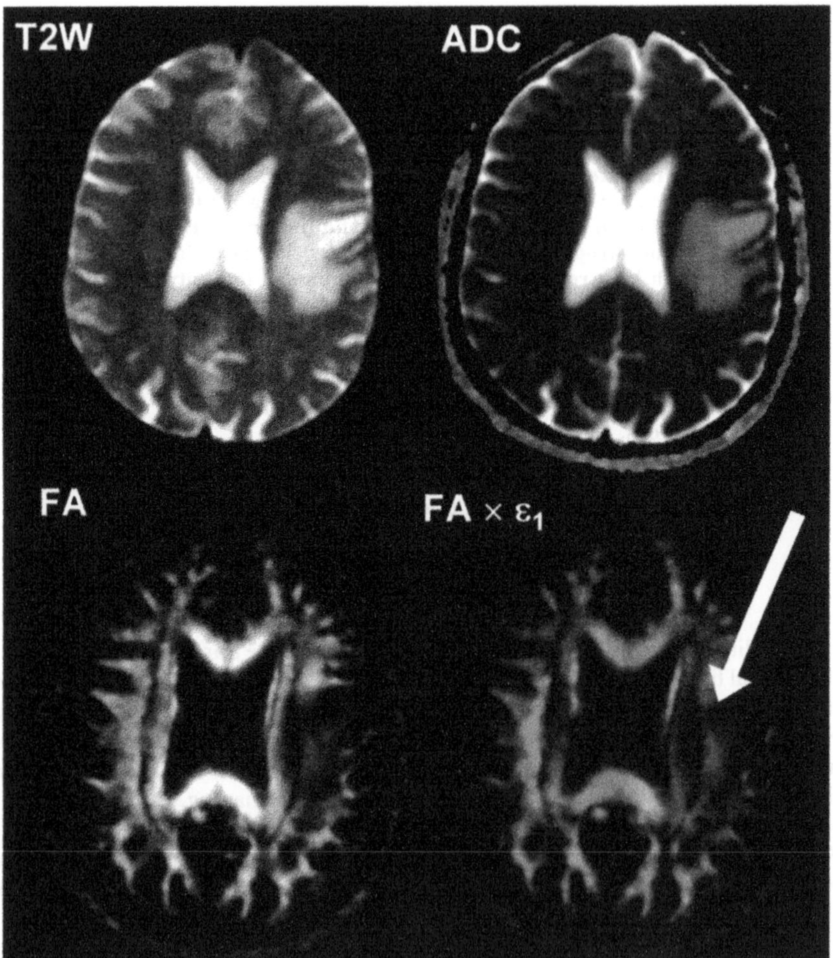

FIGURE 1. T2-weighted (T2W), ADC, FA, and FA-weighted, directionally encoded DTI color maps (FA × ε_1) in a patient with a region of bland (tumor-free) vasogenic edema involving the left corona radiata and superior longitudinal fasciculus.[16] This region shows reduced FA, but normal color hues (*arrow*).

tensor shape and orientation, opens many new possibilities for tissue characterization. For example, the major eigenvalue of the diffusion tensor has been found to be significantly lower in the WM surrounding high-grade gliomas than in the WM surrounding metastases, even when the anisotropy has shown no difference.[14] Tensor shape metrics have revealed a relative shift from linear diffusion (prolate tensors) to planar diffusion (oblate tensors) in peritumoral fiber tracts, relative to contralateral tracts.[15] Such examples in which the diffusion tensor is exploited beyond the ADC and FA for purposes of tumoral or peritumoral tissue characterization are only just beginning to be explored.

The directionally encoded color map, with hues reflecting tensor orientation and intensity weighted by FA, provides an aesthetic and informative summary of DTI features throughout the brain in an easily interpreted format. The use of these maps is becoming increasingly common in both basic and clinical research, as well as in purely clinical settings. Our group has described four distinct patterns of tumor-related pathology in WM tracts on FA-weighted, directionally encoded color maps, categorized on the basis of tensor anisotropy and orientation.[16] Unfortunately, these

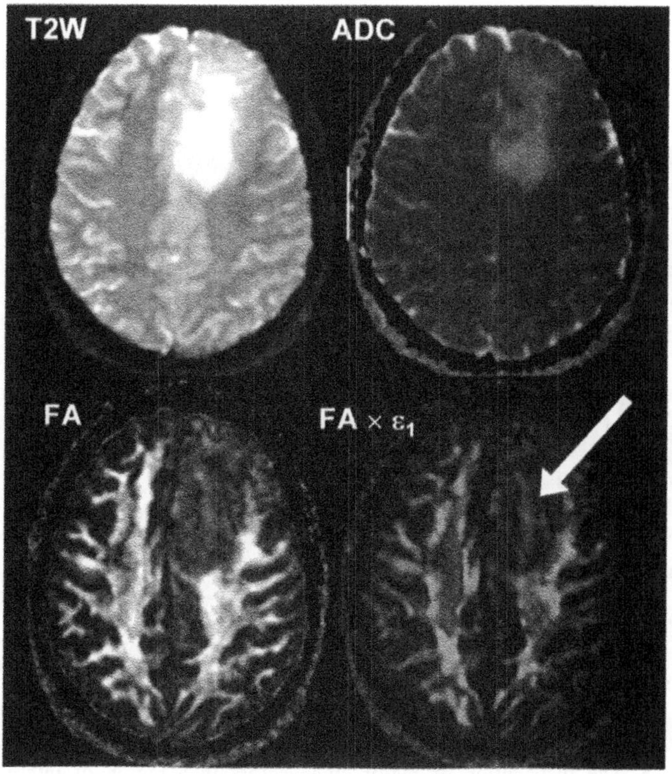

FIGURE 2. T2-weighted (T2W), ADC, FA, and FA-weighted, directionally encoded DTI color maps (FA $\times$ ε_1) in a patient with an infiltrating glioma involving the left corona radiata and subcortical U-fibers.[16] This region shows both reduced FA and abnormal color hues (*arrow*), which are not explained by any apparent tract deviation.

patterns rely on subjective assessments of color hue and intensity; although such assessments can be clinically relevant, a more objective, quantitative approach is needed for purposes of further research. Specifically, a formal methodology for analyzing regional distributions of tensor orientations, including measures of regional directional tendency and dispersion, intervoxel coherence, and interhemispheric symmetry, is needed in order to exploit DTI more reliably in the setting of cerebral neoplasm. We are developing various metrics to meet this need,[17] employing the theory and methods of directional statistics.

For example, we have observed several WM tracts infiltrated by bland edema where anisotropy is reduced with no appreciable change in hue on directionally encoded color maps, while several tracts infiltrated by gliomas have had anisotropy reductions accompanied by altered hues that were unexplained by visible tract deviations (FIGS. 1 and 2). Subjective impressions such as these must be validated quantitatively, motivating our development of directional statistical approaches.

REGIONAL METRICS FOR TENSOR ORIENTATION AND SYMMETRY

Directional Histograms ("Rose" Diagrams)

The distribution of major eigenvectors in a region of interest (ROI) can be described in terms of their polar and azimuthal angles (θ and ϕ, respectively) in 3D space.[17]

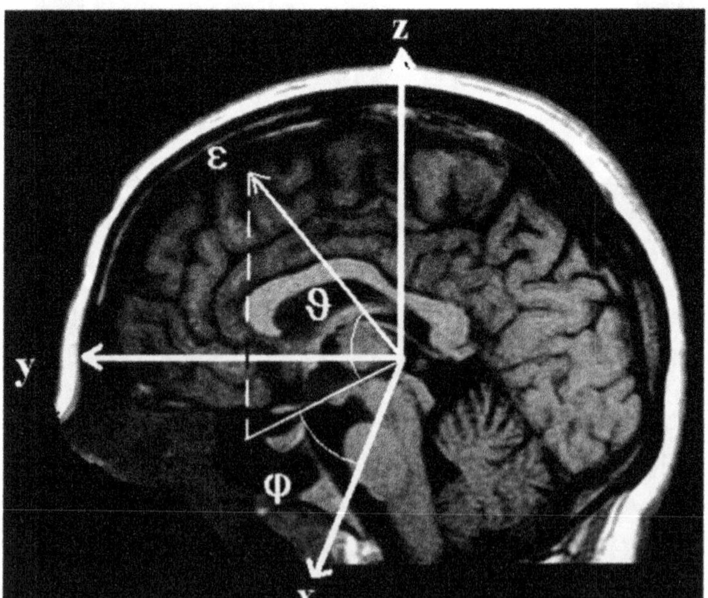

FIGURE 3. The distribution of major eigenvectors in a region of interest can be described in terms of their polar and azimuthal angles (θ and ϕ, respectively) in 3D space.[17] Specifically, the polar angle (θ) denotes the angle between the eigenvector and the x-y plane, and the azimuthal angle (ϕ) denotes the angle between the eigenvector's projection in the x-y plane and the x-axis.

Specifically, the polar angle (θ) denotes the angle between the eigenvector and the *x-y* ("axial") plane, and the azimuthal angle (ϕ) denotes the angle between the eigenvector's projection in the *x-y* plane and the *x*-axis (right-left direction) (FIG. 3). Since the WM tract orientations will be mirrored across the cerebral hemispheres, the rotation conventions of ϕ are reversed for regions in the left and the right hemispheres to facilitate symmetry comparisons. The distribution of major eigenvector directions can be calculated for any ROI and subjected to further analysis or displayed as a pair of spherical-coordinate histograms (polar and azimuthal), or "rose" diagrams. In a rose diagram, the arc subtended by each sector bin represents a range of angles, and the length of each bin reflects the frequency of occurrence for that range (FIG. 4). Note that this approach is not rotationally invariant unless θ and ϕ are referenced to anatomically standardized planes. For example, the *y-z* ("sagittal")

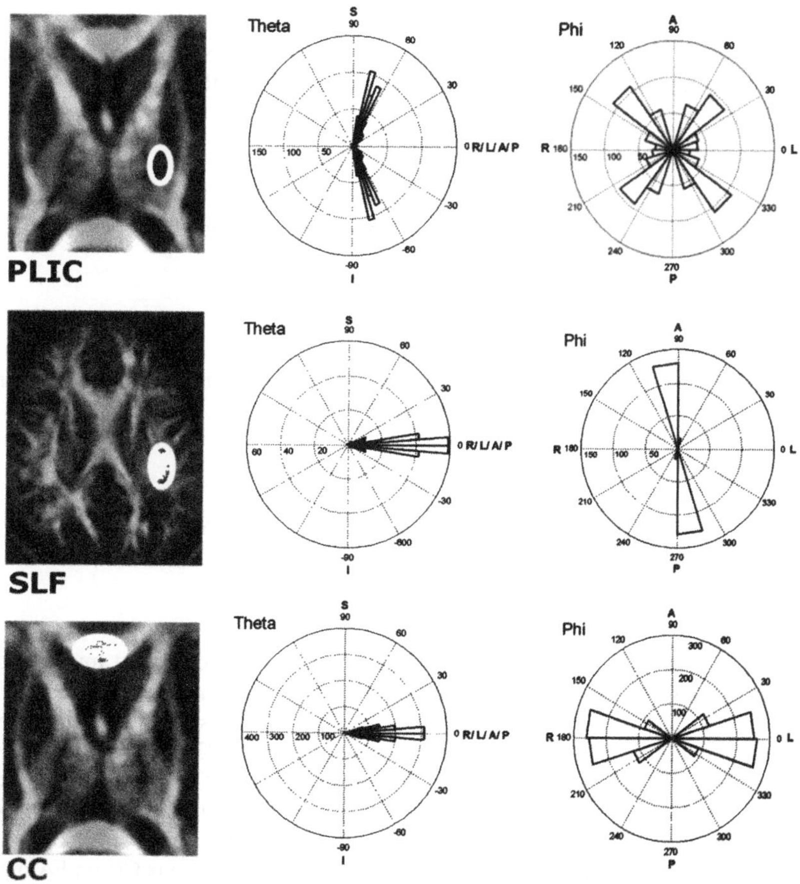

FIGURE 4. Directional histograms (rose diagrams) for posterior limb of internal capsule (PLIC), superior longitudinal fasciculus (SLF), and genu of the corpus callosum (CC) in a normal subject. Note the unique distributions characterizing each of these major tracts. These distributions can be subjected to statistical analysis seeking changes related to pathology.

plane could be defined along the interhemispheric fissure, and the *x-y* plane along the intercommissural line and perpendicular to the *y-z* plane.

Scatter Matrix

The "preferred" orientation and dispersion of major eigenvectors (analogous to mean and variance, respectively, of eigenvector orientation) in an ROI can also be summarized in terms of the scatter matrix.[17] If ε_{1i} is the major eigenvector of the diffusion tensor for the *i*-th voxel within an ROI containing *n* voxels, then the scatter matrix for the ROI is defined as

$$\boldsymbol{T} = (1/n)\sum_{i=1}^{n} \varepsilon_{1i}\varepsilon_{1i}^{t}$$

where *t* denotes the vector transpose. Like the diffusion tensors themselves, $\boldsymbol{T}$ is a 3×3 diagonally symmetric matrix with its own eigenvectors, $\boldsymbol{t}_1$, $\boldsymbol{t}_2$, and $\boldsymbol{t}_3$, which describe the orientation of the tensor distribution within the ROI.

Symmetry Index

A measure of interhemispheric symmetry for homologous ROIs in the brain may be defined as the dot product of the scatter matrix major eigenvectors for left and right ROIs:[17]

$$S = \boldsymbol{t}_{1\mathbf{L}} \cdot \boldsymbol{t}_{1\mathbf{R}}.$$

A value of *S* approaching unity indicates a high degree of interhemispheric symmetry, while *S* near zero indicates that the left and right tensor distributions are nearly orthogonal. Note that accurate *S* measurements require the *y-z* plane to be parallel to the interhemispheric fissure.

CASE EXAMPLES: VASOGENIC EDEMA AND INFILTRATING GLIOMA

Case 1: Vasogenic Edema

A patient with a small (~1 cm) brain metastasis surrounded by a much larger region of vasogenic edema (presumed tumor-free) involving the corona radiata (CR) underwent DTI at 1.5 T using a quadrature head coil and a single-shot, diffusion-weighted spin-echo EPI sequence with diffusion encoding in 23 directions ($b = 0$ and 912 s/mm^2, slice thickness = 3 mm, TR/TE = 4000/72 ms, FOV = 24 cm, matrix = 128×128 interpolated to 256×256, NEX = 4). The diffusion tensor at each voxel was estimated and diagonalized, and maps of FA were generated, using published methods.[18,19] A 3D ROI was manually traced within the edematous CR on FA maps and a homologous ROI was traced in the contralateral CR. Directional histograms for these ROIs (FIG. 5) revealed only nominal changes in tensor distributions, and the symmetry index, calculated as described above, was $S = 0.99$, indicating a high degree of interhemispheric symmetry.

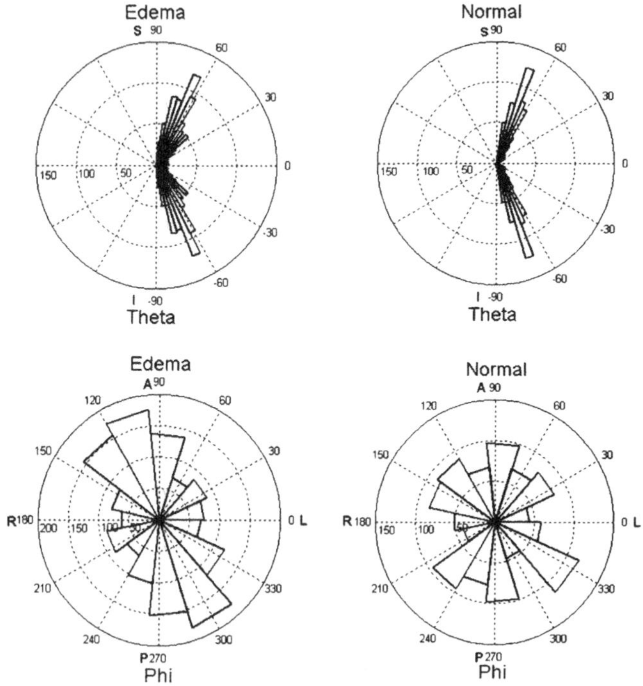

FIGURE 5. ROI rose diagrams of the corona radiata (CR) in case 1, a patient with bland vasogenic edema.[17] (*Left*) Edema ROI. (*Right*) Homologous ROI in normal contralateral CR. Only nominal changes in tensor distributions are evident. Symmetry index $S = 0.99$.

Case 2: Infiltrating Glioma

A patient with an oligodendroglioma (WHO grade II) infiltrating the superior longitudinal fasciculus (SLF) underwent DTI with postprocessing as described for case 1. Directional histograms for homologous SLF ROIs (FIG. 6) revealed more substantial changes in tensor distributions than those seen in the edema case, and the symmetry index was $S = 0.74$, indicating a lesser degree of interhemispheric symmetry.

CONCLUSIONS

These examples serve to demonstrate our approach to the quantitation of regional diffusion tensor distributions using directional statistical methods. Other approaches are certainly possible (e.g., ref. 20). Quantitation along these lines is urgently needed in order to validate the subjective impressions derived from the popular directionally encoded color maps and to enable statistical data analysis whenever relevant changes in fiber tract orientations are postulated or (subjectively) observed. The rose diagrams enable histogram-based analyses of major eigenvector distributions within ROIs;

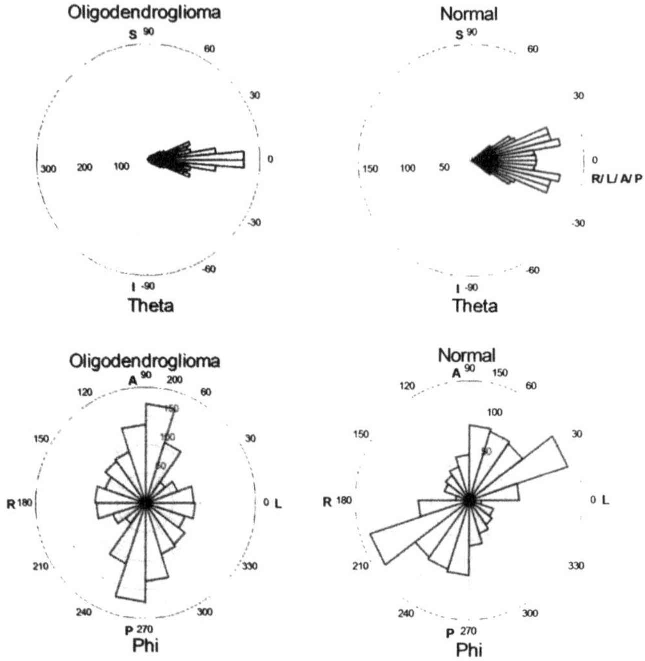

FIGURE 6. ROI rose diagrams of the superior longitudinal fasciculus (SLF) in case 2, a patient with an infiltrating oligodendroglioma.[17] (*Left*) Tumor ROI. (*Right*) Homologous ROI in normal contralateral CR. Changes in tensor distributions are more significant than for bland edema in case 1. Symmetry index $S = 0.74$.

major CNS fiber tracts are expected to have characteristic profiles on these histograms (e.g., FIG. 4), which may be tested for changes resulting from pathology. The symmetry index, derived from the scatter matrix, enables a patient whose disease is limited to one hemisphere to serve as his/her own control. The lesser degree of interhemispheric symmetry that our approach revealed in an infiltrating glioma, as compared to vasogenic edema, suggests that regional analysis of tensor orientation may have an important role to play in this and other problems of tissue characterization. Further study is needed to determine the sensitivity and specificity of our approach in the face of variations in tumor type, grade, location, prior treatment, etc.

REFERENCES

1. BRUNBERG, J.A., T.L. CHENEVERT, P.E. MCKEEVER *et al.* 1995. *In vivo* MR determination of water diffusion coefficients and diffusion anisotropy: correlation with structural alteration in gliomas of the cerebral hemispheres. AJNR Am. J. Neuroradiol. **16:** 361–371.
2. STADNIK, T.W., C. CHASKIS, A. MICHOTTE *et al.* 2001. Diffusion-weighted MR imaging of intracerebral masses: comparison with conventional MR imaging and histologic findings. AJNR Am. J. Neuroradiol. **22:** 969–976.

3. KONO, K., Y. INOUE, K. NAKAYAMA *et al.* 2001. The role of diffusion-weighted imaging in patients with brain tumors. AJNR Am. J. Neuroradiol. **22:** 1081–1088.
4. SINHA, S., M.E. BASTIN, I.R. WHITTLE & J.M. WARDLAW. 2002. Diffusion tensor MR imaging of high-grade cerebral gliomas. AJNR Am. J. Neuroradiol. **23:** 520–527.
5. BASTIN, M.E., S. SINHA, I.R. WHITTLE & J.M. WARDLAW. 2002. Measurements of water diffusion and T1 values in peritumoral oedematous brain. Neuroreport **13:** 1335–1340.
6. TIEN, R.D., G.J. FELSBERG, H. FRIEDMAN *et al.* 1994. MR imaging of high-grade cerebral gliomas: value of diffusion-weighted echo planar pulse sequences. AJR Am. J. Roentgenol. **162:** 671–677.
7. KRABBE, K., P. GIDEON, P. WAGN *et al.* 1997. MR diffusion imaging of human intracranial tumours. Neuroradiology **39:** 483–489.
8. CASTILLO, M., J.K. SMITH, L. KWOCK & K. WILBER. 2001. Apparent diffusion coefficients in the evaluation of high-grade cerebral gliomas. AJNR Am. J. Neuroradiol. **22:** 60–64.
9. LU, S., D. AHN, G. JOHNSON & S. CHA. 2003. Peritumoral diffusion tensor imaging of high-grade gliomas and metastatic brain tumors. AJNR Am. J. Neuroradiol. **24:** 937–941.
10. LU, S., D. AHN, G. JOHNSON *et al.* 2004. Diffusion-tensor MR imaging of intracranial neoplasia and associated peritumoral edema: introduction of the tumor infiltration index. Radiology **232:** 221–228.
11. PROVENZALE, J.M., P. MCGRAW, P. MHATRE *et al.* 2004. Peritumoral brain regions in gliomas and meningiomas: investigation with isotropic diffusion-weighted MR imaging and diffusion-tensor MR imaging. Radiology **232:** 451–460.
12. MAIER, S.E., P. BOGNER, G. BAJZIK *et al.* 2001. Normal brain and brain tumor: multicomponent apparent diffusion coefficient line scan imaging. Radiology **219:** 842–849.
13. FIELD, A.S., A.L. ALEXANDER, V.M. HAUGHTON *et al.* 2002. Effect of vasogenic edema on sensitivity of white matter tractography with diffusion tensor imaging. *In* ASNR 2002 Proceedings, American Society of Neuroradiology 40th Annual Meeting, Vancouver, British Columbia, Canada, May 11–17, 2002; O-269, p. 199.
14. WIEGELL, M.R., J.W. HENSON, D.S. TUCH & A.G. SORENSEN. 2003. Diffusion tensor imaging shows potential to differentiate infiltrating from non-infiltrating tumors. *In* Proceedings of the International Society for Magnetic Resonance in Medicine (ISMRM) 11th Scientific Meeting, Toronto, Ontario, Canada, July 10–16, 2003, p. 2075.
15. ZHANG, S., M.E. BASTIN, D.H. LAIDLAW *et al.* 2004. Visualization and analysis of white matter structural asymmetry in diffusion tensor MRI data. Magn. Reson. Med. **51:** 140–147.
16. FIELD, A.S., A.L. ALEXANDER, Y-C. WU *et al.* 2004. Diffusion tensor eigenvector directional color imaging patterns in the evaluation of cerebral white matter tracts altered by tumor. J. Magn. Reson. Imaging **20:** 555–562.
17. WU, Y-C., A.S. FIELD, M. CHUNG *et al.* 2004. Quantitative analysis of diffusion tensor orientation: theoretical framework. Magn. Reson. Med. **52:** 1146–1155.
18. HASAN, K.M., D.L. PARKER & A.L. ALEXANDER. 2001. Comparison of gradient encoding schemes for diffusion-tensor MRI. J. Magn. Reson. Imaging **13:** 769–780.
19. HASAN, K.M., P.J. BASSER, D.L. PARKER & A.L. ALEXANDER. 2001. Analytical computation of the eigenvalues and eigenvectors in DT-MRI. J. Magn. Reson. **152:** 41–47.
20. SCHWARTZMAN, A., R.F. DOUGHERTY & J.E. TAYLOR. 2004. Comparison of principal diffusion directions using directional statistics. *In* Proceedings of the 10th Annual Meeting of the Organization for Human Brain Mapping, Budapest, Hungary, June 13–17, 2004; WE-300.

Applications of Diffusion Tensor MR Imaging in Multiple Sclerosis

YULIN GE, MENG LAW, AND ROBERT I. GROSSMAN

Department of Radiology, New York University Medical Center, New York, New York 10016, USA

ABSTRACT: Multiple sclerosis (MS) is an inflammatory demyelinating disease of the central nervous system that is the most common cause of nontraumatic disability in young adults in the United States. In recent years, magnetic resonance imaging (MRI) has been established as an important paraclinical tool in MS for the assessment of clinical diagnosis, natural history, and treatment effects. In MS studies, there are many advantages to having a sensitive and reliable *in vivo* method for investigating the specific pathological changes of white matter and its integrity during the disease process. As a consequence, in the past decade, the application of MRI to the study of MS has been explored from conventional MRI to new advanced quantitative techniques with greater pathological specificity and sensitivity. Diffusion tensor imaging (DTI) is one of the most promising techniques with regard to MS. It quantifies the amount of nonrandom water diffusion within tissues and provides unique *in vivo* information about the pathological processes that affect water diffusion as a result of brain microstructural damage. This review outlines the current state of the art and future direction of DTI and fiber tractography in the study of MS disease.

KEYWORDS: diffusion tensor imaging (DTI); multiple sclerosis (MS); fiber tractography; lesion; white matter

INTRODUCTION

Multiple sclerosis (MS) is an inflammatory demyelinating disease of the central nervous system (CNS) that is the most common cause of nontraumatic disability in young adults in the United States.[1] Neuropathologic findings in MS include a T-cell-mediated inflammatory process, which is associated with destruction of myelin sheaths, indicating that it is mainly a white matter–presenting disease. In addition, axonal injury is found to occur in both acute inflammatory and chronic MS lesions.[2] Clinical features of MS may vary and involve the motor, sensory, cognitive, and visual pathways. Although the pathological mechanisms underlying different disease courses are still not well known, there are several phenotypes of MS with regard to distinct clinical manifestations of the disease, including relapsing-remitting (RR), secondary progressive (SP), and primary progressive (PP) MS.[3]

Address for correspondence: Meng Law, Department of Radiology, New York University Medical Center, 530 First Avenue, Basement HCC, New York, NY 10016. Voice: 212-263-3854; fax: 212-263-8186.
lawm01@med.nyu.edu

Ann. N.Y. Acad. Sci. 1064: 202–219 (2005). © 2005 New York Academy of Sciences.
doi: 10.1196/annals.1340.039

In recent years, magnetic resonance imaging (MRI) has been established as an important paraclinical tool in MS for the assessment of clinical diagnosis, natural history, and treatment effects.[4,5] In MS studies, there are many advantages to having a sensitive and reliable *in vivo* method for investigating the specific pathological changes of white matter and its integrity during the disease process. As a consequence, in the past decade, the application of MRI to the study of MS has been explored from conventional MRI to new advanced quantitative techniques with greater pathological specificity and sensitivity.[6] Diffusion tensor imaging (DTI) is one of the most promising of these techniques that has been investigated in the study of MS. It quantifies the amount of nonrandom water diffusion within tissues and provides unique *in vivo* information about the pathological processes that affect water diffusion as a result of brain microstructural damage.

DTI is utilized in an experimental and, in some instances, in a clinical setting to provide *in vivo* visualization of white matter microstructure and fiber organization that is not available with other imaging methods. The diffusion ellipsoid obtained from DTI not only provides a quantitative assessment of the magnitude of water diffusion, but also the predominant direction of water diffusion within the image voxel. In brain white matter, the motion of water molecules can be hindered by the presence of structural barriers created by the highly organized myelinated axonal fiber tracts; therefore, water diffusion is much greater along the fibers than across the fiber.[7] Such information can be quantified by deriving quantitative scalar indices, such as mean diffusivity (MD) and fractional anisotropy (FA), which have significant clinical value for the cellular level integrity of highly ordered white matter tissue. Since DTI data are sensitive to anisotropy concerning the orientation of tissue microstructure, this information also offers the basis of fiber tractography, a method to determine the pathways of anatomical connectivity of the CNS.

Unlike conventional MRI, DTI provides a unique method of imaging contrast and quantification that allows better understanding of the pathophysiology of MS-related white matter damage that would also be useful for monitoring disease progression and treatment effects. This review outlines the current state of the art and future direction of DTI and fiber tractography in the study of MS disease.

DTI STUDIES IN MS LESIONS

MS lesions are pathologically heterogeneous and appear as different patterns on MRI with variable sizes and appearance; some undergo acute inflammatory changes, while others may show extensive tissue destruction. Larsson *et al.*[8] first applied diffusion imaging in 1992 in a study of MS, and later much of the early work came from the use of diffusion-weighted MRI in MS studies.[8–14] All of these studies have shown increased MD in lesions of MS patients when compared with the normal white matter of healthy controls. The higher values of MD in MS lesions are consistent with the findings in studies[15,16] of experimental allergic encephalomyelitis (EAE), an animal model of MS, in which water diffusion was found increased in the experimental pathology, suggesting that inflammatory demyelinating changes in MS can be reflected by the water diffusion changes on diffusion-weighted imaging.

In MS lesions, the highest diffusion values appear to be found in nonenhancing T1-hypointense lesions as compared with enhancing lesions and nonenhancing T1-

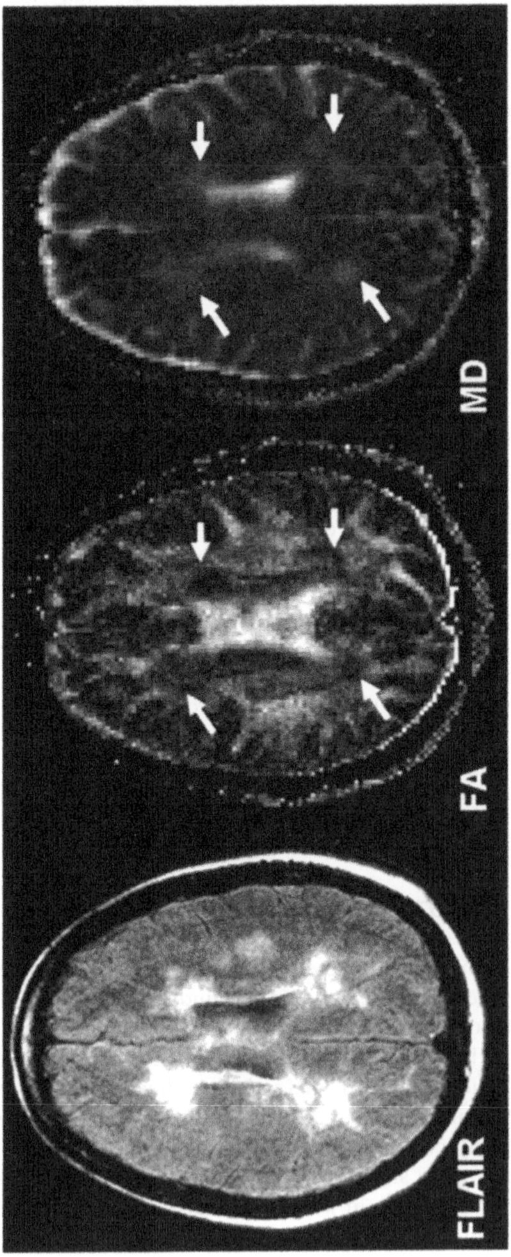

FIGURE 1. Fluid-attenuated inversion recovery imaging (FLAIR) and DTI-derived fractional anisotropy (FA) and mean diffusivity (MD) maps in a 29-year-old patient with relapsing-remitting MS showing decreased FA and increased MD in MS lesions.

isointense lesions.[11,13,14] This may be due to the long-standing destructive damage in those hypointense lesions or so-called black holes,[17] in which water diffusion is most mobile or less restricted; thus, diffusion imaging seems to be useful in assessing the severity of tissue damage in MS and holds promise to be a surrogate marker of clinical disability. Contrast-enhancing lesions, which represent early active areas of major blood-brain-barrier breakdown, are usually seen in MS disease.[18] Some studies have shown that enhancing lesions can be differentiated from nonenhancing lesions by measuring their MD values,[13,14] but others have failed to show this utility.[11,19] This discrepancy may be due to the variable degree of tissue damage during the lesion active period as reflected by their variable appearance on MRI. Contrast enhancement in MS lesions may vary in terms of shape, size, and age and usually disappears within 6 weeks or less. Although diffusion-weighted imaging cannot differentiate enhancing and nonenhancing lesions by measuring their MD, DTI studies have shown that FA is always lower in enhancing than in nonenhancing lesions, indicating that anisotropy, as an additional diffusion tensor property, is more sensitive in differentiating pathological substrates of MS lesions.[20–22]

It is now generally agreed that both MD and FA metrics should be estimated to maximize the yield from diffusion imaging and diffusion tensor imaging (see FIG. 1). This is because diffusion is inherently a three-dimensional process; in highly organized tissue, such as white matter, the diffusion may vary greatly between perpendicular and parallel direction of the major axis of axonal fibers.[23] Newly formed and visualized enhancing lesions are usually small and nodular, while old and reactive lesions may appear larger, some with ring enhancement. The FA is generally markedly reduced in ring-enhancing lesions,[19] suggesting pronounced tissue destruction of the white matter microstructure at the site of enhancement. Studies have also shown significantly increased MD values[13,19,24] in ring-enhancing lesions when compared with homogeneously enhancing lesions. This is likely due to the extracellular vasogenic edema at the enhancing rim, which is more profound than that in the enhancing region. These underlying pathologic features are not quantifiable on conventional MRI. Since lesions in SPMS are often old and chronic with extensive tissue destruction, larger degrees of changes of MD and FA were also reported in these lesions as compared with those in RRMS.[17]

DTI STUDIES IN NORMAL-APPEARING WHITE MATTER

It has become increasingly evident that MS is a disease that affects the brain globally, and normal-appearing white matter (NAWM) on conventional MRI is microscopically abnormal.[4] The disease processes do not occur only within lesion plaques seen on T2- or T1-weighted imaging, but also in white matter distant to the lesions that appears macroscopically normal. This may be part of the reason why there is not a strong correlation between the presence of T2 or T1 lesions (lesion load) on MRI and clinical status/outcome. The disease burden beyond the lesion load in NAWM is not usually determined.[6] Although conventional MRI provides a direct measure of the extent of macroscopic pathology in MS, such as lesion plaques, it suffers with little pathological specificity and lack of sensitivity in detecting occult lesions at the microscopic level. In the last decade, several nonconventional MR techniques have been developed to quantify pathologic changes not identifiable with routine MR sequences. DTI is one of these new advanced MRI techniques with the ability to improve the sensitivity in detecting microstructural architecture changes in normal-appearing brain tissues.

Studies of DTI in MS NAWM have revealed decreased FA and increased MD in different regions that appear normal on conventional MRI,[11,12,22,25–29] suggesting the presence of microscopic pathology beyond the resolution of conventional MRI. Although the DTI abnormalities seem to be quite widespread in NAWM, they tend to be more severe in the periplaque regions. Guo *et al.*[28] reported that there is a gradient of white matter abnormality extending in an outward direction from the plaque with the lowest FA, followed by lesser degrees of anisotropy decrease in the periplaque region and the least severe of FA changes in the other NAWM regions. Anisotropy measurements seem likely to be potentially more sensitive than diffusivity measurements for the detection of MS pathology.[28] The exact pathological correlations with these changes of DTI measurement in NAWM are still not fully understood, although they may be related to a low-grade inflammatory and demyelinating process found in NAWM. As with all histopathological features found in MS lesions,[30] the pathologic features in NAWM may include edema, demyelination, cellular infiltration, gliosis, and axonal loss. Among these discrete abnormalities, myelin and axonal loss are considered to contribute most to the DTI changes. Recent evidence from a quantitative postmortem study[31] demonstrated a significant reduction of axonal density or the total number of axons passing through the areas of corpus callosum that grossly appear normal in MS patients. It is possible that significant DTI changes in the NAWM are due to a subtle net loss of structural barriers to water molecular motion within the tissues.

Since the corpus callosum is commonly involved in MS disease with demyelinating lesions often located at the callosum or callososeptal interface, as well as finding callosal atrophy, a study of DTI[29] showed significant water diffusion changes in the normal-appearing corpus callosum (NACC) in a group of patients with early MS (mean disease duration: 2.7 years); however, such significant abnormalities of DTI were not observed in frontal and occipital NAWM regions. The degree of diffusion changes in the corpus callosum also correlates with cerebral lesion load in these patients. These results suggest that there is a preferential occult injury of corpus callosum in early MS, which is likely due to the accumulative bridging effects of the corpus callosum that result from Wallerian degeneration injury from distal lesion plaques. Furthermore, the diffusion abnormalities can be detected in the corpus callosum at the earliest stage of a clinically isolated syndrome suggestive of MS, before any atrophy and lesions are detected.[32] Because the corpus callosum is the largest white matter fiber tract in the brain and connects and transverses a large volume of subcortical white matter from the two hemispheres, the effects of focal occult pathology and Wallerian degeneration from distal lesions will be more severe as compared with other NAWM regions. On the other hand, since the corpus callosum is also a very well organized and densely packed fiber structure, it is likely to have much higher FA in the normal state and changes may also be more markedly manifested once pathology occurs. Thus, corpus callosum occult injury measured by DTI is potentially useful in the early stages of MS or even can serve as an early marker of primary demyelination in patients presenting with a clinically isolated syndrome (see FIG. 2).

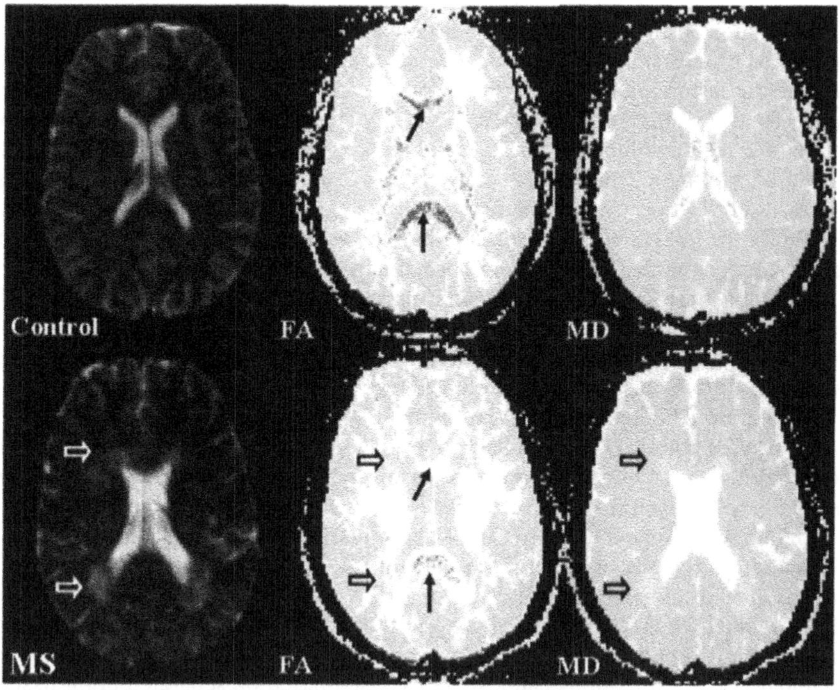

FIGURE 2. Representative images from a control subject (*top row*) and a patient (*bottom row*) at the level of corpus callosum (CC) of a DTI raw image (*left column*), color-coded FA map (*middle column*), and MD map (*right column*). CC showed the highest FA value in the healthy volunteer and was markedly reduced in the patient. Compared to other normal-appearing white matter (NAWM) regions, abnormalities of FA and MD are more apparent in normal-appearing CC regions, except in lesions in the patient.

Studies in which diffusion changes measured on DTI in NAWM have been used to precede new lesion formation indicate another clinical application of DTI in terms of studying lesion pathogenesis and natural history.[25,33] The work of Werring and colleagues, in a longitudinal study,[33] demonstrated a steady and moderate increase of MD in prelesional NAWM areas followed by a rapid and marked increase at the time of contrast enhancement of the lesion. Although this new pathological activity may develop for many months prior to focal lesion formation,[33–35] a preexisting pathological process must occur in the NAWM, which can be detected by DTI. Therefore, the degree of diffusion changes in NAWM measured by DTI may have predictive value of the subsequent lesion activity and evolution.

GLOBAL MEASUREMENTS OF DTI IN MS

In MS, methods for quantifying the disease burden are always desirable to monitor the disease progression and to assess the effect of new drugs.[36] Several methods to estimate the lesion load based on the conventional T2- and enhanced T1-weighted sequences have been developed and are typically aimed at the quantification of lesion number and lesion volume.[37–39] These quantitative methods have already been routinely used as outcome measures in clinical trials to assess both acute and chronic disease changes.[40] However, the macroscopic lesion load on conventional MRI does not account for the total microscopic disease burden. Hence, a global index that includes the microscopic burden of disease in normal-appearing brain tissue is necessary.

As demonstrated in the studies using magnetization transfer imaging,[41,42] DTI changes can also be assessed using histogram analysis. This is done by coregistering two maps of DTI metric with segmented brain tissue to generate a pixel-by-pixel histogram.[43,44] The histogram-derived DTI metrics will allow us to estimate the disease burden from both macro- and microscopic levels of the whole brain. Cercignani *et al.*[26] studied both MD and FA histograms in segmented whole-brain parenchyma and found a significant difference in the histograms between patients and controls. The quantities derived from the histogram (i.e., mean and peak location) were found to correlate with clinical disability in MS patients. The results suggest that a global DTI histogram is likely to represent the net amount of disease burden with improved correlation with clinical disability as compared with lesion load on conventional MRI. Global DTI measurement might serve as an additional method to monitor disease evolution and therapeutic response in MS (see FIG. 3).

Global DTI abnormalities can also be assessed based on separated brain tissues after the segmentation process. Although the majority of pathology in MS will be found in white matter, postmortem studies have suggested that gray matter is not spared.[45–47] The reasons that gray matter involvement is often neglected are probably due to the relatively small size of lesions and the small amount of myelin in gray matter, which often go undetected on conventional MRI. In addition, the inflammatory activity in gray matter lesions is usually less,[48] which is also likely to result in poor sensitivity of detection with conventional MRI as compared with white matter lesions. DTI is emerging as a tool in the detection of subtle gray matter abnormalities. In agreement with the postmortem findings, histogram-derived DTI metrics in the cerebral normal-appearing gray matter (NAGM) were found with greater diffusivity in MS patients than in healthy controls.[49] Patients with SPMS have higher water diffusivity than those with RRMS. Basal ganglionic (deep gray) structures also showed abnormal diffusion changes with decreased FA and increased MD.[50,51] Since gray matter and white matter differ not only functionally, but also anatomically, in general, MD is higher and FA is lower in gray matter than in white matter. This is due to findings of fewer myelinated fibers in gray matter, resulting in a decreased degree of anisotropy.

DTI histograms can be used to assess the global burden of MS pathology; histogram analysis provides a complete picture from several histogram-derived parameters about the extent and severity of pathology with or without T2-visible lesions by taking the brain as a whole. Typically, patients with MS have high average MD, high peak position, and low histogram peak height on the MD histogram, and low average FA, low peak position, and high histogram peak height on the FA histogram, when

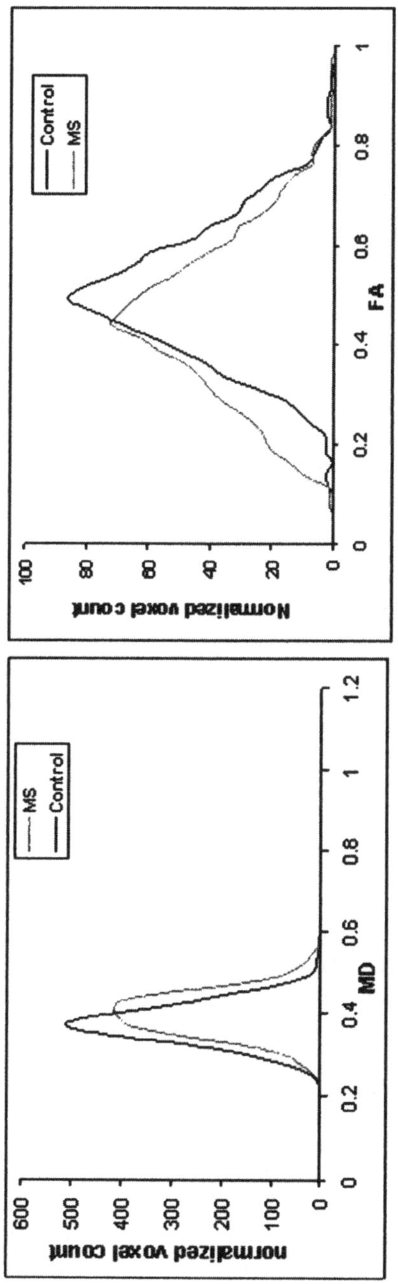

FIGURE 3. DTI-derived MD (*left*) and FA (*right*) histograms of NAWM from an MS patient and a normal control subject. Note the shift to higher value of the MD histogram and to lower value of the FA histogram in the patient as compared with the control.

compared with those in healthy controls. The histogram-derived DTI metrics have been shown to be correlated with clinical disability.

DTI STUDIES IN DIFFERENT PHENOTYPES OF MS

The most common clinical phenotypes include relapsing-remitting (RR), secondary progressive (SP), and primary progressive (PP) MS.[3] RRMS is defined as disease relapses with either full recovery or sequelae, but without disease progression during the periods between the relapses; SPMS is characterized by disease progression following an initial RRMS course with or without occasional relapses, minor remissions, and plateau; PPMS has a clinical course with a progressive accumulation of neurological deficit from onset without relapse or remission.[52] The marked discrepancy between these phenotypes of their MR activity and clinical deterioration may well reflect different mechanisms of disease evolution. One hypothesis is that neurological deficit in RRMS results from incomplete recovery from relapses due to persistent demyelination, whereas deficit in progressive disease arises from progressive axonal loss. Intrinsic changes in NAWM are also likely to contribute to disease progression, and there are known to be microscopic abnormalities in macroscopically normal white matter.[53]

Previous studies have suggested that DTI may enable differentiation of more specific distinction of clinical subgroups than conventional MRI. There is a negative correlation between MD and the degree of hypointensity of lesions on T1-weighted images,[17,24] and diffusivity is found significantly higher in SPMS lesions than in RRMS lesions,[17] suggesting the severity of microstructural damage of lesions in these two phenotypes may be different. This is in agreement with the findings of more hypointense lesions and more severe disability in SPMS as compared with RRMS. In histogram analysis studies, the whole-brain MD histogram in SPMS patients was shifted to higher abnormal values compared with RRMS patients.[43] Although patients with early RRMS may not differ from normal controls in DTI-derived measurements in normal-appearing brain tissues,[29,43,49] patients with SPMS significantly differed from the control subjects and patients with RRMS.[49,54] The correlation between DTI measures and clinical disability is also found stronger in SPMS patients, suggesting a role for DTI in monitoring advanced phases of the disease.[22]

The PPMS group has a unique clinical course and usually appears with atypical clinical and MRI features. Although, during the disease course, there are fewer lesions developing and little gadolinium enhancement,[55] the patients with PPMS usually demonstrate severe clinical disability. Several studies have investigated the DTI abnormalities in NAWM and NAGM and found small—but widespread—MD and FA changes in PPMS as compared with healthy controls.[14,22,49,56,57] Importantly, the diffuse abnormalities in normal-appearing brain tissues measured by DTI may contribute to the severe disability that patients with PPMS often develop, despite a paucity of lesions on conventional MRI. Significant correlations between lesion load and DTI abnormalities in the normal-appearing brain tissues were observed for patients with RRMS and SPMS, but not for patients with PPMS.[56,58] These microscopic changes in these patients seem to be independent of the extent of T2-visible abnormalities. However, the degree of DTI abnormality in normal-appearing brain tissues in PPMS when compared with other subtypes is still controversial and clarification is needed from further studies with larger patient samples. Even though identification of the exact pathophysiological processes involved in these different phenotypes of MS is a challenge, further studies are warranted to investigate the potential utility of DTI measures in distinguishing between axonal loss and demyelination and between demyelinating and remyelinating lesions. This may influence the commencement and choice of treatment.

DTI FIBER TRACTOGRAPHY IN MS

Recently, there has been interest in using fiber tractography obtained from DTI data sets in characterizing white matter tract directionality and integrity in relation to brain tumors.[59–63] The potential patterns of white matter tract alteration by cerebral neoplasms include tracts that are deviated, infiltrated, edematous, or destroyed. It is possible given the alterations in FA within and adjacent to lesions that similar patterns of behavior may occur with MS lesions. Some recent experimental studies[64,65] have demonstrated that lesions can be shown to transect white matter fiber tracts in a similar fashion to brain tumors (FIG. 4). Fiber tractography has a potential role in quantifying the degree of axonal loss and demyelination within

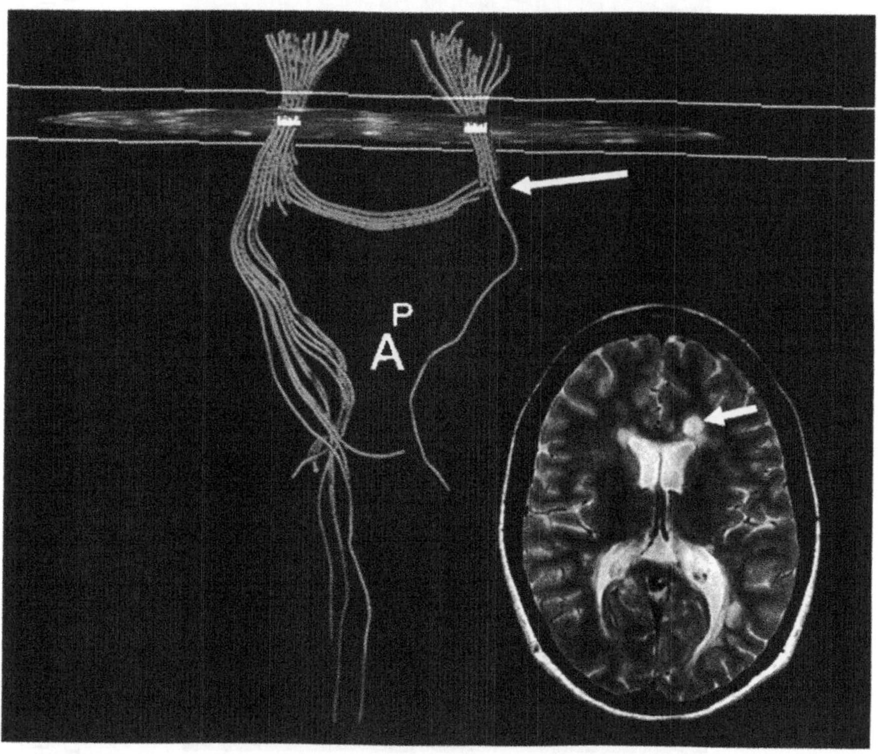

FIGURE 4. Fiber tractography in an MS patient. Fixed-size seeds are placed symmetrically in both sides of frontal white matter regions. The number of fibers in the left side is significantly decreased with a lesion (*arrow*) on the affected tracks.

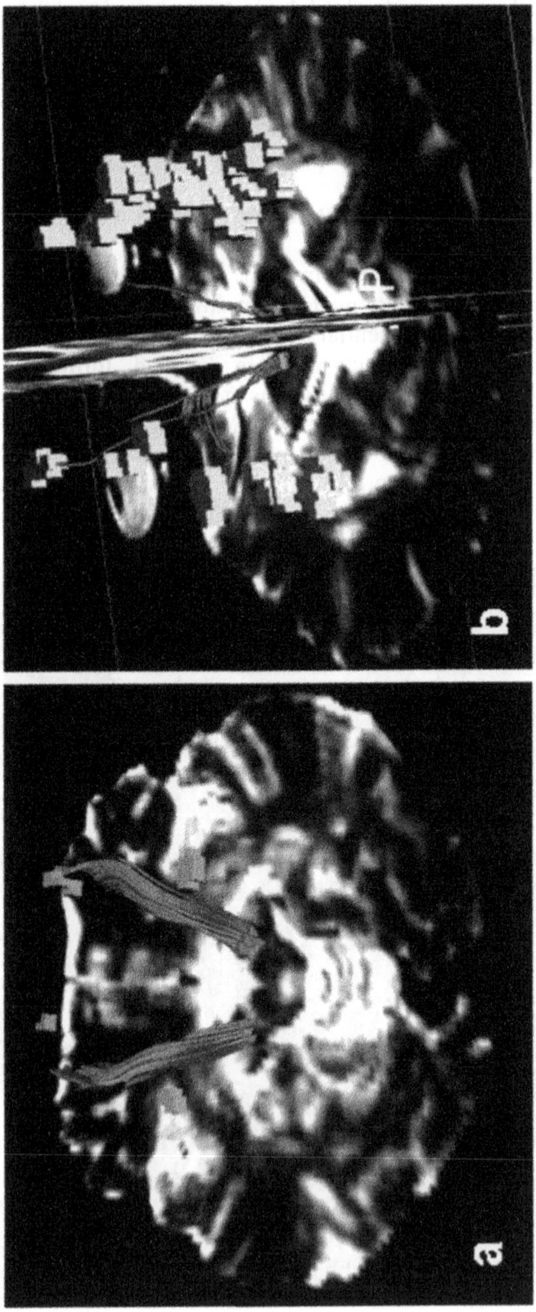

FIGURE 5. Fiber tracking in the corticospinal tract at the brain stem level in two MS patients. Fixed ROIs are placed symmetrically in both sides of the cerebral peduncle of the brain stem in one patient with lower **(a)** versus another patient with higher **(b)** lesion number/load. Note that the number of fibers of corticospinal tracts is significantly decreased in the patient with a higher lesion load, indicating Wallerian degeneration due to the remote lesions.

different lesion types and NAWM. The differences in white matter tract disruption can be directly visualized and may help to better understand the association between lesion type and location with clinical symptomatology as well as monitoring disease progression. In a review of 10 patients with RRMS, a positive correlation between FT (fiber tractography) and FA ($P < 0.0001$) and a negative correlation between FT and MD ($P < 0.0001$) were found. A negative correlation between FT and lesion size ($P = 0.0038$) was also demonstrated, suggesting that lesion size may affect the number of fibers transected.[66] Compared with NAWM (mean FT: 29.8; FA: 0.41; MD: 0.84), lesions were associated with significantly lower FT and FA, but significantly higher MD (mean FT: 21.4; FA: 0.37; MD: 1.19). Differences between lesion types were also noted: isointense lesions (mean FT: 25.2; FA: 0.39; MD: 1.10), hypointense lesions (mean FT: 19.4; FA: 0.37; MD: 1.24), and enhancing lesions (mean FT: 11.0; FA 0.19; MD: 1.33).

Measurement of lesion load by conventional MR imaging has been pursued with the aim of determining the severity of disease and response to therapy. Because the hallmark of MS is multifocal lesions characterized by inflammatory demyelinating changes that primarily involve the white matter tracts in the brain, it is likely that lesions can result in Wallerian degeneration of more remote white matter tracts and can cause microscopic involvement of distant, but contiguous, white matter tracts without evidence of abnormal signal.[2] Using fiber tracking, the degree of fiber tract loss in corticospinal tracts at the level of the brain stem in MS patients is likely to relate to the lesion load in the supratentorial brain. This suggests fiber tractography can provide a method for quantifying Wallerian degeneration and axonal transection from remote lesions in MS. Fiber tractography accurately delineated the corticospinal tract from cerebral peduncle in brain stem through internal capsule to cerebral cortical gyri. FA was significantly lower (FA = 0.52) in patients with higher lesion load (volume > 1360 mm^3) as compared with those patients (FA = 0.63) with lower lesion load ($P = 0.03$). Correspondingly, there were fewer fibers generated in the corticospinal tracts in patients with higher lesion load (FIG. 5). Also, patients with lesions in the corticospinal tracts showed lower FA values and lower number of fiber tracts as compared with those who did not have lesions on this pathway. Thus, fiber tracking may be a method for characterization of pathology within the brain stem remote from lesions in the supratentorial brain.

With faster alternative techniques being developed for acquiring DTI data sets such as multishot EPI, diffusion-weighted PROPELLER, spin-echo navigator spiral DTI, and parallel imaging methods like SENSE, acquisition times have been reduced to allow data to be acquired from structures like the human spinal cord. The spinal cord has intrinsic cord motion and CSF pulsation, making acquisition of DTI metrics difficult unless faster sequences are utilized. DTI in the spinal cord has demonstrated significantly decreased FA in the lateral and posterior tracts in the cervical spinal cord of MS patients, in the absence of spinal cord signal abnormality at conventional MR examination. These findings may be clinically more sensitive than conventional MR in differentiating primary from secondary causes of demyelination, such as cervical spondylosis, which seems to have a propensity for involving the posterior and central regions of the spinal cord. Furthermore, it may allow characterization of different spinal cord pathways within the spinal cord, such as the anterior and posterior column pathways, which again will have diagnostic, prognostic, and therapeutic implications (FIG. 6).

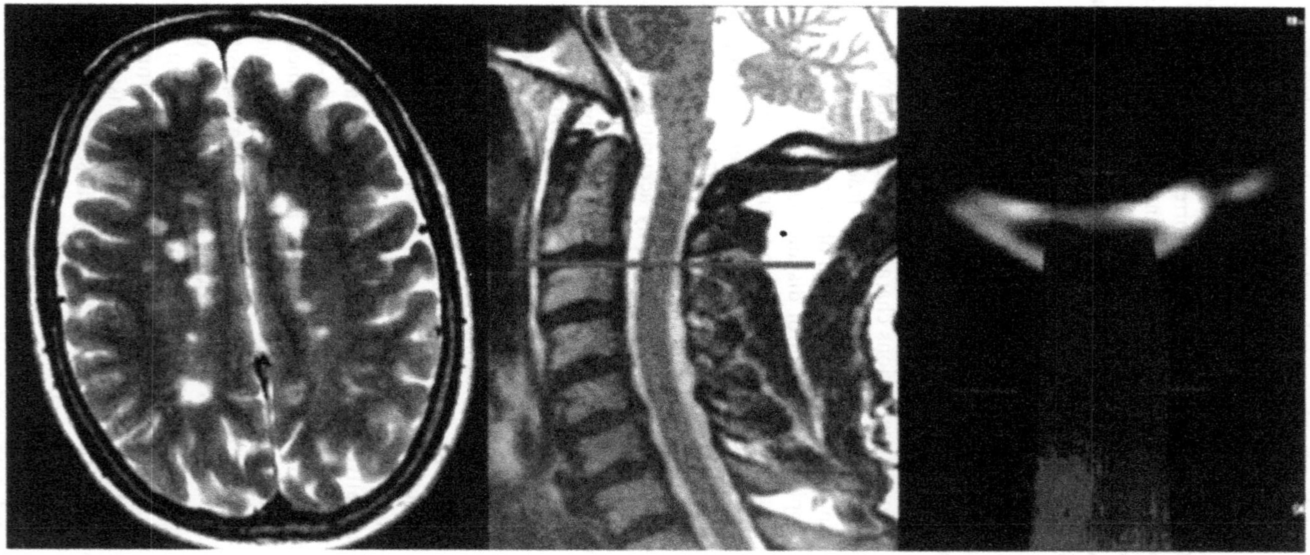

FIGURE 6. **(Left)** Axial T2-weighted MR image of a patient with relapsing-remitting MS. **(Center)** Sagittal image demonstrates the absence of cervical spinal cord signal abnormality. **(Right)** Axial $b = 0$ diffusion-weighted image at the C2–C3 level (red line, center image) with superimposed three-dimensional diffusion fiber tractography.

FUTURE DIRECTION

In recent years, extensive and impressive advances in the use of DTI have been made to develop a quick and reliable clinical tool in the assessment of diseases of the CNS. Several key elements will emerge in the future study of MS using DTI. First, since highly ordered white matter fiber tracts have components of both axon and myelin, and DTI can encode properties that vary with space, shape, and direction of the compartment that is accessible to the water molecule, DTI has the potential to decode these properties in order to differentiate the origin of tissue injury to see whether only myelin or both myelin and axonal injury occurs. This has been demonstrated in some experimental work by decoding axial and radial diffusivity of the optic nerve in animal studies. Application in the human brain will have clinically important significance in terms of characterizing the severity of tissue injury and differentiating axonal injury, demyelination, and remyelination during lesion evolution. Second, the newly emerging technique of fiber tractography from DTI data has a major impact on visualizing and quantifying axonal fibers *in vivo*. Future research in MS using fiber tracking technology can provide direct evidence not only for clinical manifestations caused by the primary site where axonal injury occurs, but also for secondary degenerative processes due to axonal pathway transection from a distant site. Third, histopathological correlation of DTI metrics in MS is challenging, but will help resolve many questions regarding the specific characterization of tissue damage at different stages of the disease process. Comparing DTI metrics with advanced MRI techniques will also help to define the value of DTI in MS. Lastly, in the meantime, it is important to have sufficient validation of DTI techniques for their incorporation into future clinical trials, including the assessment of inherent variation in both cross-sectional and longitudinal fashions, as well as determination of the most optimal imaging sequence, technique, and postprocessing methodologies.

CONCLUSIONS

The ability of DTI in measuring the diffusion of water molecules and their interaction with the cellular structures provides a unique tool to looking into the nature and extent of pathological damage of microstructural architecture that occurs in the diseases of the CNS. MS is one of these CNS diseases that is frequently investigated with DTI. DTI metrics and fiber tracking of different MS lesions allow more accurate characterization of intrinsic tissue damage in MS. Together with perfusion MR imaging, magnetization transfer imaging, and MR spectroscopy, DTI has made a significant contribution to the evaluation of "invisible" disease burden or occult lesions in normal-appearing brain tissues (i.e., NAWM, NAGM) in MS and yielded increased degree of pathological specificity of lesion development. The quantitative nature of DTI may play a role in assessing the outcome of clinical trials as an additional surrogate marker in monitoring the therapeutic response. Future longitudinal studies and histopathological correlative data are crucially important for improving our understanding of DTI quantities in monitoring pathophysiological changes in MS.

ACKNOWLEDGMENTS

This work was partially supported in part by Grant Nos. R37 NS 29029-11 from the National Institutes of Health and NCRR M01 RR00096 (GCRC).

REFERENCES

1. BAUM, H.M. & B.B. ROTHSCHILD. 1981. The incidence and prevalence of reported multiple sclerosis. Ann. Neurol. **10**(5): 420–428.
2. TRAPP, B.D., J. PETERSON, R.M. RANSOHOFF *et al.* 1998. Axonal transection in the lesions of multiple sclerosis. N. Engl. J. Med. **338**(5): 278–285.
3. LUBLIN, F.D. & S.C. REINGOLD. 1996. Defining the clinical course of multiple sclerosis: results of an international survey—National Multiple Sclerosis Society (USA) Advisory Committee on Clinical Trials of New Agents in Multiple Sclerosis. Neurology **46**(4): 907–911.
4. MILLER, D.H., R.I. GROSSMAN, S.C. REINGOLD & H.F. MCFARLAND. 1998. The role of magnetic resonance techniques in understanding and managing multiple sclerosis. Brain **121**(part 1): 3–24.
5. MILLER, D.H., A.J. THOMPSON & M. FILIPPI. 2003. Magnetic resonance studies of abnormalities in the normal appearing white matter and grey matter in multiple sclerosis. J. Neurol. **250**(12): 1407–1419.
6. FILIPPI, M. 2001. Non-conventional MR techniques to monitor the evolution of multiple sclerosis. Neurol. Sci. **22**(2): 195–200.
7. BEAULIEU, C. & P.S. ALLEN. 1994. Determinants of anisotropic water diffusion in nerves. Magn. Reson. Med. **31**(4): 394–400.
8. LARSSON, H.B., C. THOMSEN, J. FREDERIKSEN *et al.* 1992. *In vivo* magnetic resonance diffusion measurement in the brain of patients with multiple sclerosis. Magn. Reson. Imaging **10**(1): 7–12.
9. CHRISTIANSEN, P., P. GIDEON, C. THOMSEN *et al.* 1993. Increased water self-diffusion in chronic plaques and in apparently normal white matter in patients with multiple sclerosis. Acta Neurol. Scand. **87**(3): 195–199.
10. HORSFIELD, M.A., M. LAI, S.L. WEBB *et al.* 1996. Apparent diffusion coefficients in benign and secondary progressive multiple sclerosis by nuclear magnetic resonance. Magn. Reson. Med. **36**(3): 393–400.
11. FILIPPI, M., G. IANNUCCI, M. CERCIGNANI *et al.* 2000. A quantitative study of water diffusion in multiple sclerosis lesions and normal-appearing white matter using echo-planar imaging. Arch. Neurol. **57**(7): 1017–1021.
12. CERCIGNANI, M., G. IANNUCCI, M.A. ROCCA *et al.* 2000. Pathologic damage in MS assessed by diffusion-weighted and magnetization transfer MRI. Neurology **54**(5): 1139–1144.
13. ROYCHOWDHURY, S., J.A. MALDJIAN & R.I. GROSSMAN. 2000. Multiple sclerosis: comparison of trace apparent diffusion coefficients with MR enhancement pattern of lesions. AJNR Am. J. Neuroradiol. **21**(5): 869–874.
14. DROOGAN, A.G., C.A. CLARK, D.J. WERRING *et al.* 1999. Comparison of multiple sclerosis clinical subgroups using navigated spin echo diffusion-weighted imaging. Magn. Reson. Imaging **17**(5): 653–661.
15. HEIDE, A.C., T.L. RICHARDS, E.C. ALVORD, JR., *et al.* 1993. Diffusion imaging of experimental allergic encephalomyelitis. Magn. Reson. Med. **29**(4): 478–484.
16. RICHARDS, T.L., E.C. ALVORD, JR., Y. HE *et al.* 1995. Experimental allergic encephalomyelitis in non-human primates: diffusion imaging of acute and chronic brain lesions. Multiple Sclerosis **1**(2): 109–117.
17. CASTRIOTA-SCANDERBEG, A., F. TOMAIUOLO, U. SABATINI *et al.* 2000. Demyelinating plaques in relapsing-remitting and secondary-progressive multiple sclerosis: assessment with diffusion MR imaging. AJNR Am. J. Neuroradiol. **21**(5): 862–868.
18. GROSSMAN, R.I. & J.C. MCGOWAN. 1998. Perspectives on multiple sclerosis. AJNR Am. J. Neuroradiol. **19**(7): 1251–1265.

19. BAMMER, R., M. AUGUSTIN, S. STRASSER-FUCHS *et al.* 2000. Magnetic resonance diffusion tensor imaging for characterizing diffuse and focal white matter abnormalities in multiple sclerosis. Magn. Reson. Med. **44**(4): 583–591.
20. WERRING, D.J., C.A. CLARK, G.J. BARKER *et al.* 1999. Diffusion tensor imaging of lesions and normal-appearing white matter in multiple sclerosis. Neurology **52**(8): 1626–1632.
21. CASTRIOTA-SCANDERBEG, A., F. FASANO, G. HAGBERG *et al.* 2003. Coefficient D(av) is more sensitive than fractional anisotropy in monitoring progression of irreversible tissue damage in focal nonactive multiple sclerosis lesions. AJNR Am. J. Neuroradiol. **24**(4): 663–670.
22. FILIPPI, M., M. CERCIGNANI, M. INGLESE *et al.* 2001. Diffusion tensor magnetic resonance imaging in multiple sclerosis. Neurology **56**(3): 304–311.
23. MORI, S., R. ITOH, J. ZHANG *et al.* 2001. Diffusion tensor imaging of the developing mouse brain. Magn. Reson. Med. **46**(1): 18–23.
24. NUSBAUM, A.O., D. LU, C.Y. TANG & S.W. ATLAS. 2000. Quantitative diffusion measurements in focal multiple sclerosis lesions: correlations with appearance on TI-weighted MR images. AJR Am. J. Roentgenol. **175**(3): 821–825.
25. ROCCA, M.A., M. CERCIGNANI, G. IANNUCCI *et al.* 2000. Weekly diffusion-weighted imaging of normal-appearing white matter in MS. Neurology **55**(6): 882–884.
26. CERCIGNANI, M., M. BOZZALI, G. IANNUCCI *et al.* 2001. Magnetisation transfer ratio and mean diffusivity of normal appearing white and grey matter from patients with multiple sclerosis. J. Neurol. Neurosurg. Psychiatry **70**(3): 311–317.
27. GUO, A.C., V.L. JEWELLS & J.M. PROVENZALE. 2001. Analysis of normal-appearing white matter in multiple sclerosis: comparison of diffusion tensor MR imaging and magnetization transfer imaging. AJNR Am. J. Neuroradiol. **22**(10): 1893–1900.
28. GUO, A.C., J.R. MACFALL & J.M. PROVENZALE. 2002. Multiple sclerosis: diffusion tensor MR imaging for evaluation of normal-appearing white matter. Radiology **222**(3): 729–736.
29. GE, Y., M. LAW, G. JOHNSON *et al.* 2004. Preferential occult injury of corpus callosum in multiple sclerosis measured by diffusion tensor imaging. J. Magn. Reson. Imaging **20**(1): 1–7.
30. DE GROOT, C.J., E. BERGERS, W. KAMPHORST *et al.* 2001. Post-mortem MRI-guided sampling of multiple sclerosis brain lesions: increased yield of active demyelinating and (p)reactive lesions. Brain **124**(part 8): 1635–1645.
31. EVANGELOU, N., M.M. ESIRI, S. SMITH *et al.* 2000. Quantitative pathological evidence for axonal loss in normal appearing white matter in multiple sclerosis. Ann. Neurol. **47**(3): 391–395.
32. RANJEVA, J.P., J. PELLETIER, S. CONFORT-GOUNY *et al.* 2003. MRI/MRS of corpus callosum in patients with clinically isolated syndrome suggestive of multiple sclerosis. Multiple Sclerosis **9**(6): 554–565.
33. WERRING, D.J., D. BRASSAT, A.G. DROOGAN *et al.* 2000. The pathogenesis of lesions and normal-appearing white matter changes in multiple sclerosis: a serial diffusion MRI study. Brain **123**(part 8): 1667–1676.
34. FILIPPI, M., M.A. ROCCA, G. MARTINO *et al.* 1998. Magnetization transfer changes in the normal appearing white matter precede the appearance of enhancing lesions in patients with multiple sclerosis. Ann. Neurol. **43**(6): 809–814.
35. PIKE, G.B., N. DE STEFANO, S. NARAYANAN *et al.* 2000. Multiple sclerosis: magnetization transfer MR imaging of white matter before lesion appearance on T2-weighted images. Radiology **215**(3): 824–830.
36. MILLER, D.H., P.S. ALBERT, F. BARKHOF *et al.* 1996. Guidelines for the use of magnetic resonance techniques in monitoring the treatment of multiple sclerosis: US National MS Society Task Force. Ann. Neurol. **39**(1): 6–16.
37. UDUPA, J.K., L.G. NYUL, Y. GE & R.I. GROSSMAN. 2001. Multiprotocol MR image segmentation in multiple sclerosis: experience with over 1,000 studies. Acad. Radiol. **8**(11): 1116–1126.
38. SIMON, J.H., A. SCHERZINGER, U. RAFF & X. LI. 1997. Computerized method of lesion volume quantitation in multiple sclerosis: error of serial studies. AJNR Am. J. Neuroradiol. **18**(3): 580–582.

39. Filippi, M. & R.I. Grossman. 2002. MRI techniques to monitor MS evolution: the present and the future. Neurology **58**(8): 1147–1153.
40. Evans, A.C., J.A. Frank, J. Antel & D.H. Miller. 1997. The role of MRI in clinical trials of multiple sclerosis: comparison of image processing techniques. Ann. Neurol. **41**(1): 125–132.
41. van Buchem, M.A., J.K. Udupa, J.K. McGowan *et al.* 1997. Global volumetric estimation of disease burden in multiple sclerosis based on magnetization transfer imaging. AJNR Am. J. Neuroradiol. **18**(7): 1287–1290.
42. Ge, Y., R.I. Grossman, J.K. Udupa *et al.* 2002. Magnetization transfer ratio histogram analysis of normal-appearing gray matter and normal-appearing white matter in multiple sclerosis. J. Comput. Assisted Tomogr. **26**(1): 62–68.
43. Nusbaum, A.O., C.Y. Tang, T. Wei *et al.* 2000. Whole-brain diffusion MR histograms differ between MS subtypes. Neurology **54**(7): 1421–1427.
44. Rashid, W., A. Hadjiprocopis, C.M. Griffin *et al.* 2004. Diffusion tensor imaging of early relapsing-remitting multiple sclerosis with histogram analysis using automated segmentation and brain volume correction. Multiple Sclerosis **10**(1): 9–15.
45. Kidd, D., F. Barkhof, R. McConnell *et al.* 1999. Cortical lesions in multiple sclerosis. Brain **122**(part 1): 17–26.
46. Peterson, J.W., L. Bo, S. Mork *et al.* 2001. Transected neurites, apoptotic neurons, and reduced inflammation in cortical multiple sclerosis lesions. Ann. Neurol. **50**(3): 389–400.
47. Brownell, B. & J.T. Hughes. 1962. The distribution of plaques in the cerebrum in multiple sclerosis. J. Neurol. Neurosurg. Psychiatry **25:** 315–320.
48. Bo, L., C.A. Vedeler, H. Nyland *et al.* 2003. Intracortical multiple sclerosis lesions are not associated with increased lymphocyte infiltration. Multiple Sclerosis **9**(4): 323–331.
49. Bozzali, M., M. Cercignani, M.P. Sormani *et al.* 2002. Quantification of brain gray matter damage in different MS phenotypes by use of diffusion tensor MR imaging. AJNR Am. J. Neuroradiol. **23**(6): 985–988.
50. Fabiano, A.J., J. Sharma, B. Weinstock-Guttman *et al.* 2003. Thalamic involvement in multiple sclerosis: a diffusion-weighted magnetic resonance imaging study. J. Neuroimaging **13**(4): 307–314.
51. Ciccarelli, O., D.J. Werring, C.A. Wheeler-Kingshott *et al.* 2001. Investigation of MS normal-appearing brain using diffusion tensor MRI with clinical correlations. Neurology **56**(7): 926–933.
52. Thompson, A.J., C.H. Polman, D.H. Miller *et al.* 1997. Primary progressive multiple sclerosis. Brain **120**(part 6): 1085–1096.
53. Allen, I.V. & S.R. McKeown. 1979. A histological, histochemical, and biochemical study of the macroscopically normal white matter in multiple sclerosis. J. Neurol. Sci. **41**(1): 81–91.
54. Nusbaum, A.O. 2002. Diffusion tensor MR imaging of gray matter in different multiple sclerosis phenotypes. AJNR Am. J. Neuroradiol. **23**(6): 899–900.
55. Thompson, A.J., A.G. Kermode, D. Wicks *et al.* 1991. Major differences in the dynamics of primary and secondary progressive multiple sclerosis. Ann. Neurol. **29**(1): 53–62.
56. Rocca, M.A., G. Iannucci, M. Rovaris *et al.* 2003. Occult tissue damage in patients with primary progressive multiple sclerosis is independent of T2-visible lesions—a diffusion tensor MR study. J. Neurol. **250**(4): 456–460.
57. Rovaris, M., M. Bozzali, G. Iannucci *et al.* 2002. Assessment of normal-appearing white and gray matter in patients with primary progressive multiple sclerosis: a diffusion-tensor magnetic resonance imaging study. Arch. Neurol. **59**(9): 1406–1412.
58. Oh, J., R.G. Henry, C. Genain *et al.* 2004. Mechanisms of normal appearing corpus callosum injury related to pericallosal T1 lesions in multiple sclerosis using directional diffusion tensor and 1H MRS imaging. J. Neurol. Neurosurg. Psychiatry **75**(9): 1281–1286.
59. Field, A.S. & A.L. Alexander. 2004. Diffusion tensor imaging in cerebral tumor diagnosis and therapy. Top. Magn. Reson. Imaging **15**(5): 315–324.

60. FIELD, A.S., A.L. ALEXANDER, Y.C. WU *et al.* 2004. Diffusion tensor eigenvector directional color imaging patterns in the evaluation of cerebral white matter tracts altered by tumor. J. Magn. Reson. Imaging **20**(4): 555–562.
61. JELLISON, B.J., A.S. FIELD, J. MEDOW *et al.* 2004. Diffusion tensor imaging of cerebral white matter: a pictorial review of physics, fiber tract anatomy, and tumor imaging patterns. AJNR Am. J. Neuroradiol. **25**(3): 356–369.
62. NIMSKY, C., O. GANSLANDT, P. HASTREITER *et al.* 2005. Intraoperative diffusion-tensor MR imaging: shifting of white matter tracts during neurosurgical procedures—initial experience. Radiology **234**(1): 218–225.
63. NIMSKY, C., O. GANSLANDT, P. HASTREITER *et al.* 2005. Preoperative and intraoperative diffusion tensor imaging–based fiber tracking in glioma surgery. Neurosurgery **56**(1): 130–138.
64. GE, Y., K. TUVIA, M. LAW *et al.* 2005. Corticospinal tract degeneration in brainstem in patients with multiple sclerosis: evaluation with diffusion tensor tractography. *In* Proceedings: Int. Soc. Magn. Reson. Med. (ISMRM); Miami.
65. CAMPI, A., P. STAEMPFLI, T. JAERMANN *et al.* 2004. Preliminary study with high resolution SENSE DTI at 3 T: DTI and fiber tracking study in multiple sclerosis. *In* Proceedings of the 42nd Annual Meeting of the American Society of Neuroradiology, Seattle, p. 7.
66. SOOHOO, S., Y. GE, M. LAW *et al.* 2005. Lesional fractional anisotropy, diffusivity, and fiber tractography in patients with multiple sclerosis with DTI at 3 tesla. *In* Proceedings: Int. Soc. Magn. Reson. Med. (ISMRM); Miami.

Quantitative Analysis of Diffusion Tensor Imaging Data in Serial Assessment of Krabbe Disease

JAMES M. PROVENZALE,[a] MARIA ESCOLAR,[b] AND JOANNE KURTZBERG[c]

[a]*Department of Radiology, Duke University Medical Center, Durham, North Carolina, USA*

[b]*Department of Pediatrics, University of North Carolina, Chapel Hill, North Carolina, USA*

[c]*Department of Pediatrics, Duke University Medical Center, Durham, North Carolina, USA*

ABSTRACT: **Krabbe disease is a rare autosomal recessive pediatric white matter (WM) disorder that is due to deficiency of a specific enzyme, β-galactocerebrosidase. This report reviews our experience with use of diffusion tensor imaging (DTI) in serial assessment of WM changes in Krabbe disease following stem cell transplantation. DTI appears to be a sensitive means to monitor effects of stem cell transplantation on WM development in Krabbe disease. The group of early transplantation infants was clearly distinguishable from the group of late transplantation infants based on anisotropy measurements. Good correlation also was seen between neurodevelopmental scores and anisotropy measurements. The work described here in Krabbe disease may serve as a model for application of DTI to other therapies in various WM disorders such as multiple sclerosis and dysmyelinating disorders of childhood.**

KEYWORDS: **Krabbe disease; diffusion tensor imaging (DTI); white matter (WM); transplantation; stem cell; anisotropy; children**

Krabbe disease (globoid cell leukodystrophy) is a rare autosomal recessive pediatric white matter (WM) disorder that is due to deficiency of a specific enzyme, β-galactocerebrosidase.[1] These substances accumulate within macrophages, and one of these substances (galactosylsphingosine) is toxic to the cells responsible for myelin formation (oligodendroglia). As a result, infants with the early-onset (infantile) form of Krabbe disease fail to develop normal myelin, resulting in development of severe deficits in the central and peripheral nervous systems.[2] Without treatment, children with early infantile Krabbe disease typically follow a progressive course of neurological deterioration characterized by a vegetative state and death within the first 2–4 years of life. Unfortunately, in the past, no curative or temporizing therapy existed. Children with infantile Krabbe disease typically are seen to have hyperintense lesions within WM on T2-weighted MR images and, on occasion, within gray matter struc-

Address for correspondence: James M. Provenzale, M.D., Department of Radiology, Duke University Medical Center, Box 3808, Durham, NC 27710. Voice: 919-684-7218; fax: 919-684-7157. prove001@mc.duke.edu

Ann. N.Y. Acad. Sci. 1064: 220–229 (2005). © 2005 New York Academy of Sciences. doi: 10.1196/annals.1340.040

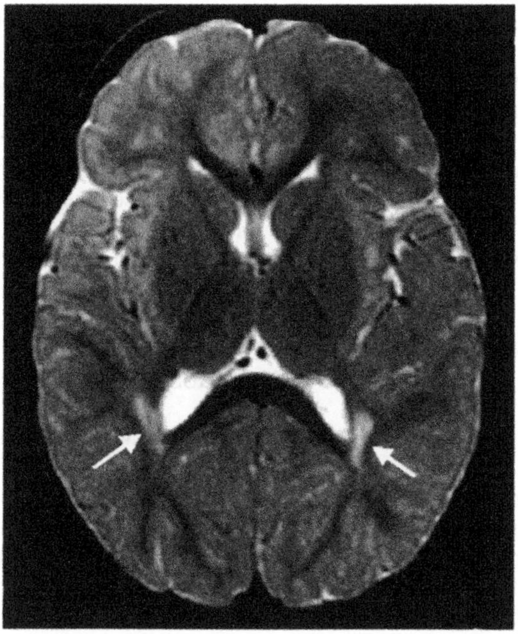

FIGURE 1. Axial T2-weighted MR image in an infant with Krabbe disease shows regions of abnormal hyperintense signal (*arrows*) consistent with areas of dysmyelination. (Reprinted from ref. 3 with permission.)

tures (FIG. 1). A recent report noted pyramidal tract involvement in 90% of early-onset Krabbe children, cerebellar WM involvement in 80%, deep gray matter (i.e., dentate nucleus, basal ganglia, or thalamus) involvement in 70%, posterior corpus callosal involvement in 60%, and parieto-occipital WM involvement in 50%.[2]

This report reviews our experience with use of diffusion tensor imaging in serial assessment of WM changes in Krabbe disease following stem cell transplantation. For full details of methodology and results, the reader is directed to a previous publication (see ref. 3).

INDICATIONS FOR STEM CELL TRANSPLANTATION THERAPY

Recently, hematopoietic stem cell transplantation has shown promise as therapy for Krabbe disease based on the fact that donor leukocytes can provide the deficient enzyme to cells in the peripheral and central nervous system. In a previous study, 5 children with Krabbe disease treated with hematopoietic stem cell transplantation were found to have reversal or lack of progression of clinical and radiological findings with resumption of leukocyte β-galactocerebrosidase activity.[3] As a result of such promising initial studies, Duke University Medical Center has become a major treatment center for hematopoietic stem cell transplantation for pediatric WM disorders such as Krabbe disease. Workers in the Pediatric Bone Marrow and Stem Cell Transplant Program at our institution have developed the use of umbilical cord blood, a widely available entity, as a reliable source of stem cells.[4] As a result of the need to assess a relatively large number of patients with pediatric WM disorders undergoing stem cell transplantation, it was necessary to develop a means to reproducibly assess the degree of the lack of myelination in such patients. Conventional MR imaging is sensitive to changes in brain development, such as myelination, because progression of myelination during childhood can be depicted as development of hypointense signal within WM on T2-weighted images. However, assessment of myelination using conventional MR imaging is based solely on visual inspection and, because it is not quantitative, it is subject to interobserver variability and open to reader bias.

ROLE FOR DIFFUSION TENSOR IMAGING IN KRABBE DISEASE ASSESSMENT

The progression in anisotropy seen within the WM of normal children during development, while not synonymous with myelination, is thought to reflect, at least in part, changes reflecting myelination. In the remainder of this chapter, we will speak of increases in anisotropy values as showing evidence of myelination; however, the reader must keep in mind that the exact changes underlying increases in anisotropy are a matter of active debate and do not solely reflect myelination, but also likely such changes as increases in numbers of axons and other microstructural changes indicating increasing tissue organization. With this fact in mind, we submit that development of a quantitative means for assessing WM myelination that is reproducible and not subject to reader bias would be a true advance in evaluation of WM disorders.

Diffusion tensor imaging (DTI) is a noninvasive imaging technique that offers a sensitive means to assess patterns of brain WM development. Because water motion in myelinated WM matter is anisotropic (i.e., has a tendency to diffuse along one direction rather than in all directions) and the MR signal using DTI is sensitized to microscopic movement of water molecules, myelinated WM is seen to have higher anisotropy values on DTI-derived anisotropy maps. On anisotropy maps, anisotropy values can be seen to be higher in more compact WM structures (e.g., posterior limb of internal capsule) compared to less compact WM structures (e.g., subcortical WM) and gray matter.[5] Anisotropy values are seen to increase in various WM regions throughout childhood.[5] The increase in anisotropy values within WM with age is presumed to represent the effects of progression of myelination as well as further growth of axons.[6] One immediate advantage of use of anisotropy maps is that they provide a quantitative, reproducible means of assessing WM integrity, which is not available from conventional MR images. Because anisotropy values of WM structures in normal children are known to increase with age during myelination, it appears appropriate that DTI could serve as a sensitive means to quantify failure of myelination in pediatric WM disorders and to monitor myelination in response to therapeutic interventions. Application of DTI to Krabbe disease could prove further valuable to the medical community because DTI as a means of assessing therapeutic response is a relatively underexplored topic. This report details the methods by which we have used DTI to assess WM changes in Krabbe patients and represents one of the first reports in the medical literature to use DTI to assess for responses to a therapy aimed at brain white matter.

EARLY STUDIES USING DIFFUSION TENSOR IMAGING FOR DISEASE ASSESSMENT

In a previous study, we compared DTI findings in a small group of Krabbe patients who were treated with hematopoietic stem cell transplantation and untreated children with Krabbe disease.[7] That study assessed Krabbe patients at a single time point after transplantation rather than on serial imaging. We found that mean fractional anisotropy (FA) values in WM structures in treated Krabbe children were intermediate between values in untreated Krabbe children and values in normal children. These findings suggested to us that a possibly beneficial treatment effect was present and that DTI might be a reliable means of assessing treatment response.

One factor that became apparent early in the assessment of Krabbe patients is that increases in anisotropy can only be assessed (a) using age-matched controls that are within a relatively narrow age range relative to the index patient and (b) by comparison of specific, well-identified WM structures. Comparison to age-matched controls is needed because of the relatively rapid changes that occur in various WM structures during the first few years of life. Inappropriate comparison of Krabbe infants against younger children could result in falsely elevated anisotropy ratios in patients. Measurements within well-identified WM structures are needed because the regional variations between FA values in WM sites would make for inaccurate comparisons if FA values within a WM structure in which WM fibers are not compact (e.g., subcortical WM) were compared with FA values within a WM structure with compact fibers (e.g., corpus callosum). In our experience, comparison with controls is most easily understood as a ratio of FA values within a specific structure in a Krabbe patient to FA values within the same structure in a group of normal control subjects.

Our early experience with the first group of Krabbe patients that we studied led us to postulate two hypotheses. Our first hypothesis was based on the fact that, although Krabbe infants typically have no detectable abnormality on routine physical examination at birth, on close neurological examination subtle neurological deficits can be seen in the first few years of life. Further evidence of early involvement of the central nervous system in Krabbe newborn infants is found in the fact that subtle demyelinating lesions have been reported on MR imaging in an affected fetus.[8] Based on these facts, our first hypothesis was that newborn infants with Krabbe disease will have decreased anisotropy values compared to normal neonates. Verification of this hypothesis would potentially prove to be important because assessment of myelination on conventional (i.e., spin echo) images of the newborn infant is difficult because of the lack of identifiable myelination milestones in the newborn. Therefore, detection of delayed myelination in Krabbe newborn infants on conventional MR images is difficult. Finding differences in anisotropy values of newborn Krabbe infants and normal newborns would be an important advance in MR imaging assessment of these patients and would be evidence that DTI provides information that is not available on conventional MR images.

SUBSEQUENT STUDIES TO INVESTIGATE DISEASE PROGRESSION

Early clinical experience with stem cell transplantation in Krabbe patients had shown us that the greatest neurological improvement was seen in infants who underwent transplantation in the first few months of life. This opinion is supported by occasional case reports in the medical literature.[9] On this basis, we hypothesized that infants treated with hematopoietic stem cell transplantation at a very early age would have a greater increase of anisotropy values following therapy compared to infants who were treated at a later age.[3] To test this hypothesis, we compared anisotropy values in various WM structures on serial imaging studies in a small group of 3 Krabbe infants who underwent stem cell transplantation in the first of life and 4 infants who underwent transplantation at a later stage (5–8 months of life). The two groups differed from one another in a number of ways. First, the early transplantation group infants were suspected as having Krabbe disease on the basis of an affected sibling. These infants were identified as having Krabbe disease at birth or *in utero* on the basis of abnormal serum β-galactocerebrosidase levels, before clinical symptoms and signs were evident. The average age at time of pretreatment imaging in these 3 infants was 13 days. The second group of infants differed from the first group because they were diagnosed based on the presence of neurological symptoms. The average age at time of the pretreatment MR imaging examination was 5.5 months. The full details of this study have been previously published.[3]

All infants underwent allogeneic hematopoietic stem cell transplantation at our institution between August 1999 and January 2002. Six infants were transplanted with unrelated, banked, partially HLA-mismatched umbilical cord blood donors and 1 infant received HLA-matched bone marrow from a carrier sibling. To be included in the study, infants had to have undergone MR imaging using DTI at the time of initial diagnosis (i.e., just prior to transplantation) and at least one subsequent MR imaging examination with DTI performed at least 10 months after transplantation.

The comparison population consisted of children who underwent clinical imaging that included DTI, were found to have a normal clinical MR examination, and were never subsequently diagnosed with a neurological disease. The most common reasons for MR imaging in the comparison group were growth delay, developmental delay, and correlation with suspected abnormalities seen on CT imaging. All children underwent imaging for clinical purposes rather than as individuals who entered a prospective trial after a normal screening neurological examination. Therefore, the comparison children are presumed, rather than proven, to be normal. The Institutional Review Board at our medical center granted a waiver of informed consent for inclusion of these comparison patients.

MR scanning following transplantation was not performed at set intervals, but instead was performed at the time of routine clinical follow-up examination at our institution. As a result, subsequent MR imaging was performed at various times relative to transplantation. It is known that myelination proceeds most rapidly in the first year of life and anisotropy values accordingly have the fastest rate of increase during that time period.[4] For this reason, it was important that the age difference between Krabbe infants in the first year of life and control infants be minimized. Therefore, for patients who were less than 1 year of age, we compared Krabbe infants solely with control infants who were within 1 month of age of Krabbe patients. After the first year of life, rate of myelination is thought to be slower than in the first year. Accordingly, anisotropy values increase at a slower rate after the first year compared to the first year of life. For this reason, we deemed it acceptable to allow a 2-month time interval between the age of Krabbe patients at the time of scanning and the age of control children. For each child, we chose posttransplantation MR scans that were performed closest to yearly intervals following transplantation.

In order to correlate DTI findings with clinical features, all Krabbe children underwent periodic neurodevelopmental evaluations after transplantation. We assessed solely the most recent neurological evaluation for the following neurodevelopmental categories relative to normal age-matched children: expressive language, receptive language, gross motor skills, fine motor skills, and cognitive ability.[10] We assigned standard scores (i.e., scores relative to the general population; a score of 100 represented the average score for the normal population) for each parameter. Patients were then assigned a single score from a range of 1–5 in each category depending on their standard score. Those children with a standard score of >100 were assigned a score of 5, those with a score between 70 and 100 were assigned a score of 4, those with a score between 40 and 70 were assigned a score of 3, those with a score between 10 and 40 were assigned a score of 2, and those with a score of ≤10 were assigned a score of 1. Five neurodevelopmental categories were assessed and the score for each category ranged between 5 and 1; the maximal functional score was 25 and the minimal functional score was 5. Mean scores for developmental age-adjusted parameters in the early transplantation group were as follows: expressive language, 3.33; receptive language, 3.67; gross motor skills, 2.67; fine motor skills, 3.67; and cognitive ability, 3.33. The mean cumulative score for the five developmental categories for the group was 16.6. All children in the late transplantation group had developmental scores of 1 in every category. Therefore, the mean cumulative score for the five developmental categories for the group was 5.00.

All imaging was performed on a 1.5-T clinical MR imaging scanner (Signa; GE Healthcare, Milwaukee, WI) using a standard head coil. All Krabbe patients underwent DTI using four excitations (using a pulse sequence that was designed to optimize signal-to-noise, but had twice the DTI imaging time of that used for routine clinical imaging), and all children in the comparison group underwent DTI using two excitations. To assess DTI images, a single observer placed region-of-interest measurements (ROIs) on FA maps in the following structures: peripheral frontal lobe WM, genu of the corpus callosum, splenium of the corpus callosum, and posterior limb of the internal capsule (hereafter referred to as the internal capsule). The reader is referred to a previous publication for the exact scanning protocol details and for details of ROI placement.[3]

An attempt was made to draw the ROIs for FA measurement at consistent locations from scan to scan. We drew ROIs for the internal capsule and the genu and splenium of the corpus callosum within the midportion of these structures (FIGS. 2 and 3). Consistent positions were somewhat more difficult in the other regions. We placed frontal WM ROIs in the WM of the superior frontal gyri on the image one or two slices cephalad to the roof of the lateral ventricle, within the most peripheral portion of the WM that could still accommodate a standard-sized ROI. The observer then chose the ROI that showed highest anisotropy values within that small WM region, which meant that the ROI placement differed to some extent from study to study within the same patient. Again, these ROIs were almost exclusively drawn in regions that had normal signal on T2-weighted images.

Early transplantation infants had only mild hyperintense signal abnormalities on pretransplantation T2-weighted images, which were primarily located in the internal capsule and dentate nuclei of the cerebellum. On pretransplantation DTI studies, all 3 early transplantation infants had mean FA ratios for all 4 WM regions between 97% and 117% (FIG. 4). The reader is referred to a previous publication for the exact values.[3]

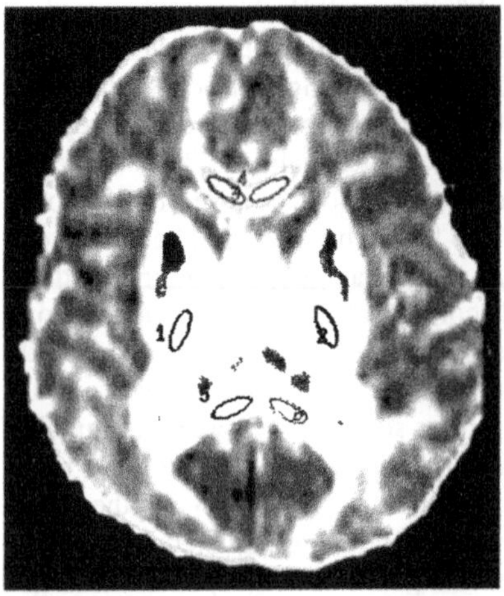

FIGURE 2. Axial image from color-coded fractional anisotropy map (in which high anisotropy values are depicted in yellow, intermediate values are depicted in green, and low values are depicted in blue) shows method for placement of regions of interest (ROIs) for measuring anisotropy values. ROIs 1 and 2 are located in the posterior limb of internal capsule, ROIs 3 and 4 are located in the genu of the corpus callosum, and ROIs 5 and 6 are located in the splenium of the corpus callosum. (Reprinted from ref. 3 with permission.)

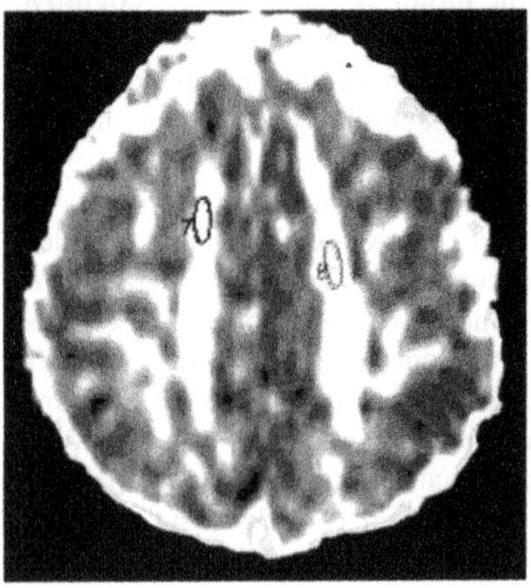

FIGURE 3. Axial image from color-coded fractional anisotropy map shows method for placement of ROIs for measuring anisotropy values in frontal white matter. Because the regions of highest anisotropy values in each hemisphere were chosen, the ROI in the right hemisphere (ROI 7) is located at a different region than the ROI in the left hemisphere (ROI 8). (Reprinted from ref. 3 with permission.)

In the late transplantation group, pretransplantation conventional MR images typically showed more diffuse, larger regions of signal abnormality than at the same stage in the early transplantation group. The periventricular WM of the frontal and parietal lobes, centrum semiovale, internal capsule, and dentate nucleus of the cerebellum were the most frequently affected regions. Pretransplantation mean FA ratios across all 4 regions in the late transplantation group ranged between 55% and 74%. The reader is referred to a previous publication for the exact values.[3] In all 4 infants, the values in the genu of the corpus callosum were substantially higher than in other locations.

As the results outlined above indicate, notable differences were seen between findings on conventional MR scans performed prior to transplantation in the two groups of infants. Pretransplantation conventional MR scans in the early transplantation (i.e., asymptomatic) infants either were normal or showed only subtle abnormalities. However, pretransplantation MR scans in the symptomatic, late transplantation infants were always abnormal and typically showed more widespread and

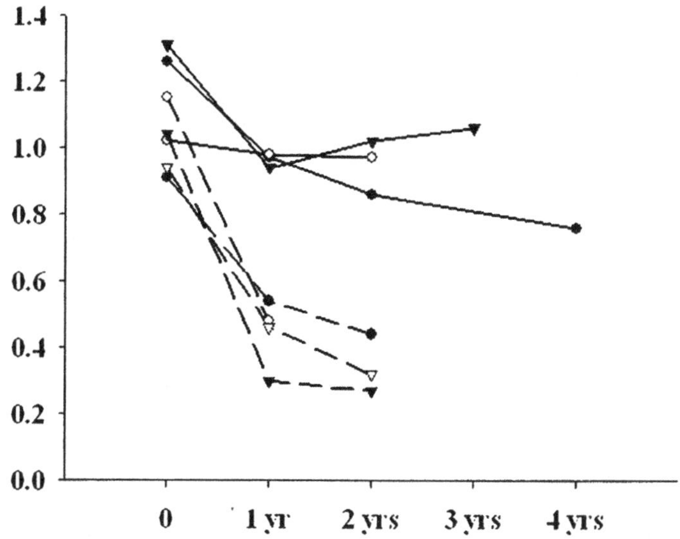

FIGURE 4. Line plot showing serial changes in fractional anisotropy (FA) ratios (i.e., FA value of Krabbe infant relative to mean FA values of 5 normal infants) in genu of the internal capsule in early treatment and late treatment Krabbe infants. The study consisted of 3 early transplantation infants and 4 late transplantation infants. Time after transplantation is shown on the x-axis and FA ratios are shown on the y-axis. The early treatment infants are signified by *solid lines*, and late treatment infants by *dashed lines*. Although the FA ratios are similar at the time of transplantation for the two groups, following transplantation FA ratios remain high for early transplantation infants, whereas the ratios markedly decrease for the late transplantation infants. Similar patterns were seen for other WM regions studied. (Reprinted from ref. 3 with permission.)

larger regions of abnormal signal. Lesions in the late transplantation infants were generally seen in the WM of the parietal and occipital lobes, centrum semiovale, and corpus callosum. The differences between the two groups likely simply reflect the fact that the late transplantation infants were symptomatic and further along in their disease course at the time of initial imaging.

Marked decreases in anisotropy values were seen in late transplantation children. However, only mild or no changes were seen for anisotropy values in the early transplantation infants at this stage. Our first hypothesis was that baseline FA ratios would be decreased for Krabbe patients. This hypothesis was clearly verified for late transplantation infants, but not for early transplantation infants. It appears that Krabbe infants in the first month of life have relatively normal myelination and axonal integrity as measured on DTI. It is also quite possible that dysmyelination and failure of myelination are present at the time of birth in these infants, but that they are too subtle to be reflected on DTI. However, by the time that late transplantation infants had become symptomatic, their abnormalities were detectable on both conventional MR scans and DTI.

On conventional MR images at 1 year posttransplantation, early transplantation children were seen to have a mild increase in the extent of hyperintense signal abnormalities in the central and periventricular WM regions of the frontal, parietal, and occipital lobes, the centrum semiovale, the internal capsule, and the dentate nuclei. The anisotropy ratios in these regions continued to be only mildly abnormal. The reader is referred to a previous publication for the exact values.[3] Thus, these children continued to maintain only mildly decreased anisotropy ratios, which appears to indicate that myelination and axonal integrity were progressing. At 2 years following transplantation, no substantial disease progression was seen on conventional MR images in any early transplantation child compared to the 1-year posttransplantation scans. The reader is referred to a previous publication for the exact values.[3] In general, FA ratios tended to be above 80% and lowest ratios were in the internal capsule. Thus, similar to the DTI pattern seen at 1 year, anisotropy values rose and anisotropy ratios were relatively well maintained at the 2-year follow-up study.

At 1 year following transplantation in the late transplantation children, conventional MR images generally showed substantial increase in regions of abnormal signal, with the corpus callosum and pons being preferentially affected in the interval. Mean FA ratios in all regions studied were now even further decreased, especially in the genu of the corpus callosum. The reader is referred to a previous publication for the exact values.[3] At 2 years following transplantation, conventional MR imaging typically showed enlargement of preexistent regions of abnormal signal and development of abnormal signal in the thalamus. Mean FA ratios generally continued to slightly decrease, but mild increases were seen in the internal capsule in 3 children. The reader is referred to a previous publication for the exact values.[3] As these data indicate, in the late transplantation children, heterogeneous changes were seen with decreases in FA ratios seen in many locations studied. Marked decreases in FA ratios in the genu of the corpus callosum were uniformly seen. Results in the splenium of the corpus callosum and frontal WM were heterogeneous. However, 3 children had moderate increases in FA values in the internal capsule, which is thought to represent a mild degree of interval myelination and establishment of intact axonal pathways.

Our second hypothesis, that early transplantation infants would have less pronounced abnormalities as measured by DTI than late transplantation infants, was verified. Over the period of a few years, differences in anisotropy between the two groups became more pronounced, that is, differences between the groups were even more apparent on posttransplantation imaging than on pretransplantation scans. These differences were also reflected on conventional imaging studies, in which the late transplantation group much more frequently developed abnormalities in the corpus callosum, central and subcortical WM regions, cerebellar WM, and (in severe cases) the pons. The differences in findings in the two groups were also reflected in differences in the clinical neurodevelopmental examination scores.

We found marked differences on follow-up DTI examinations in early transplantation children and late transplantation children. However, late transplantation infants often had no increase in anisotropy values or a decrease in anisotropy values and decreases in FA ratios. Therefore, our second hypothesis regarding effects of stem cell transplantation on DTI studies in the two groups of infants appears to have been verified. Nonetheless, prolonged survival in our late transplantation children is notable because the early infantile form of Krabbe disease has a uniformly fatal prognosis, with death usually occurring in the first year of life.

In summary, DTI appears to be a sensitive means to monitor effects of stem cell transplantation on WM development in Krabbe disease. The group of early transplantation infants was clearly distinguishable from the group of late transplantation infants based on anisotropy measurements. In addition, good correlation was seen between neurodevelopmental scores and anisotropy measurements. The work described here in Krabbe disease may serve as a model for application of DTI to other therapies in various WM disorders such as multiple sclerosis and dysmyelinating disorders of childhood.

REFERENCES

1. SUZUKI, K. 2003. Globoid cell leukodystrophy (Krabbe's disease): update. J. Child Neurol. **1:** 595–603.
2. LOES, D.J., C. PETERS & W. KRIVIT. 1999. Globoid cell leukodystrophy: distinguishing early-onset from late-onset disease using a brain MR imaging scoring method. AJNR **20:** 316–323.
3. MCGRAW, P., L. LIANG, M. ESCOLAR *et al.* 2005. Serial assessment of anisotropy measurements in Krabbe disease patients treated with hematopoietic stem cell transplantation—initial experience. Radiology **236:** 221–230.
4. KRIVIT, W., E.G. SHAPIRO, C. PETERS *et al.* 1998. Hematopoietic stem-cell transplantation in globoid-cell leukodystrophy. N. Engl. J. Med. **338:** 1119–1127.
5. MCGRAW, P., L. LIANG & J.M. PROVENZALE. 2002. Evaluation of normal age-related changes in anisotropy during childhood with diffusion-tensor imaging. AJR **179:** 1515–1522.
6. NEIL, J.J., S.I. SHIRAN, R.C. MCKINSTRY *et al.* 1998. Normal brain in human newborns: apparent diffusion coefficient and diffusion anisotropy measured by using diffusion tensor MR imaging. Radiology **209:** 57–66.
7. GUO, A., J.R. PETRELLA, J. KURTZBERG & J.M. PROVENZALE. 2001. Evaluation of white matter anisotropy in Krabbe disease using diffusion tensor MR imaging. Radiology **218:** 809–815.
8. OKEDA, R., Y. SUZUKI, S. HORIGUCHI & T. FUJII. 1979. Fetal globoid cell leukodystrophy in one of twins. Acta Neuropathol. **47:** 151–154.
9. CANIGLIA, M., I. RANA, R.M. PINTO *et al.* 2002. Allogeneic bone marrow transplantation for infantile globoid-cell leukodystrophy (Krabbe's disease). Pediatr. Transplant. **6:** 427–431.
10. MULLEN, E.M. 1995. The Mullen Scales of Early Learning: AGS Edition. American Guidance Service. Circle Pines, MN.

Index of Contributors